T0205393

AI-Enabled Threat Detection and Security Analysis for Industrial IoT

Hadis Karimipour • Farnaz Derakhshan
Editors

AI-Enabled Threat Detection and Security Analysis for Industrial IoT

Editors
Hadis Karimipour
Department of Electrical and Software Engineering
University of Calgary
Calgary, AB, Canada

Farnaz Derakhshan
North York, ON, Canada

ISBN 978-3-030-76615-3 ISBN 978-3-030-76613-9 (eBook)
https://doi.org/10.1007/978-3-030-76613-9

This Springer imprint is published by the registered company Springer Nature Switzerland AG
The registered company address is: Gewerbestrasse 11, 6330 Cham, Switzerland

Preface

The pervasive deployment of Internet of Things (IoT) devices facilitates data exchange of modern objects across every aspect of our lives, which allows new use cases in smart cities, smart homes, smart grid, health, cyber-physical systems (CPS), and industrial IoT (IIoT). Specifically, the advent of Industry 4.0 takes advantage of the IoT developments by making use of intelligent, interconnected cyber-physical systems to automate all phases of industrial operations. Although IIoT improves efficiency, reliability, and economical benefits through IoT-enabled CPS, it brings new challenges for the security community by increasing the attack landscape. The increasing number of cybersecurity incidents in CPS and IIoT, additionally stress the need to strengthen cyber resilience.

Cyber attacks in the CPS and IIoT can result in anomalous behavior in the system that may compromise physical security, cause production downtimes, damaging equipment as well as ensuring financial and reputational losses. With the evolution of adversarial techniques, current threats become even more complicated. Therefore, self-learning/cognitive approaches are required to protect the infrastructure from malicious network attacks and unauthorized access. Cognitive ability detection and Artificial Intelligence (AI) techniques, such as deep learning, machine learning, and reinforcement learning, have proven to be beneficial in learning the anomalous pattern from data to detect cyber attacks and reduce the workload of analysts.

To inform current IoT engineers and the future IoT players, materials summarizing the state-of-the-art techniques and possible challenges in IoT security are required. This handbook focuses on cutting-edge research from both academia and industry, with an emphasis on the scientific foundations and engineering techniques for securing critical IoT-enabled infrastructure and their underlying computing and communicating systems. Utility networks, transportation systems, and wireless communication and sensor networks are examples of such infrastructures. Providing a fundamental and theoretical background with a clear, comprehensive overview of security issues in IoT is useful for students in the fields of information technology, computer science/engineering, or electrical engineering.

This book will cover the anomaly detection and cybersecurity concepts in cutting-edge technologies, i.e., IoT, IIoT, and cyber-physical systems (CPS) and inform the reader of anomaly detection and defensive mechanisms in critical IoT-enabled industries. It also addresses the technical challenges associated with the design of secure IIoT and Industry 4.0 architecture by providing real-world problems and solutions from a wide variety of attack scenarios to provide intelligent automated IoT-enabled CPSs against cyber attack.

Calgary, AB, Canada Hadis Karimipour
North York, ON, Canada Farnaz Derakhshan

Contents

Artificial Intelligence for Threat Detection and Analysis in Industrial IoT: Applications and Challenges

Hadis Karimipour and Franaz Derakhshan

1 Introduction

The exponential usage of the Internet of Things (IoT) worldwide is transforming businesses and industries by providing real-time visibility into the flow of materials, products, and information. Many firms are investing in IoT technologies to redesign their workflows, improve process tracking, and optimize costs [1–3]. As such, IoT is widely recognized as a vital field in technology, paving the way for a more connected future. Among the vast applications of IoT technologies is its integration into manufacturing, automation, and critical infrastructure. Coined with the term Industrial IoT (IIoT), this domain of technology focuses on communication among industrial machines to enhance the efficiency of automation processes [4, 5].

While advantageous in various ways, the vast and growing integration of IIoT into all major industries also raises serious concerns and vulnerabilities regarding security threats [6]. Therefore, the use of IIoT in critical infrastructures, like hospitals, transportation systems, and power grids, increases the potential damage caused by cyber-attacks. This damage can extend beyond violating the privacy of individuals to potentially sabotaging an entire community [7]. IIoT devices may store an individual's health data and medical records, shopping behaviour, finances, and even location history. On an enterprise level, IIoT devices may store inventory quantities, business orders, and other information that may be sensitive to a

H. Karimipour (✉)
Department of Electrical and Software Engineering, University of Calgary, Calgary, AB, Canada
e-mail: hadis.karimipour@ucalgary.ca

F. Derakhshan
Faculty of Electrical and Computer Engineering, University of Tabriz, Tabriz, Iran
e-mail: derakhshan@tabrizu.ac.ir

H. Karimipour, F. Derakhshan (eds.), *AI-Enabled Threat Detection and Security Analysis for Industrial IoT*, https://doi.org/10.1007/978-3-030-76613-9_1

company [8, 9]. As such, security is a critical concern that surrounds the development of IIoT technologies and their integration into society.

A notable application of IIoT is in Industrial Control Systems (ICS) [10, 11]. Nowadays, ICS, which are one type of Operational Technology (OT), are deployed in a wide range of critical infrastructures such as smart grids, oil and gas refinery, transportation systems, Unmanned Aerial Vehicles (UAVs), nuclear power generation, water and gas distribution networks, and advanced communication systems [12–17]. The rapid integration of the IIoT in ICS leads to the widespread use of sensors, networked devices, and data acquisition systems, which is prone to various cyber and/or physical security threats and challenges such as privacy, access control, secure communication, and secure storage of data [18–22]. Traditionally, the ICS security was provided by physical obscurity, or a so-called air gap by keeping these systems on isolated communication networks [23]. By introducing Industry 4.0 and IIoT, ICS information is routed to sophisticated applications across enterprises through the local area network and the internet; and this is where security by obscurity is no longer a valid security solution to protect the system [24].

While there were concerns expressed about the security of critical CPSs and IIoT as early as 2000s, it was not until the 2010 Stuxnet attack [25, 26] that the security of CPSs entered into public and government discourse. In Stuxnet, zero-day exploits were mounted on a USB drive and injected malicious code into Siemens Programmable Logic Controller (PLC) to spin centrifuges at their natural frequencies, causing their wear rates to be much higher than expected. Another major cyber-attack that happened in 2015 against power substations in Ukraine resulted in power outages affecting 225,000 people [27]. The Black-Energy malware was used to target the power grids in Ukraine, causing an industrial power outage, which affected thousands of citizens [28, 29]. A more recent case was reported in April 2018 by three U.S. gas pipeline suffering a shutdown of electronic customer communication systems for several days [30].

The significant number and diversity of nodes in IIoT environments result in a large and complex attack surface. These environments (e.g., ICS, oil and gas, smart grids, transportation system) must be continuously monitored and protected [31]. This is very challenging, especially as IIoT devices are distributed across rural and remote areas. These characteristics are limiting applications of traditional security solutions such as endpoint security methods, firewalls, and security information and event management systems (SIEMs) in protecting IIoTs [32, 33].

Artificial Intelligence (AI) is proven to be a useful technique in the security analysis of IIoT. AI algorithms can be trained to detect intrusion and attacks in different layers of IIoT in a timely and reliable manner [34–37]. AI techniques can easily analyze the pattern and can attain the knowledge to find countermeasures to avoid cyber-attacks. Given the maturity and complexity of modern cyber threats and the current cybersecurity skills shortage, AI is seen as game-changers in assisting human analysts to detect, secure, and mitigate modern attacks. Although the application of AI in security comes with tremendous advantages, it also brings up certain challenges related to the verification of AI systems, their trustworthiness, computational burden, and their security issues, which should be thoroughly addressed to ensure the efficiency of the AI-based solutions.

2 Book Outline

This book presents an overview of AI-based security solution and in IIoT by examining the state of the art security measures. Furthermore, the book proposes various defence strategies, including intelligent cyber-attack and anomaly detection algorithms for different IIoT applications. This book is comprised of 13 chapters. The next chapter, chapter "Complementing IIoT Services Through AI: Feasibility and Suitability", presents a brief overview of the opportunities and challenges of realizing the AI in IIoT environments [38]. Chapter "Data Security and Privacy in Industrial IoT" will touch on issues related to data security and privacy in the IIoT environment [39]. It reviews various proposed countermeasures for security at different surfaces in IIoT, including authentication techniques, key establishment techniques, and intrusion detection techniques. Chapter "Blockchain Applications in the Industrial Internet of Things" discusses common issues and challenges related to the blockchain-based approaches on IIoT structures [40]. One of the important use cases of IIoT is on smart grids, which is discussed in chapter "Application of Deep Learning on IoT-Enabled Smart Grid Monitoring" [41]. This chapter gives an overview of the application of AI in smart grid state estimation, which has a key role in monitoring and control of these systems. Chapter "Cyber Security of Smart Manufacturing Execution Systems: A Bibliometric Analysis" aims to present a bibliographic analysis of the smart manufacturing execution systems and their integration with IIoT [42]. Chapter "The Role of Machine Learning in IIoT Through FPGAs" try to cover challenges faced in IIoT and developed ML-based solutions, which can address some of them. It also discusses the important role of FPGAs in implementing ML solutions [43].

The book then examines more advanced and specific topics in AI-based solutions developed for IIoT environments. Chapter "Deep Representation Learning for Cyber-Attack Detection in Industrial IoT" propose an unsupervised deep representation learning to handle the imbalanced IIoT data. The new representation is evaluated using seven IIoT datasets and compared with six other ML techniques in accuracy, precision, recall, and f-measure [44]. Analysis of various intelligent mining techniques for efficient anomaly detection in IIoT systems is proposed in chapter "Classification and Intelligent Mining of Anomalies in Industrial IoT" [45]. In this respect, the authors reviewed existing studies highlighting their main features. They also discuss the remaining open problems that need to be solved to shed light for future research in the field. Chapter "A Snapshot Ensemble Deep Neural Network Model for Attack Detection in Industrial Internet of Things" proposed a Snapshot Ensemble Deep Neural Network (SEDNN) model to detect cyber-attacks on IIoT environment [46]. A privacy-preserving federated learning solution for the security of industrial cyber-physical systems is also proposed in chapter "Privacy Preserving Federated Learning Solution for Security of Industrial Cyber Physical Systems" [47]. Chapter "A Multi-stage Machine Learning Model for Security Analysis in Industrial Control System" developed a multi-stage machine learning model for cyber-attack detection and identification in ICS [48]. The proposed model was tested and evaluated on samples from the data set for the water storage tank and gas pipeline. Finally, chapter "A Recurrent Attention Model for Cyber Attack Classification" proposed a recurrent attention model for cyber-attack classification in IIoT environments [49]. The proposed approach utilizes visualization to highlight regions of importance and learn feature representation for detecting polymorphic malware, which is an extremely difficult task due to its dynamic nature.

References

1. I. Lee, K. Lee, "The Internet of Things (IoT): Applications, Investments, and Challenges for Enterprises", Business Horizons, vol. 58, 431–440, 2015.
2. H. HaddadPajouh, A. Dehghantanha, R. M. Parizi, M. Aledhari, and H. Karimipour, "A Survey on Internet of Things Security: Requirements, Challenges, and Solutions", Internet of Things, pp. 100–129, 2019.
3. Z. K. Aldein Mohammed and E. S. Ali Ahmed, "Internet of Things Applications, Challenges and Related Future Technologies," World Sci. News, vol. 67, no. 2, pp. 126–148, 2017.
4. J. Sakhnini, H. Karimipour, A. Dehghantanha, R. M. Parizi, and G. Srivastava, "Security Aspects of Internet of Things Aided Smart Grids: a bibliometric survey", Internet of Things, pp. 100–111, 2020.
5. H. Karimipour and H. Leung, "Relaxation-based anomaly detection in cyber-physical systems using ensemble kalman filter," IET Cyber-Physical Syst. Theory Appl., 2020.
6. S. M. Tahsien, H. Karimipour, P. Spachos, "Machine Learning Based Solutions for Security of Internet of Things (IoT): A survey", Journal of Network and Computer Applications, vol. 161, 1–13, 2020.
7. Y. Guan and X. Ge, "Distributed Attack Detection and Secure Estimation of Networked Cyber-Physical Systems Against False Data Injection Attacks and Jamming Attacks," IEEE Transactions on Signal and Information Processing over Networks, vol. 4, no. 1, pp. 48–59, 2018.
8. H. Karimipour, A. Dehghantanha, R. M. Parizi, K. R. Choo, H. Leung, "A Deep and Scalable Unsupervised Machine Learning System for Cyber-Attack Detection in Large-Scale Smart Grids", IEEE Access, vol. 7, pp. 80778–80788, 2019.
9. S. Mohammadi, H. Mirvaziri, M. Ghazizadeh-Ahsaee, H. Karimipour, "Cyber Intrusion Detection by Combined Feature Selection Algorithm", Journal of Information Security and Applications, vol. 44, pp. 80–88, 2019.
10. K. Paridari, N. O'Mahony, A. El-Din Mady, R. Chabukswar, M. Boubekeur and H. Sandberg, "A Framework for Attack-Resilient Industrial Control Systems: Attack Detection and Controller Reconfiguration," Proceedings of the IEEE, vol. 106, no. 1, pp. 113–128, 2018.
11. A. Alabasi, H. Karimipour, A. Dehghantanha, "An Ensemble Deep Learning-based Cyber-Attack Detection in Industrial Control System", IEEE Access, vol. 8, pp. 83965-83973, April. 2020.
12. J. P. A. Yaacoub, O. Salman, H. N. Noura, N. Kaaniche, A. Chehab, and M. Malli, "Cyber-physical Systems Security: Limitations, Issues and Future Trends," Microprocess. Microsyst., vol. 77, p. 103201, Sep. 2020.
13. F. Darbandi, A. Jafari, H. Karimipour, A. Dehghantanha, F. Derakhshan, and K. K. R. Choo, "Real-time Stability Assessment in Smart Cyber-physical Grids: A Deep Learning Approach," IET Smart Grid, vol. 3, no. 4. pp. 454–461, 2020.
14. A. Yazdinejad, R. M. Parizi, A. Dehghantanha, H. Karimipour, G. Srivastava, and M. Aledhari, "Enabling Drones in the Internet of Things with Decentralized Blockchain-based Security," IEEE Internet of Things Journal, 2020.

15. H. HaddadPajouh, R. Khayami, A. Dehghantanha, K.-K. R. Choo, and R. M. Parizi, "AI4SAFE-IoT: An AIpowered secure architecture for edge layer of Internet of things," *Neural Computing and Applications,* vol. 32, no. 20, pp. 16119–16133, 2020
16. A. Yazdinejad, R. M. Parizi, A. Dehghantanha, and K.-K. R. Choo, "Blockchain-enabled authentication handover with efficient privacy protection in SDN-based 5G networks," *IEEE Transactions on Network Science and Engineering*, 2019.
17. M. Saharkhizan, A. Azmoodeh, A. Dehghantanha, K. -K. R. Choo and R. M. Parizi, "An Ensemble of Deep Recurrent Neural Networks for Detecting IoT Cyber Attacks Using Network Traffic," in *IEEE Internet of Things Journal*, vol. 7, no. 9, pp. 8852–8859, Sept. 2020, doi: https://doi.org/10.1109/JIOT.2020.2996425.
18. A. Yazdinejad, R. M. Parizi, A. Dehghantanha, Q. Zhang, and K.-K. R. Choo, "An energy-efficient SDN controller architecture for IoT networks with blockchain-based security," *IEEE Transactions on Services Computing,* vol. 13, no. 4, pp. 625–638, 2020
19. M. Aledhari, R. M. Parizi, A. Dehghantanha and K. R. Choo, "A Hybrid RSA Algorithm in Support of IoT Greenhouse Applications," *2019 IEEE International Conference on Industrial Internet (ICII)*, Orlando, FL, USA, 2019, pp. 233–240, https://doi.org/10.1109/ICII.2019.00049.
20. A. Yazdinejad, R. M. Parizi, G. Srivastava, A. Dehghantanha, and K.-K. R. Choo, "Energy efficient decentralized authentication in internet of underwater things using blockchain," in *2019 IEEE Globecom Workshops (GC Wkshps)*, 2019, pp. 1–6: IEEE
21. F. Kargl, R. W. van der Heijden, H. König, A. Valdes and M. C. Dacier, "Insights on the Security and Dependability of Industrial Control Systems," in IEEE Security & Privacy, vol. 12, no. 6, pp. 75–78, Nov.-Dec. 2014
22. Namavar Jahromi A, Hashemi S, Dehghantanha A, et al "An Improved Two-hidden-layer Extreme Learning Machine for Malware Hunting" Computer and Security, vol. 89, 1–22, 2019.
23. M. Zhang, J. Chen, S. He, L. Yang, X. Gong and J. Zhang, "Privacy-Preserving Database Assisted Spectrum Access for Industrial Internet of Things: A Distributed Learning Approach," IEEE Transactions on Industrial Electronics, vol. 67, no. 8, pp. 7094–7103, 2020.
24. A. N. Jahromi, J. Sakhnini, H. Karimipour, A. Dehghantanha, "A Deep Unsupervised Representation Learning Approach for Effective Cyber-physical Attack Detection and Identification on Highly Imbalanced Data", 29th Annual International Conf. on Computer Science and Software Engineering, pp.1–10, Toronto, Canada, Nov. 2019.
25. L. J. Trautman and P. C. Ormerod, "Industrial Cyber Vulnerabilities: Lessons from Stuxnet and the Internet of Things." Accessed: Dec. 15, 2020. [Online]. Available: http://www.whitehouse.gov/the-press-office/2013/02/12/presidential-.
26. T. M. Chen and S. Abu-Nimeh, "Lessons from Stuxnet," Computer (Long. Beach. Calif)., vol. 44, no. 4, pp. 91–93, Apr. 2011.
27. G. Falco, C. Caldera, and H. Shrobe, "IIoT Cybersecurity Risk Modeling for SCADA Systems," IEEE Internet Things J., vol. 5, no. 6, pp. 4486–4495, Dec. 2018.
28. CISA, "Cyber-attack against Ukrainian critical infrastructure," 2016. [Online]. Available: https://www.us-cert.gov/ics/alerts/IR-ALERT-H-16-056-01
29. A. Al-Abassi, J. Sakhnini and H. Karimipour, "Unsupervised Stacked Autoencoders for Anomaly Detection on Smart Cyber-physical Grids," 2020 IEEE International Conference on Systems, Man, and Cybernetics (SMC), pp. 3123–3129, 2020.
30. K. J. Higgins, "Security Incidents Rise in Industrial Control Systems", 2010. [Online]. Available: https://www.darkreading.com/attacks-breaches/security-incidents-rise-inindustrial-control-systems-/d/d-id/1133388?
31. Y. Li, Y. Xiao, Y. Li, and J. Wu, "Which Targets to Protect in Critical Infrastructures-a Game-theoretic Solution from a Network Science Perspective," IEEE Access, vol. 6, pp. 56214–56221, 2018.
32. H. Karimipour, P. Srikantha, J. Wei-Kocsis, "Security of Cyber-Physical Systems: Vulnerability and Impact", Springer Books, Aug. 2020.

33. H. Karimipour, V. Dinavahi, "Robust Massively Parallel Dynamic State Estimation of Power Systems Against Cyber-Attack", IEEE Access, vol. 6, pp. 2984–2995, Dec. 2017.
34. E. M. Dovom, A. Azmoodeh, A. Dehghantanha, D. E. Newton, R. M. Parizi, H. Karimipour, "Fuzzy Pattern Tree for Edge Malware Detection and Categorization in IoT", Journal of Systems Architecture, vol. 97, pp. 1–7, 2019.
35. X. Qiu, Z. Du and X. Sun, "Artificial Intelligence-Based Security Authentication: Applications in Wireless Multimedia Networks," IEEE Access, vol. 7, pp. 172004–172011, 2019
36. S. Singh, H. Karimipour, H. HaddadPajouh, and A. Dehghantanha, "Artificial Intelligence and Security of Industrial Control Systems," in Handbook of Big Data Privacy, Springer International Publishing, 2020, pp. 121–164.
37. G. Li, K. Ota, M. Dong, J. Wu and J. Li, "DeSVig: Decentralized Swift Vigilance Against Adversarial Attacks in Industrial Artificial Intelligence Systems", IEEE Transactions on Industrial Informatics, vol. 16, no. 5, pp. 3267–3277, 2020.
38. F.Banaie, and M. Hashemzadeh, "Complementing IIoT Services through AI: Feasibility and Suitability", AI-Enabled Threat Detection and Security Analysis for Industrial IoT, Springer Cham, pp. 1–16, 2021.
39. N. Sharghivand, F. Derakhshan, "Data Security and Privacy in Industrial IoT", AI-Enabled Threat Detection and Security Analysis for Industrial IoT, Springer Cham, pp. 1-21, 2021.
40. S. N. Ghabela, Sh. Yousefib, H. Karimipourc, "Blockchain Applications in the Industrial Internet of Things", AI-Enabled Threat Detection and Security Analysis for Industrial IoT, Springer Cham, pp. 1-40, 2021.
41. I. Al-Omari, Sh. Hadayeghparast, H. Karimipour, "Application of Deep Learning on IoT-enabled Smart Grid Monitoring", AI-Enabled Threat Detection and Security Analysis for Industrial IoT, Springer Cham, pp. 1-29, 2021.
42. A. H.Bahrami, H. M. Rouzbahani, "Cyber Security of Smart Manufacturing Execution Systems: A Bibliometric Analysis", AI-Enabled Threat Detection and Security Analysis for Industrial IoT, Springer Cham, pp. 1-21, 2021.
43. B. Joudat and M. Z. Lighvan, "The Role of Machine Learning in IIoT Through FPGAs", AI-Enabled Threat Detection and Security Analysis for Industrial IoT, Springer Cham, pp. 1–21, 2021.
44. A. N. Jahromi, H. Karimipour, A. Dehghantanha, R. M. Parizi, "Deep Representation Learning for Cyber-Attack Detection in Industrial IoT", AI-Enabled Threat Detection and Security Analysis for Industrial IoT, Springer Cham, pp. 1–29, 2021.
45. N. Sharghivand, F. Derakhshan, "Classification and Intelligent Mining of Anomalies in Industrial IoT", AI-Enabled Threat Detection and Security Analysis for Industrial IoT, Springer Cham, pp. 1–19, 2021.
46. H. M.Rouzbahani, A. H. Bahrami, H. Karimipour, "A Snapshot Ensemble Deep Neural Network Model for Attack Detection in Industrial Internet of Things", AI-Enabled Threat Detection and Security Analysis for Industrial IoT, Springer Cham, pp. 1–18, 2021.
47. S. H. Majidi, H.Asharioun, "Privacy Preserving Federated Learning Solution for Security of Industrial Cyber Physical Systems", AI-Enabled Threat Detection and Security Analysis for Industrial IoT, Springer Cham, pp. 1–18, 2021.
48. P. Semwal, "A Multi-Stage Machine Learning Model for Security Analysis in Industrial Control System", AI-Enabled Threat Detection and Security Analysis for Industrial IoT, Springer Cham, pp. 1–35, 2021.
49. N. Alsadi, H. Karimipour, A. Dehghantanha, G. Srivastava, "A Recurrent Attention Model for Cyber Attack Classification", AI-Enabled Threat Detection and Security Analysis for Industrial IoT, Springer Cham, pp. 1–35, 2021.

Complementing IIoT Services Through AI: Feasibility and Suitability

Fatemeh Banaie and Mahdi Hashemzadeh

1 Introduction

Internet of Things (IoT) has emerged as a new technology paradigm envisioned to enable the interoperable interactions of machines and devices over the internet. The structure of IoT entails the use of millions of smart devices which are able to efficiently share and process data among each other, thereby providing reliable monitoring and controlling systems [1]. IoT offers the ability to learn and interact with environmental indicators in realizing the automated real-time decision-making processes. This paves the way for enabling efficient productions and manufacturing processes in the industrial domain. Applying IoT to industrial applications has raised a new research area called Industrial IoT (IIoT) that enables the industry to analyze the acquired data from industrial assets and systems. This is a notable feature in the context of the fourth industrial revolution, known as industry 4.0 [2].

This infrastructure can lead to a significant improvement in performance, energy efficiency, and reduced response time of the devices [3]. However, IIoT-enabled multi-source manufacturing data generated by industrial devices are required to be analyzed in real-time to achieve the operation optimization and strategic decision-making [4]. The large quantity of data in resource-constrained devices and growing concerns of data privacy are preventing IIoT solutions to achieve the desired quality of services. Therefore, the realization of IIoT requirements in terms of the network reliability, real-time processing and transmission, and industrial information security can be met with the new technologies such as AI and edge computing techniques [5, 6].

F. Banaie · M. Hashemzadeh (✉)
Faculty of Information Technology and Computer Engineering, Artificial Intelligence and Machine Learning Research Laboratory, Azarbaijan Shahid Madani University, Tabriz, Iran
e-mail: hashemzadeh@azaruniv.ac.ir

H. Karimipour, F. Derakhshan (eds.), *AI-Enabled Threat Detection and Security Analysis for Industrial IoT*, https://doi.org/10.1007/978-3-030-76613-9_2

Edge computing refers to the edge-processing model that provides a flexible and efficient edge network for heterogeneous industrial devices. It leverages edge nodes with sufficient computing resources for implementing the local pre-processing of real-time industrial data, which is essential for accommodating the growing computational demands [7]. It can potentially reduce the communication bandwidth and overall delay of the system, thereby improve the overall performance of the system. More importantly, it also allows enterprises to build the effective solutions for protecting the security and privacy risks [7, 8]. In this context, edge computing incorporates AI technologies for data mining and analysis process [9].

The IIoT enables the successful cooperating of AI and big data techniques. AI-assisted data analysis framework requires proximate and prompt cloud resources for manufacturing data processing. Therefore, integrating AI into edge computing is a promising solution for deploying efficient distributed computing services [10, 11], known as Edge Intelligence (EI). The realization of edge intelligence in IIoT can be further reinforced by integrating ML methods. In particular, model training and model evaluation in data analysis and prediction processes can be carried out locally in edge devices called ML as a Service (MLaaS) [12]. EI provides some benefits in terms of *personalization*, *responsiveness*, and *privacy issues*. Notably, it enables not only accurate services through customizing AI models, but also provides fast and adaptive services for time-varying industrial process. Moreover, information processing at the network edge ensures the private services. The rest of this chapter discusses the opportunities and essential issues of the paradigm where ML models are executed locally in the industrial manufacturing network.

2 IIoT with Edge Intelligence

The IIoT provides an efficient computational platform that is able to monitor and control the manufacturing processes with the aid of information technologies. In this platform, the acquired data from industrial assets and devices can be efficiently processed and analyzed by incorporating the AI technologies. In smart manufacturing, data analysis is a critical feature in realizing the automation and intelligence of IIoT systems. Mainly, ML is a popular modeling technique that can be applied to data-driven applications. Learning techniques typically utilize a sufficient amount of data for training the model in different areas such as regression, classification, clustering, and forecasting. Thus, intelligent manufacturing requires cloud-assisted service for processing and analyzing the industrial big data. For this purpose, utilizing edge devices can be a promising development trend that provides computational power and service accuracy on the edge servers. The relationship between AI, EI, Edge Computing (EC), and IIoT is shown in Fig. 1. Given these concepts and the relationships between them, in this section, we investigate the preliminaries of an intelligent computing framework for IIoT.

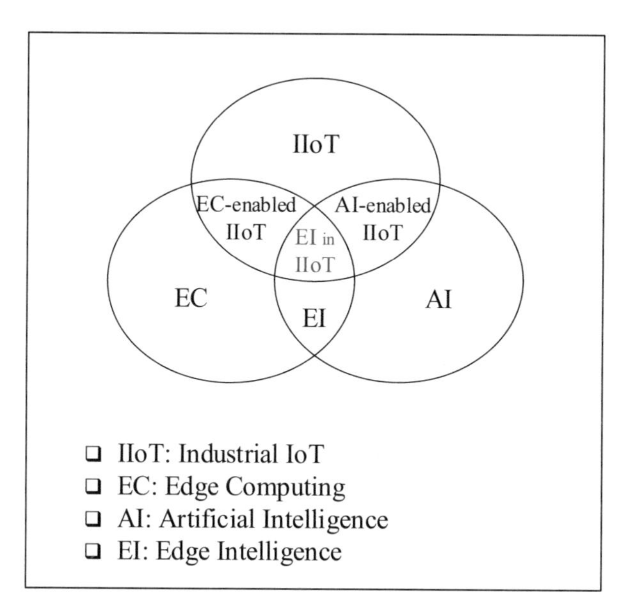

Fig. 1 AI, EC, EI, and IIoT in venn diagram

2.1 *The Challenges of EC in IIoT*

EC is an innovative paradigm for industrial devices that utilizes sensing, computational resources, and data processing techniques to provide an efficient manufacturing infrastructure. However, processing a large quantity of data in edge nodes may bring some new challenges in terms of the confidentiality of information and performance-related issues that should be addressed in design and implementation of a system. The following focuses on the detailed discussion of the design issues.

Data Processing and Analysis Big industrial data are mainly driven by millions of smart devices and business processes. To develop an efficient edge service in IIoT, it is critical to design the appropriate approaches for data capturing and storing the vast amount of heterogeneous data in distributed edge servers. Since the data quantity is huge and grows rapidly, the solution must not only be able to store the industrial data effectively, but also support scalability and flexibility. This is the requirement of applications envisaged in IIoT vision [13, 14]. In addition, data analysis schemes should be considered to meet the requirements of data processing and management. AI-enabled technologies are proved as an effective solution in providing a real-time data management platform in terms of accuracy, adaptability, and the security of data [15, 16].

Security and Privacy Integrating EC technologies with IIoT enhances the security and privacy of the produced data, as it decreases the data transmission in the network. However, traditional security solutions cannot guarantee the full requirements for edge services. Different kinds of malicious attacks can threat confidentiality due to the ubiquitous network environment [17, 18]. Moreover, existing cybersecurity frameworks are not applicable for industrial systems, as the unique characteristics of IIoT in providing the strict performance and reliability requirements for supporting the critical functionality [19]. AI technologies are the potent methods for investigating the normal/abnormal behavior of the manufacturing components and devices in the IIoT environment. In particular, these methods can assist in developing the security-based intelligent systems [20].

Energy Consumption In general, the total energy consumption consists of the amount of energy consumed by industrial devices for collecting and processing the industrial information, and the energy consumed for data transmission among these devices. Although edge-enabled computing enhances the energy expenditure of the sensory devices, big data processing, and association imposes the higher energy levels on the edge devices that need to be considered in design stages [21].

Resource Management and Task Scheduling An edge-enabled computing model provides flexible computations and storage services for intelligent industrial systems. However, the heterogeneity of this platform in terms of the higher real-time task requests, terminal assets and devices, and edge nodes necessitates the creation of an efficient task scheduling scheme. The problem lies in determining the rules about how to perform the data transmission and task scheduling among edge nodes to minimize the delay and energy consumption, whilst enhance the overall performance of the manufacturing systems. Moreover, the scheduling strategy should assign the tasks among edge nodes to guarantee the load-balancing and prolong the lifecycle of the system [7, 22].

2.2 *Classifications of AI Techniques*

Nowadays, AI has become an integral part of our daily lives. Motivated by the recent advancement in AI techniques and the impacts on a wide variety of domains, ranging from automatic face-focus to robotics, a set of intelligent applications have quickly ascended to the spotlight in the industrial field. In particular, AI is a generic term, which involves various techniques summarized in Fig. 2.

Among them, ML is an effective method that has received greater attention in recent years owing to the achieved accuracy [9]. As shown in Fig. 2, ML-based approaches are generally divided into supervised, unsupervised, and Reinforcement Learning (RL) methods. These methods can improve the performance of the system through training the machine using gathered data from the real world. Besides ML techniques that utilize neural networks for learning, the deep representation of data, achieve remarkable results in a broad spectrum of fields. In contrast to the ML approaches that require a feature extractor for transmitting raw data into proper representation, the Deep Learning (DL) method develops its own representations

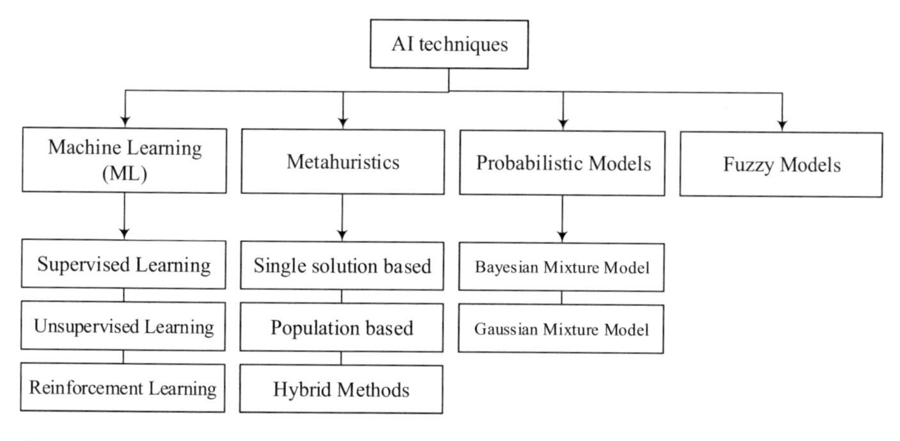

Fig. 2 Summary of AI technologies [23]

for pattern recognition. In this manner, systems can successfully learn more complex functions with unprecedented and undefined conditions.

2.3 Machine Learning Techniques in IIoT

ML is one of the popular AI-based techniques that provide the ability to learn with data and enhancing the performance of computer systems in decision-making processes, without having explicit programs [23, 24]. The aid of these techniques is to model the concepts from perceptions. According to these perceptions, learning techniques can be in three categories: *supervised learning,* in which labeled data are applied in classification or regression tasks, *unsupervised learning* with unlabeled data in clusters, and *reinforcement learning* approach that utilizes the concept of agents for maximizing the cumulative reward.

Nowadays, ML-based schemes have found their applications in upcoming decentralized and intelligent information and networking systems. They have significant potential in improving the deployment of communications and networking systems, as they are able to extract features from a large amount of data. The application of ML-based algorithms in networking is twofold. First, it can help in optimal decisions with learning network patterns, such as routing decisions in traffic patterns [25, 26]. Second, the performance and resource usage optimization in a network can be solved by the intelligent task allocation and scheduling schemes [27, 28].

The heterogeneous manufacturing data are analyzed through a ML-assisted approaches in IIoT. For example, the authors in [29] proposed an ML-aided information management system to enhance the user request service. They leveraged some indexing techniques for achieving the effective data management, then an ML algorithm is applied to improve the accuracy of the request processing. The authors in [10] investigated the tradeoff between the service delivery latency and energy expenditure in IIoT. Generally, cloud resources are used for data processing in these approaches. However, it imposes much cost in terms of network bandwidth and service latency, while transmitting data to the cloud servers; so, it is beneficial to apply the distributed computing services.

2.4 Machine Learning Techniques in Edge Computing

AI-aided edge computation services could efficiently empower the Manufacturing Devices (MD) with low latency computing capability. Typically, edge resources cannot afford the complex AI tasks. Thus, distributed AI services can be performed with multiple edge devices to provide an efficient service provisioning. For example, an AI-based privacy-preserving service division is presented in [30] that conducts the service composition on encrypted data using a homomorphic encryption algorithm. In [31], a federated deep reinforcement learning-based framework is proposed that improves latency by applying the edge caching technique. This framework utilizes the local training parameters of the base stations as the initial input for the global training in the next stage. In [32], a QoE-based computation offloading model is presented that improves the service latency, energy consumptions, and task success rate using a deep reinforcement learning algorithm. It could improve the QoE performance, besides achieving the better instability and faster convergence.

2.5 Edge Intelligent IIoT

The recent proliferation of the computation-intensive manufacturing applications generates a large volume of the industrial data at the network edge. This incurs an urgent need for AI techniques at the network edge to release the informative potential of big data. Big data has a crucial role in AI development that has recently moved from datacenters toward the growing widespread devices, e.g., IIoT devices. The emerging paradigm that moves computing tasks and services from core to the network edge has led to a promising area of EI. EI combines the edge advantages (e.g., reduced latency and network traffic) with AI strategies that result in further benefits in the following aspects [9].

1. *Big data analysis at the network edge*: the growing number of smart devices and assets leads to the large volume of industrial data in IIoT. In this context, decision-making processes can be accelerated through AI strategies in data analytic and information extraction. Among them, deep learning is a strong approach that can meet the requirements of big data analytic. Deep learning models have also achieved remarkable results in automatically identifying patterns and anomaly detection in data. Then, the extracted information is used for real-time predictive decision-making in industrial environments.

2. *Efficient data processing using edge resources*: it is already proved that data has a vital effect on the development of AI models. Traditionally, the acquired data from IoT devices and industrial assets are sent and stored in the cloud data centers. This incurs the higher latency and wastage of bandwidth resources. To address these challenges, computing tasks and services are moved to the network edge in recent years. In this way, the generated data can be processed locally to achieve low-latency responses in real-time manufacturing systems.
3. *Ubiquitous AI services*: AI has been recognized as an essential solution in a variety of application domains that influence our everyday lives [33–38]. The potential of AI in improving the smart products and services imposes the need for bringing AI closer for every person and device [39]. Clearly, edge computing can assist in achieving this goal by enabling ubiquitous AI at the network edge.

3 AI-Enhanced Cooperative Computing Architecture

This section introduces the architecture of smart manufacturing resources based on the cooperative edge computing in IIoT. It is consists of three operational layers named manufacturing assets, edge-devices, and remote cloud resources, as depicted in Fig. 3. Multi-source manufacturing data is collected and delivered to the base stations (BS) or edge devices. Real-time manufacturing services include self-monitoring, production and logistic status, fault detection, and service management [10, 40].

Edge layer provides a lightweight smart service by processing the real-time data and perception events, and transforming them into dynamic behaviors in manufacturing systems. Computing service of edge devices may be different, depending on the application and service accuracy. However, cooperative computing service through edge and cloud accommodates both prompt and comprehensive computing services. Owing to the additional cost of communications between edge and cloud servers, the deployment of computing services between them has a significant impact on performance. Additionally, a task scheduling strategy is required for assigning the computing tasks to the heterogeneous edge servers according to their specifications and quality of service (QoS) requirements.

4 Potential Advantages of Learning Techniques in Edge Intelligent IIoT

The applications of learning techniques in IIoT would enable the further extraction of the information, and the deployment of the innovative applications in the intelligent manufacturing domain. More specifically, applying information technologies in the industrial field can provide a flexible infrastructure for smart manufacturing

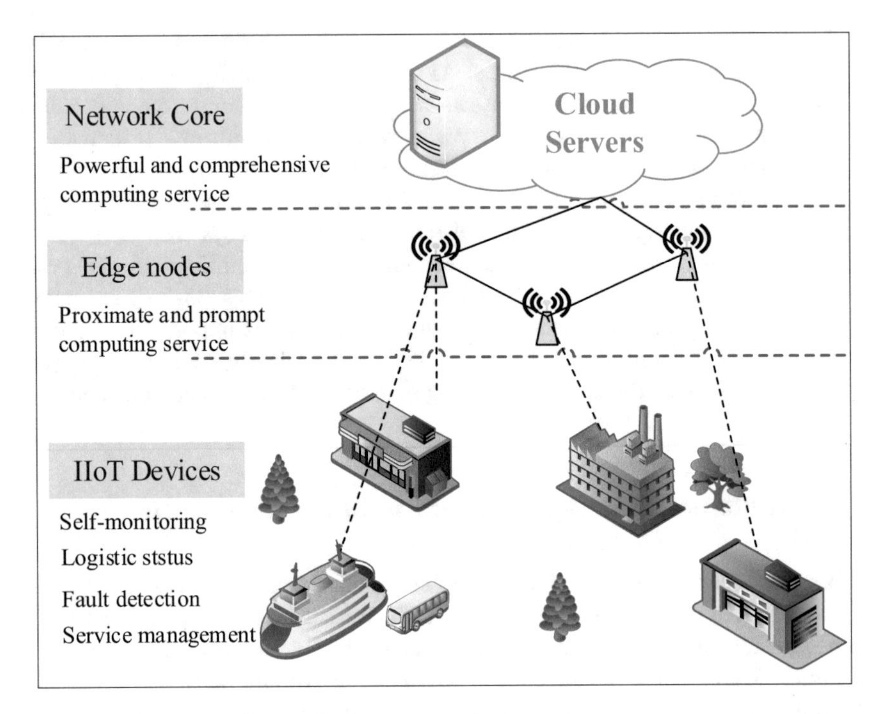

Fig. 3 Three-layered architecture for IIoT

services. This allows the industrial assets and devices to have a seamless autonomous communication in the service delivery process. For example, Prognostic and Health Management (PHM) is a novel paradigm that utilizes the collected operational information of the system for predicting and taking suitable decisions before system failure [41]. A large number of sensors are utilized to provide a real-time monitoring system, which is essential in IIoT [42]. This system has the ability to detect and predict the system failures by integrating the autonomous support system with the information system. The initial processing of data is performed at the network edge that enhances the latency of the emergency decisions.

In the context of big data, learning can be used for the efficient extracting and mining of the features in exact classifications or autonomous decision-making processes. Moreover, an intelligent edge-computing platform would be capable of learning the surrounding conditions, mapping out the events, monitoring and tracking manufacturing faults, effective action prediction, and providing a fast response to real-time changes.

5 Practical Limitation and Open Issues

Although there is a growing interest in EI, the study of EI is still in the early stage. This section discusses some challenges and open issues in EI-based IIoT, including software platforms and middleware, load balancing, EI model design, and security issues.

5.1 Software Platforms and Middleware

Recently, cloud-based AI service provisioning systems have received a lot of attentions in industrial field. In this regard, some companies, such as Amazon's Greengrass, Microsoft Azure, and Google Cloud IoT Edge try to deliver software platforms and middleware for the edge services. On the other hand, the growing number of AI-assisted computation-intensive applications leads to the development of the pervasive EI platform and middleware.

In order to realize the potential of EI services, several key challenges should be addressed in terms of the compatibility, portability, and programing issues. Owing to the diverse and heterogeneous EI services, a middleware should be developed for providing seamless and smooth services. This platform should support the portability between different AI programing frameworks, such as Tensor flow and Torch. Moreover, it should provide a lightweight virtualization and computing service [9].

5.2 Task Offloading and Load Balancing

Pervasive computing in EI is a distributed system paradigm that has variety of computing resources. Therefore, it is required to have an effective task offloading and load balancing scheme for task dissemination among the resources and servers. Particularly, data offloading schemes aid to balance the overall load of the system among the limited computing resources. This can help in improving the overall latency of the service delivery in system, which is necessary for manufacturing applications. In this context, machine learning models can be used to set up the efficient balancing schemes.

5.3 EI Model Design

AI models are usually resource-intensive and require powerful computing capability. To address this problem, model compression techniques can be applied to resize the AI models. To this end, model simplification is used for adapting the model to the edge resources, which includes weight pruning and data quantification. In weight pruning method, the removal of neurons with small contribution makes the model smaller. Data quantization utilizes a small data format with fewer bits in representing the input/output that improves the operational speed of the instructions [7].

5.4 Security Issues

Owing to the open nature of the pervasive computing, privacy and security problems is one of the key challenges of IIoT. Manufacturing assets and devices produce a large amount of data that may contain sensitive information about location, activity records, production process, and manufacturing information. Therefore, designing an appropriate distributed security mechanism is critical to guarantee the user privacy and data integrity for industrial applications. Distributed learning models are a feasible solution for the privacy-friendly local data training schemes.

6 Conclusion

Recent advances on ubiquitous computing play a crucial role in boosting AI techniques in a resource-constrained environment. Moving the AI frontier from the remote cloud to the network edge can pave the way for computation-intensive AI applications. This can help to tackle the limitations of the bandwidth and latency in computation-intensive decision-making processes.

This chapter reviewed the novel paradigm of edge intelligence, motivations for pushing artificial intelligence frontier to the network edge, and the reference architecture of edge intelligence in industrial IoT (IIoT). Specifically, we discussed the emerging learning models in the industrial field for training and perceiving manufacturing data and processes at the network edge. This review could be a good step for motivating researchers to make more attention to the industry development.

References

1. H. H. Pajouh, A. Dehghantanha, R. Parizi, H. Karimipour, "A Survey on Internet of Things Security: Requirements, Challenges, and Solutions", Internet of Things Journal, pp. 1–16, Oct. 2019. https://doi.org/10.1016/j.iot.2019.100129
2. H. Haddadpajouh, A. Mohtadi, A. Dehghantanaha, H. Karimipour, X. Lin and K. -K. R. Choo, "A Multi-Kernel and Meta-heuristic Feature Selection Approach for IoT Malware Threat Hunting in the Edge Layer," in IEEE Internet of Things Journal, https://doi.org/10.1109/JIOT.2020.3026660.
3. Y. Lu, "Industry 4.0: A survey on technologies, applications and open research issues," *Journal of Industrial Information Integration*, vol. 6, pp. 1–10, 2017, doi: https://doi.org/10.1016/j.jii.2017.04.005.
4. Y. Wang, S. Wang, B. Yang, L. Zhu, and F. Liu, "Big data driven Hierarchical Digital Twin Predictive Remanufacturing paradigm: Architecture, control mechanism, application scenario and benefits," *Journal of Cleaner Production,* vol. 248, p. 119299, 2020, doi: https://doi.org/10.1016/j.jclepro.2019.119299.
5. H. Karimipour, A. Dehghantanha, R. M. Parizi, K. R. Choo and H. Leung, "A Deep and Scalable Unsupervised Machine Learning System for Cyber-Attack Detection in Large-Scale

Smart Grids," in IEEE Access, vol. 7, pp. 80778–80788, 2019, doi: https://doi.org/10.1109/ACCESS.2019.2920326.
6. A. Al-Abassi, J. Sakhnini, H. Karimipour, "Unsupervised Stacked Autoencoders for Anomaly Detection on Smart Cyber-physical Grids", IEEE System, Men, Cybernetic (IEEE SMC), pp. 1–5, Aug. 2020. Accepted
7. T. Qiu, J. Chi, X. Zhou, Z. Ning, M. Atiquzzaman, and D. O. Wu, "Edge Computing in Industrial Internet of Things: Architecture, Advances and Challenges," *IEEE Communications Surveys & Tutorials,* pp. 1–1, 2020, doi: https://doi.org/10.1109/COMST.2020.3009103.
8. F. Banaie, J. Misic, V. B. Misic, M. H. Y. Moghaddam, and S. A. H. Seno, "Performance Analysis of Multithreaded IoT Gateway," *IEEE Internet of Things Journal,* vol. 6, no. 2, pp. 3143–3155, 2019, doi: https://doi.org/10.1109/JIOT.2018.2879467.
9. Z. Zhou, X. Chen, E. Li, L. Zeng, K. Luo, and J. Zhang, "Edge Intelligence: Paving the Last Mile of Artificial Intelligence With Edge Computing," *Proceedings of the IEEE,* vol. 107, no. 8, pp. 1738–1762, 2019, doi: https://doi.org/10.1109/JPROC.2019.2918951.
10. J. Sakhnini, H. Karimipour, "AI and Security of Cyber Physical Systems: Opportunities and Challenges", Handbook of Security of Cyber-Physical Systems: Vulnerability and Impact, Springer Books, March. 2020. https://doi.org/10.1007/978-3-030-45541-5_1
11. S. Yousefi, F. Derakhshan, H. S. Aghdasi, and H. Karimipour, "An energy-efficient artificial bee colony-based clustering in the internet of things," *Computers & Electrical Engineering,* vol. 86, p. 106733, 2020, doi: https://doi.org/10.1016/j.compeleceng.2020.106733.
12. B. Qolomany, I. Mohammed, A. Al-Fuqaha, M. Guizani, and J. Qadir, "Trust-Based Cloud Machine Learning Model Selection For Industrial IoT and Smart City Services," *IEEE Internet of Things Journal*, pp. 1-1, 2020, doi: https://doi.org/10.1109/JIOT.2020.3022323.
13. L. Jiang, L. D. Xu, H. Cai, Z. Jiang, F. Bu, and B. Xu, "An IoT-Oriented Data Storage Framework in Cloud Computing Platform," *IEEE Transactions on Industrial Informatics,* vol. 10, no. 2, pp. 1443-1451, 2014, doi: https://doi.org/10.1109/TII.2014.2306384.
14. S. M. Tahsien, H. Karimipour, P. Spachos, "Machine Learning Based Solutions for Security of Internet of Things (IoT): A Survey", Journal of Network and Computer Applications, vol. 161, pp. 1–18, April. 2020. https://doi.org/10.1016/j.jnca.2020.102630
15. T. Zhang, Z. Shen, J. Jin, A. Tagami, X. Zheng, and Y. Yang, "ESDA: An Energy-Saving Data Analytics Fog Service Platform," Cham, 2019: Springer International Publishing, in Service-Oriented Computing, pp. 171–185.
16. Y. Han, B. Park, and J. Jeong, Fog Based IIoT Architecture Based on Big Data Analytics for 5G-networked Smart Factory, Cham, 2019: Springer International Publishing, in Computational Science and Its Applications—ICCSA 2019, pp. 44–52.
17. T. Qiu, J. Liu, W. Si, and D. O. Wu, "Robustness Optimization Scheme With Multi-Population Co-Evolution for Scale-Free Wireless Sensor Networks," *IEEE/ACM Transactions on Networking,* vol. 27, no. 3, pp. 1028–1042, 2019, doi: https://doi.org/10.1109/TNET.2019.2907243.
18. A. Al-Abassi, H. Karimipour, A. Dehghantanha, and R. M. Parizi, "An Ensemble Deep Learning-Based Cyber-Attack Detection in Industrial Control System," *IEEE Access,* vol. 8, pp. 83965–83973, 2020, doi: https://doi.org/10.1109/ACCESS.2020.2992249.
19. A. Hassanzadeh, S. Modi, and S. Mulchandani, "Towards effective security control assignment in the Industrial Internet of Things," in *2015 IEEE 2nd World Forum on Internet of Things (WF-IoT)*, 14–16 Dec. 2015 2015, pp. 795–800, https://doi.org/10.1109/WF-IoT.2015.7389155.
20. A. S. Lalos, A. P. Kalogeras, C. Koulamas, C. Tselios, C. Alexakos, and D. Serpanos, "Secure and Safe IIoT Systems via Machine and Deep Learning Approaches," in *Security and Quality in Cyber-Physical Systems Engineering: With Forewords by Robert M. Lee and Tom Gilb,* S. Biffl, M. Eckhart, A. Lüder, and E. Weippl Eds. Cham: Springer International Publishing, 2019, pp. 443–470.
21. S. Sharma and H. Saini, "Fog assisted task allocation and secure deduplication using 2FBO2 and MoWo in cluster-based industrial IoT (IIoT)," *Computer Communications,* vol. 152, pp. 187–199, 2020. https://doi.org/10.1016/j.comcom.2020.01.042.

22. L. Yin, J. Luo, and H. Luo, "Tasks Scheduling and Resource Allocation in Fog Computing Based on Containers for Smart Manufacturing," *IEEE Transactions on Industrial Informatics,* vol. 14, no. 10, pp. 4712–4721, 2018, doi: https://doi.org/10.1109/TII.2018.2851241.
23. G. Zhu, J. Zan, Y. Yang, and X. Qi, "A Supervised Learning Based QoS Assurance Architecture for 5G Networks," *IEEE Access,* vol. 7, pp. 43598–43606, 2019, doi: https://doi.org/10.1109/ACCESS.2019.2907142.
24. S. Yousefi, F. Derakhshan, and H. Karimipour, "Applications of Big Data Analytics and Machine Learning in the Internet of Things," in *Handbook of Big Data Privacy*, K.-K. R. Choo and A. Dehghantanha Eds. Cham: Springer International Publishing, 2020, pp. 77-108.
25. F. Tang *et al.*, "On Removing Routing Protocol from Future Wireless Networks: A Real-time Deep Learning Approach for Intelligent Traffic Control," *IEEE Wireless Communications,* vol. 25, no. 1, pp. 154–160, 2018, doi: https://doi.org/10.1109/MWC.2017.1700244.
26. B. Mao *et al.*, "A Novel Non-Supervised Deep-Learning-Based Network Traffic Control Method for Software Defined Wireless Networks," *IEEE Wireless Communications,* vol. 25, no. 4, pp. 74–81, 2018, doi: https://doi.org/10.1109/MWC.2018.1700417.
27. L. Liu, B. Yin, S. Zhang, X. Cao, and Y. Cheng, "Deep Learning Meets Wireless Network Optimization: Identify Critical Links," *IEEE Transactions on Network Science and Engineering,* vol. 7, no. 1, pp. 167–180, 2020, doi: https://doi.org/10.1109/TNSE.2018.2827997.
28. Y. He, F. R. Yu, N. Zhao, V. C. M. Leung, and H. Yin, "Software-Defined Networks with Mobile Edge Computing and Caching for Smart Cities: A Big Data Deep Reinforcement Learning Approach," *IEEE Communications Magazine,* vol. 55, no. 12, pp. 31–37, 2017, doi: https://doi.org/10.1109/MCOM.2017.1700246.
29. G. Manogaran *et al.*, "Machine Learning Assisted Information Management Scheme in Service Concentrated IoT," *IEEE Transactions on Industrial Informatics,* pp. 1–1, 2020, doi: https://doi.org/10.1109/TII.2020.3012759.
30. M. S. Rahman, I. Khalil, M. Atiquzzaman, and X. Yi, "Towards privacy preserving AI based composition framework in edge networks using fully homomorphic encryption," Engineering Applications of Artificial Intelligence, vol. 94, p. 103737, 2020, doi: https://doi.org/10.1016/j.engappai.2020.103737.
31. X. Wang, C. Wang, X. Li, V. C. M. Leung, and T. Taleb, "Federated Deep Reinforcement Learning for Internet of Things With Decentralized Cooperative Edge Caching," *IEEE Internet of Things Journal,* vol. 7, no. 10, pp. 9441–9455, 2020, doi: https://doi.org/10.1109/JIOT.2020.2986803.
32. H. Lu, X. He, M. Du, X. Ruan, Y. Sun, and K. Wang, "Edge QoE: Computation Offloading With Deep Reinforcement Learning for Internet of Things," *IEEE Internet of Things Journal,* vol. 7, no. 10, pp. 9255–9265, 2020, doi: https://doi.org/10.1109/JIOT.2020.2981557.
33. M. Hashemzadeh and A. Zademehdi, "Fire detection for video surveillance applications using ICA K-medoids-based color model and efficient spatio-temporal visual features," *Expert Systems with Applications,* vol. 130, pp. 60-78, 2019, doi: https://doi.org/10.1016/j.eswa.2019.04.019.
34. M. Hashemzadeh, B. Asheghi, and N. Farajzadeh, "Content-aware image resizing: An improved and shadow-preserving seam carving method," *Signal Processing,* vol. 155, pp. 233-246, 2019, doi: https://doi.org/10.1016/j.sigpro.2018.09.037.
35. M. Hashemzadeh and B. Adlpour Azar, "Retinal blood vessel extraction employing effective image features and combination of supervised and unsupervised machine learning methods," *Artificial Intelligence in Medicine,* vol. 95, pp. 1-15, 2019, doi: https://doi.org/10.1016/j.artmed.2019.03.001.
36. N. Farajzadeh and M. Hashemzadeh, "Exemplar-based facial expression recognition," *Information Sciences,* vol. 460–461, pp. 318-330, 2018, doi: https://doi.org/10.1016/j.ins.2018.05.057.
37. M. Hashemzadeh and N. Farajzadeh, "A machine vision system for detecting fertile eggs in the incubation industry," *International Journal of Computational Intelligence Systems,* vol. 9, no. 5, pp. 850-862, 2016.

38. M. Hashemzadeh and N. Farajzadeh, "Combining keypoint-based and segment-based features for counting people in crowded scenes," *Information Sciences,* vol. 345, pp. 199–216, 2016, doi: https://doi.org/10.1016/j.ins.2016.01.060.
39. Y. Xiao, G. Shi, and M. Krunz, "Towards Ubiquitous AI in 6G with Federated Learning," *arXiv preprint arXiv:2004.13563,* 2020.
40. Y. Zhang, Z. Guo, J. Lv, and Y. Liu, "A Framework for Smart Production-Logistics Systems Based on CPS and Industrial IoT," *IEEE Transactions on Industrial Informatics,* vol. 14, no. 9, pp. 4019–4032, 2018, doi: https://doi.org/10.1109/TII.2018.2845683.
41. X. Yi, Y. Chen, P. Hou, and Q. Wang, "A Survey on Prognostic and Health Management for Special Vehicles," in *2018 Prognostics and System Health Management Conference (PHM-Chongqing),* 26–28 Oct. 2018, pp. 201–208, https://doi.org/10.1109/PHM-Chongqing.2018.00041.
42. A. L. Ellefsen, V. Æsøy, S. Ushakov, and H. Zhang, "A Comprehensive Survey of Prognostics and Health Management Based on Deep Learning for Autonomous Ships," *IEEE Transactions on Reliability,* vol. 68, no. 2, pp. 720–740, 2019, doi: https://doi.org/10.1109/TR.2019.2907402.

Data Security and Privacy in Industrial IoT

Nafiseh Sharghivand and Farnaz Derakhshan

1 Introduction

The Industrial Internet of Things (IIoT) is considered as one of the most promising revolutionary technologies to prompt Industry 4.0. However, the massive volume of generated data by the connected devices in the IIoT paradigm, and the existing data sharing between them lead to new security and privacy concerns (e.g. data leakage [1, 2]). For this reason, security and privacy concerns are assumed as major obstacles where things are responsible to control sensitive machinery and controlling systems in industries [3]. Indeed, according to Gartner forecast, information security is considered as one of the top concerns among enterprises adopting IoT [4].

The leakage of private data in the IIoT paradigm can lead to critical issues that may be far beyond only financial loss, such as human death and injuries [5]. Data leakage may occur during data storage, data transmission, or data sharing. Therefore, for providing a completely secure IIoT system, a holistic cybersecurity framework is required covering all abstraction layers of the system.

Hence, in the first place, the storage of IIoT devices should be protected against potential adversaries via different approaches such as encryption. If there is a central curator for data storage in the IIoT infrastructure, the increasing risks for data leakage, and the vulnerability of the curator to single point failure by malicious attacks such as DDoS needs to be addressed. This is while allowing the authorized users to access their required data. Hence, the IIoT infrastructure needs efficient identification and authentication mechanisms as well. Furthermore, if the generated data is outsourced to a cloud infrastructure, other security and privacy challenges

N. Sharghivand (✉) · F. Derakhshan
Computer Engineering Department, Faculty of Electrical and Computer Engineering, University of Tabriz, Tabriz, Iran
e-mail: n.sharghivand@tabrizu.ac.ir

H. Karimipour, F. Derakhshan (eds.), *AI-Enabled Threat Detection and Security Analysis for Industrial IoT*, https://doi.org/10.1007/978-3-030-76613-9_3

such as the authenticity of data, untrustworthiness of third parties, and robustness of cloud infrastructure should be considered.

To enhance security in the next layers, i.e. data transmission and sharing, the collaboration and data sharing among IIoT devices should be limited to authenticated devices. To do so, the communication channels between the IIoT devices should be kept secured in the first place so that the IIoT devices can easily share their data without compromising confidentiality or integrity. Then, efficient identification and authorization mechanisms should be tailored, so that only the authorized IIoT devices can access the data.

Finally, if the IIoT system is attacked for any reason, this should be detected in the fastest way before any serious damage happens. For example, anomaly detection approaches can be used to identify any anomalous behavior in the system due to cyberattacks. After this, appropriate actions must be taken to return the system to normal operation.

Figure 1 illustrates a Cyber-physical production system (CPPS) architecture and its existing attack surfaces as an example. Smart factories consist of several CPPSs, which are comprised of electronics (e.g., CPU and RAM) and monitors that are used to control physical processes through sensors and actuators [6]. At the lowest level, the deployed electronics are vulnerable to different invasive hardware and physical attacks, such as side-channel attacks and reverse-engineering attacks [7]. On the higher layer, the deployed software on the electronics can be compromised via different Trojans, viruses, or other types of malicious code. Then, the

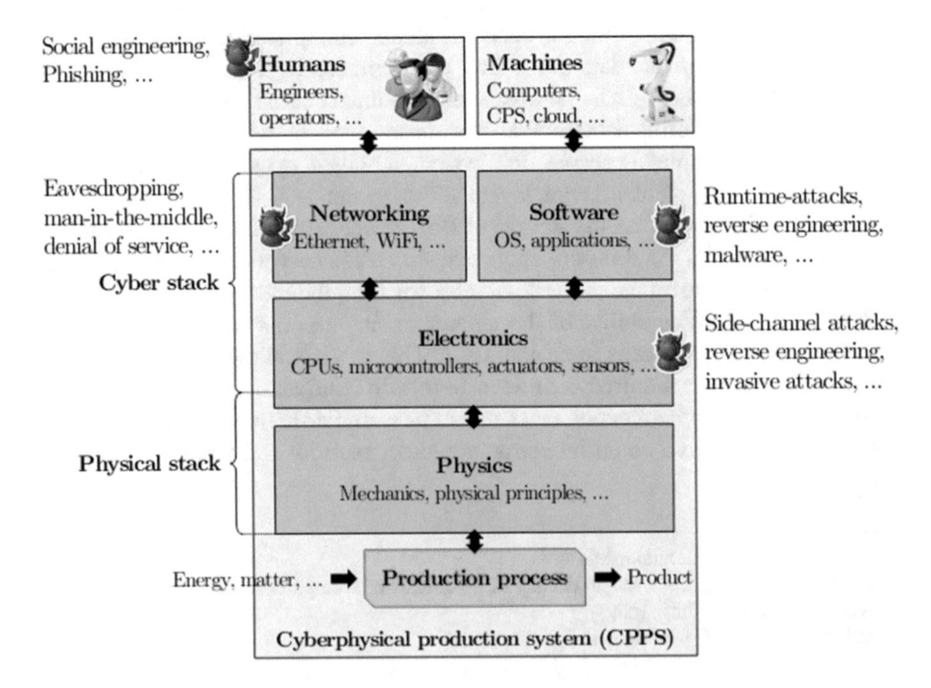

Fig. 1 Different attack surfaces in CPPS as an example [6]

networking infrastructure and the employed communication protocols can be subject to other attacks such as man-in-the-middle and denial-of-service attacks [6]. Even the human operators may intentionally or accidentally compromise the security of the system [8]. For example, a careless human user may leave his desktop logged in and then let unauthorized people use his user account, or he may intentionally attack the system for example via phishing or social engineering [6].

However, addressing all the aforementioned security and privacy issues is assumed to be very challenging due to the existing complexity and resource limitations in most of the IIoT systems. Indeed, IIoT systems usually consist of a large-scale network of heterogeneous IIoT devices and cyber-physical systems, with constrained power, computing, and communication resources of IIoT devices [9, 10].

Nonetheless, many researchers have addressed the security and privacy concerns in industrial IoT systems from different aspects in recent years. They have developed different approaches based on secure engineering, security and privacy management, identity management, industrial rights management, platform security, and communication security and privacy [7].

In this chapter, we discuss three countermeasures that are very common in IIoT systems to overcome security and privacy threats. These countermeasures include authentication techniques, key establishment techniques, and intrusion detection techniques. We also review the proposed methods in the literature which are developed based on one of these techniques. Furthermore, we review the existing real testbeds which are developed so that the proposed security mechanisms can be evaluated properly before they are used in practice.

The remainder of this chapter is organized as follows. In Sect. 2, we review the proposed techniques for intrusion detection in IIoT. In Sect. 3, we discuss and review the proposed authentication techniques in IIoT. Furthermore, the proposed key establishment techniques in IIoT are discussed and reviewed in Sect. 4. In Sect. 5, we explain the necessity of using real testbeds for security research evaluation in IIoT. Finally, we conclude the chapter in Sect. 6.

2 Intrusion Detection in IIoT

A network intrusion is described as an attempt to damage the confidentiality, integrity, or availability of the host and network [11]. It is often assumed as one of the most common threats in cyberspace.

To overcome any intrusion attack and avoid subsequent damages in any system, the first step is to detect them on time. Then, appropriate actions can be taken to thwart the attack. Here, our main focus is on the first step, i.e. the timely and efficient detection of intrusions in industrial IoT [12]. In general, the part of the system that is responsible to do this work is often referred to as the intrusion detection system (IDS). More precisely, an IDS is defined as a network security device that

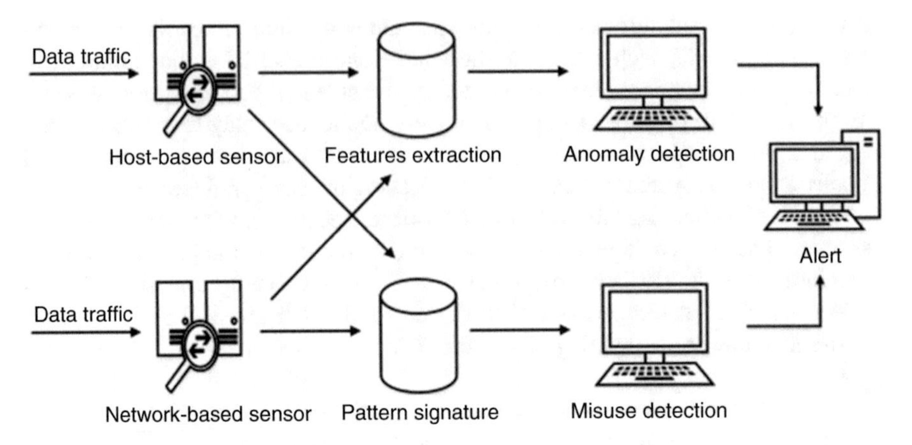

Fig. 2 An illustration of IDS operations can be divided into the monitoring, analysis, and detection stages [15]

monitors network traffic to discover suspicious transmissions on time so that taking suitable proactive actions against the existing intrusion could be possible [13, 14].

Figure 2 illustrates the existing operations in a general IDS, including the monitoring, analysis, and detection stages. As it can be seen, the first stage relies on network-based or host-based sensors. Then, the second stage relies on feature extraction methods or pattern identification methods. Finally, the last stage relies on anomaly or misuse intrusion detection [15].

An IDS is often modeled as a binary classification problem that aims to discriminate the normal network traffic behavior from anomalous. It is also sometimes implemented as a multiclass classification model that determines the type of network attack as well [16].

In the following, we explain two of the very recent works for intrusion detection in IIoT.

Li et al. [17] propose both a method for processing one-dimensional weakly three correlated feature data and a deep learning approach for intrusion detection using a multi-convolutional neural network (multi-CNN) fusion method. The authors believe that the processed data have a better training result for deep learning. In this respect, they first apply their processing method on the benchmark NSL-KDD dataset provided by [18]. According to the correlation, the feature data are divided into four parts. Then the one-dimensional feature data are converted into a grayscale graph. Next, they use it to propose their multi-CNN fusion algorithm for intrusion detection and the best of the four results emerge.

In [19] a novel federated deep learning scheme is proposed to detect cyber threats against industrial cyber-physical systems (CPSs). Namely, they first design a deep learning-based intrusion detection model for industrial CPSs, by making use of a convolutional neural network and a gated recurrent unit. Then, they develop a federated learning framework to build a comprehensive intrusion detection model by using data from multiple industrial CPSs. This framework also provides privacy preservation of data resources by allowing data processing at each industrial CPS's premise. Finally, a Paillier cryptosystem based secure communication protocol is built for the developed federated learning framework so that the security and privacy of model parameters through the training process is preserved.

However, despite the current significant advances in the field of intrusion detection in IIoT, there still exists a big room for further improvements in developing efficient IDSs in IIoT. For example, IIoT systems usually consist of a large set of devices that are distributed in the industrial environment. Hence, there is a great need to develop novel distributed IDS suitable for IIoT [20, 21]. Specifically, distributed IDS can overwhelm the limited computing resources of most IIoT devices which decreases the efficiency of centralized IDSs. Another important issue for future research is how to optimize intrusion detection algorithms automatically during running. Since IIoT systems often need to work continuously and the system parameters (e.g. steady system states, security requirements, and system constrains) may change over time. Therefore, the intrusion detection algorithm needs to be optimized automatically concerning the undergoing changes to maintain a satisfying detection accuracy [22].

3 Authentication Techniques

In a real IIoT system, often a massive amount of data is generated, processed, and exchanged by a variety of IoT devices. Each IoT device has a unique identity and is capable of sharing its sensed or processed data with other communicated devices. However, in all data communications, each IoT device should be assured that the identity of the other party is the one that it claims to be [23]. In other words, besides secure data acquisition and communication in an IIoT system, it should be also assured that the data is always accessed only by authenticated (or authorized) parties.

Therefore, an authentication process is needed before any kind of communication is established between two IoT devices. In this respect, in a two-party communication, authentication is used by an IoT device that shares its data when it needs to know exactly who is accessing its information. It is also used by the IoT device which receives the data to be sure that the data is sent by the device that must be sent. In other words, both sides need to prove their identity to the other side [24, 25].

Hence, as it can be seen, authentication can be considered as the first line of defense among various security mechanisms. It involves the basis of access control to sensitive data in industrial environments. Each user shall be first verified before being allowed to access the application data. Hence, whenever an IoT device intends to log in to a remote service provider (which may be a powerful server or a lightweight sensor node) and access the desired data/services, both the user and the service provider must validate the authenticity of the corresponding party by the acquirement of corroborative evidence [26].

To provide mutual authentication in device-to-device communications in IIoT, a great number of authentication schemes have been proposed. In the following, we review some of the salient works in this area.

Esfahani et al. [25] believe that conventional authentication schemes that have been proposed to ensure security in IoT or Machine-to-Machine (M2M) applications, such as [27–31], cannot be directly employed in IIoT. Because manufacturing machines are often limited by computation power and/or communication bandwidth. So, authentication techniques with high computing and communication resource requirements are not applicable in industrial IoT environments.

In this regard, they propose a lightweight authentication scheme to authenticate these resource-constrained types of machinery in [25] to address the above issue. More specifically, they propose a lightweight authentication mechanism that is based only on hash and XOR operations, for M2M communications in the industrial IoT environment. By M2M they mean a resource-constrained industrial device (e.g., smart sensor) including a Secure Element (SE) and a router including a Trusted Platform Module (TPM). The proposed mechanism is inspired by [32] and includes two procedures; the registration procedure and the authentication procedure.

In the registration procedure, the sensor is registered to the Authentication Server (AS) which is performed as follows. First, the sensor sends its unique ID through a secure channel to the AS. Then, the AS calculates three secret authentication parameters for the sensor and sends two of them back. These parameters are saved in the SE of the sensor. Then the registration process is complete.

In the authentication procedure, mutual authentication between the sensor and the router is achieved. It should be noted that the sensor never uses its real identity for authentication. This way, the smart sensor's ID cannot be eavesdropped by a malicious entity during the authentication process. The authentication procedure consists of the following steps. The smart sensor first generates a random number and then computes several parameters based on the generated random number and its ID. Then it sends an authentication request to the router based on the produced parameters. Once the authentication request is received by the router, it first validates the obtained message using the pre-shared key. If the message is invalid, then it is rejected by the router. However, if the message is valid, the router sends back the authentication response to the sensor. The router also computes the session key. When the sensor received the response message, it calculates the session key using the parameters in the message and then sends a message to the router. Upon receiving this final message by the router, it uses the message to verify that the smart sensor holds the legitimate session key or not. Once the verification is complete and it is proved that the session key is valid, the authentication process is finished successfully.

In their work, they have also shown that the proposed mechanism involves low computational cost and low communication and storage overhead. This is while it can achieve mutual authentication, session key agreement, and device's identity confidentiality. Moreover, the proposed mechanism is resistant against the replay, man-in-the-middle, impersonation, and modification attacks. However, the drawback of their work is that the proposed authentication technique is only applicable between a machine and a network element and thereby it cannot be generalized to other types of machines.

Zhu et al. [33] have used a directed graph structure to represent an IIoT control system. Because the existing relationships among different devices can be simply represented by the directed graph. Then, they have developed a dynamically updatable privacy-preserving authentication scheme for the general directed graphs.

It should be mentioned that several works in the literature have developed authentication schemes for directed graphs. However, they have often assumed a specific type of graph in their work and have developed their proposed authentication scheme based on that structure. For example, [34–36] have proposed authentication techniques for trees, while several other works have developed authentication schemes for directed acyclic graphs and directed cyclic graphs such as [37, 38].

Therefore, the main novelty of the proposed authentication scheme in [33] compared to the previous works in the literature is that it considers general directed graphs and not a specific type of graph. In this respect, the authors in [33] propose an authentication scheme called privacy-preserving authentication for general directed graphs (PPAG). The proposed PPAG scheme is based on the cryptographic accumulator and an underlying standard digital signature scheme. For the sake of its security, the input domain for the accumulator must be prime numbers. Thereby, to authenticate arbitrary numbers via this accumulator, a special hash function is used to transform arbitrary numbers into prime numbers. The proposed scheme supports the addition of nodes/edges in the graph, and thus the graph is dynamically updatable. Furthermore, a feasible solution for nodes/edge deletion operations is provided. The security of PPAG under the adaptive chosen-message attacks is proved. The proposed scheme can also protect the privacy of nodes and edges from leakage.

In [39] the existing security and privacy challenges in device-to-device communications have been addressed. Because two different devices that belong to two different administrative domains may have to collaborate to complete the same task. The authors claim that the previously proposed authentication approaches in the literature may result in heavy key management overhead or rely on a trusted third party. Hence, they have tried to overcome this problem by proposing an efficient blockchain-assisted secure device authentication mechanism BASA for cross-domain industrial IoT. Precisely, consortium blockchain is introduced to construct trust among different domains by enabling devices in different domains to authenticate each other. In this respect, an identity-based signature is exploited during the authentication process. Moreover, an identity management mechanism is designed to preserve the privacy of devices. This mechanism can realize that devices being authenticated remain anonymous. Furthermore, session keys between two parties are negotiated, which can secure the subsequent communications.

4 Key Establishment Techniques

In this section, we review the proposed key establishment protocols for industrial IoT in the literature. First, we review the conventional key establishment protocols which are designed at higher layers and the physical layer. Then, we review proposed cross-layer key establishment protocols which have been proposed to address the shortcomings of conventional key establishment protocols.

Figure 3 illustrates an overview of the surveyed key establishment protocols. In the following, we explain the fundamentals of each category, besides reviewing the proposed methods in each category.

4.1 *Key Establishment Protocols at Higher Layers*

At higher layers, the key establishment protocols are designed in either symmetric or asymmetric key settings [7]. In the following, we review symmetric and asymmetric key establishment protocols respectively.

4.1.1 Symmetric Key Establishment Protocols

Symmetric key establishment protocols are considered lightweight protocols since the IIoT devices can establish the communication keys with less energy consumption compared to asymmetric key establishment protocols. Different symmetric key establishment protocols have been proposed in the literature, including entity-based, probabilistic-based, polynomial-based, and matrix-based protocols [7, 40].

In entity-based key establishment protocols, such as in [41, 42], a trusted third-party is in charge of establishing communication keys. However, a critical issue in these methods is the privacy of the master key. Since, once the master key is discovered by an adversary, the security and privacy of the IIoT system will be in danger. Hence, [42] has tried to overcome this problem by enforcing IIoT devices to delete the master key after establishing the communication keys.

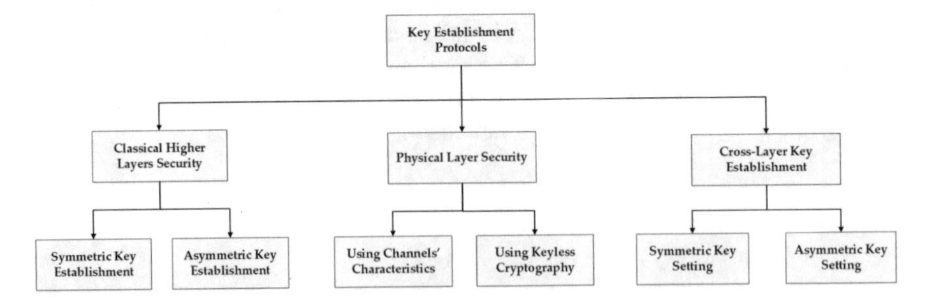

Fig. 3 An overview of key establishment protocols ADDIN

A probabilistic key distribution protocol, such as in [43, 44], often consists of three phases, including the key pre-distribution phase, the shared key discovery phase, and the path key establishment phase. In the key pre-distribution phase, a key ring is generated based on the selected random keys for each IIoT device. Then, the key ring, as well as the keys' identifiers, are loaded into the device's memory. In the next phase, the IIoT devices broadcast the loaded key identifiers such that they can find the shared keys with their neighbor devices. This way, neighbor IIoT devices can establish a communication key using the shared keys. Finally, in the last phase, any two IIoT devices can establish a path key using these devices, located at two different links between them.

Liu et al. [45] develop a general framework for establishing pairwise keys between sensor nodes using bivariate polynomials. They propose two efficient instantiations of the general framework, namely a random subset assignment key pre-distribution scheme, and a hypercube-based key pre-distribution scheme. They also present an optimization technique for polynomial evaluation which is used to compute pairwise keys.

Finally, in [46, 47], a matrix-based key distribution protocol is developed based on [48]. The basic idea of the proposed protocol in [48] is that any two IIoT devices can directly establish a communication key as long as they pre-load secrets from the same matrix space.

In [46] a new design of matrix G [48] is provided. Their proposed protocol exhibits a threshold property which ensures that when the number of compromised nodes is less than the threshold, the probability that communications between any additional nodes are compromised is close to zero. This way the adversary will have to attack a large fraction of the network before it can achieve any significant gain.

In [47], the proposed protocol in [46] is improved using deployment knowledge and avoiding unnecessary key assignments. There are also several other matrix-based key distribution protocols in the literature such as [49, 50].

4.1.2 Asymmetric Key Establishment Protocols

Asymmetric key establishment protocols also have been used widely by researchers. For example, Two-Party Password Authenticated Key Exchange (2PAKE) protocols facilitate any two IIoT devices to establish a communication key using a pre-shared short password [7].

The 2PAKE protocol was initially proposed in [51] in which it is assumed that some public parameters are generated by the system authority and made accessible to IIoT devices. Then, a communication key can be established between two devices by running the improved Cramer-Shoup encryption algorithm.

Inspired by [51], several 2PAKE protocols have been proposed in the literature such as [52, 53]. Furthermore, several Password Authenticated Group Key Exchange (GPAKE) protocols have been proposed, extending the idea of 2PAKE protocols [54–56].

4.2 Key Establishment Protocols at the Physical Layer

At the physical layer, several key establishment protocols have been developed using either the characteristics of the channels or keyless cryptography [7]. In the following, we review the proposed protocols using each of these methods.

4.2.1 Key Establishment Protocols Using channel's Characteristics

In the real world, each wireless channel between two IIoT devices undergoes time-varying and stochastic fading between the exchanged wireless signals. The fading has the property that it is invariant within the channel coherence time no matter what the direction of the signals are in the channel [7]. Moreover, the channel coherence time is a statistical measurement of time duration over which the channel impulse response is essentially invariant. Finally, the fading decorrelates over the distances of the order of half a wavelength. Hence, if the adversary is at a longer distance than this from the participants, it will be unable to extract any useful information [7].

Existing physical layer key extraction protocols based on channel characteristics generally consist of three phases, including quantization phase, reconciliation, and privacy amplification phases respectively. In the first phase, the IIoT devices sample the transmitted signals and then quantize them using predefined thresholds. At the end of this phase, the two IIoT devices obtain two initial binary bit sequences, which may demonstrate minor differences due to the effect of imperfect reciprocity and noise. In the following phases, these devices can remove the mismatch bits and make the bit sequences have sufficient entropy [7].

In this respect, several key establishment protocols at the physical layer [57] have been proposed in the literature using the characteristics of the channel such as [58–61].

In [58], an information-theoretic secret key generation (SKG) method for time division duplexing (TDD)-based orthogonal frequency-division multiplexing (OFDM) systems over multipath fading channels is proposed. The proposed SKG method is based on the physical layer properties of the wireless medium, which aims to maximize the number of secret bits given a target secret key disagreement ratio (SKDR).

Premnath et al. [59] evaluate the effectiveness of secret key extraction using the received signal strength (RSS) variations on the wireless channel between two devices. We use real world measurements of RSS in a variety of environments and settings. Based on their obtained results from real world experiments, in certain environments the extracted bits have very low entropy which make these bits unsuitable for a secret key. Indeed, an adversary can cause predictable key generation in static environments. In contrast, as the mobility of the devices increases, the existing entropy in the extracted bits grows quickly. In this respect, the authors develop an environment adaptive secret key generation scheme that uses an adaptive lossy quantizer in conjunction with Cascade-based information reconciliation and privacy amplification.

In [60] the security of a wireless channel between any two vehicles is addressed. The authors propose a scheme for message encryption by allowing two cars to extract a shared secret from RSSI (Received Signal Strength Indicator) values. The generated key is secure against an adversary with unlimited computing power, i.e., it is information-theoretically secure. Moreover, the proposed scheme can be used in the noisy vehicular environments. The unique channel conditions in vehicular networks make existing solutions of key extraction in the indoor or low-speed environments inapplicable in vehicular environments. Finally, the authors propose an online parameter learning mechanism to adapt to different channel conditions.

4.2.2 Using Keyless Cryptography

Keyless cryptography techniques at the physical layer also have been employed by the researchers in their proposed key establishment protocols. These protocols use the characteristics of anonymous channels, i.e. those wireless channels that the source of the transmitted signals remain unidentifiable by the adversary [7].

In [62], an over-the-air key establishment protocol has been designed without using asymmetric key cryptography and pre-shared secrets. The proposed protocol enables two wireless devices to establish a secret key by directly sending random signals to each other. The proposed scheme for key establishment is low cost in terms of both energy consumption and execution time.

Pietro et al. [63], present a probabilistic protocol to allow two wireless devices to commit over-the-air on a shared secret, even in the presence of a globally eavesdropping adversary. The proposed scheme is only based on plaintext messages exchange and thus does not leverage any crypto.

4.3 Cross-Layer Key Establishment Protocols

The proposed key establishment protocols at higher layers generally assume that passwords are pre-shared among IIoT devices. However, in the real world, IIoT devices are often produced by different factories and thus it is impractical to assume that these devices are released from the factory with pre-loaded passwords. Furthermore, it has been shown that devices can extract secrets using the wireless fading channel [64]. The proposed key establishment protocols at the physical layer however do not require pre-loaded secrets in IIoT devices, but they suffer from a slow key extraction rate [7].

Hence, cross-layer key establishment protocols have gained attention in recent years to address the above problems. These protocols employ the characteristics of both higher layers and the physical layer. In the following we review the proposed cross-layer key establishment protocols in the literature which are designed in either asymmetric or symmetric key settings [7].

4.3.1 Cross-Layer Key Establishment Protocols Based on Asymmetric Key Setting

In this section, we review the proposed cross-layer key establishment protocols that have been proposed in the literature based on the asymmetric key setting.

In [64] a variant of password-authenticated key exchange (vPAKE) protocol has been proposed in which the password sharing assumption has been evoked. The proposed protocol is indeed a cross-layer design in which two IIoT devices extract short secrets at the physical layer to shorten the key extraction time. The extracted short secrets are used as passwords by the users to establish a communication key at higher layers.

In some scenarios, data sharing may be required among a group of devices over the public and unreliable networks. In this respect, in [65] a group password-authenticated group key exchange (GPAKE) protocol has been proposed without the password sharing assumption. Similar to the above work, first wireless devices are used to extract short secrets at the physical layer. However, then the users establish a group key at higher layers using the extracted secrets. The proposed protocol is a cross-layer design, which is also a compiler. Indeed, the proposed protocol can transform any provably secure 2PAKE protocol into a GPAKE protocol with only one more round of communications.

4.3.2 Cross-Layer Key Establishment Protocols Based on Symmetric Key Setting

In this section, we review the proposed cross-layer key establishment protocols that have been proposed in the literature based on the symmetric key setting.

In smart homes scenarios, IIoT devices send their collected data to the home gateway. Then, the home gateway sends back suitable commands to the devices based on the analyzed data. Clearly, the data and commands are transmitted through wireless links making them vulnerable to different cyber-attacks such as eavesdropping. Hence, in order to address the security and privacy of smart home networks, Zhang et al. [66] propose a matrix-based cross-layer key establishment protocol without using any pre-shared secrets. More precisely, based on their proposed scheme, home IIoT appliances extract master keys at the physical layer using the wireless fading channels. Then, the home gateway distributes key seeds for the devices using the extracted master keys. This way, any two IIoT devices can directly establish a secret session key at higher layers. The proposed protocol consists of four phases, including the initialization, the master key extraction, the key seed distribution, and the session key establishment phases.

The problem of key establishment among smart home devices also has been addressed in [67]. In their proposed model, smart home devices first extract short random keys at the physical layer. Then, the Merkle puzzle is employed to establish secret communication keys at higher layers. In their proposed model, secret keys can be established without the secret sharing assumption or existing of an online trusted third party.

In [68], a cross-layer key establishment model for heterogeneous wireless devices in the Cyber-Physical Systems (CPS) is presented. In the proposed model, wireless devices extract master keys at the physical layer using ambient wireless signals. These master keys are shared with the system authority. Then, the authority use these master keys to distribute secrets for devices. To do so, the model converts an existing symmetric key establishment protocol into cross-layer key establishment protocols. This way, the IIoT devices will be able to establish communication keys at higher layers by calling the employed key establishment protocol and without the secret sharing assumption.

5 Real IIoT Security Testbeds

In this section, we discuss the necessity of developing and employing real IIoT testbeds for evaluating new security mechanisms before they are actually used on a real infrastructure.

As mentioned earlier, conventional industrial protocols often lack sufficient security mechanisms with the assumption that anyone on the local network is trusted. Moreover, many legacy devices in many IIoT infrastructures exhibit a long time life cycle. IIoT environments often include a heterogeneous set of these devices which are designed without suitable security mechanisms in the first place. Hence, all these issues set hurdles for fast, riskless, and cost-efficient setup of new protocols [69].

More precisely, many existing legacy industrial systems such as Programmable Logic Controller (PLC), Remote Terminal Unit (RTU), Supervisory Control and Data Acquisition (SCADA), Input/Output (I/O) devices, etc., interoperate with IoT technologies. This is while these systems, as well as many other IoT devices used in energy, water, buildings, roads, and factories, have been initially designed to provide a long time service without any concerns about their connectivity and security requirements in the future. Hence, replacing them with new gadgets and devices designed to be secure from scratch brings new challenges in terms of both technical and economic issues [70].

Moreover, the current security solutions are not often applicable in the context of IIoT because they are usually IT-centric and do not take into account a system's safety, resilience, and reliability [70].

Hence, as we thoroughly discussed in this chapter, novel security solutions are needed to be designed according to the IIoT system's requirements to fill the existing security-gap. In this respect, many researchers have addressed the existing security challenges in IIoT in recent years [71, 72]. In the previous sections, we reviewed some of these works which have been proposed for intrusion detection, authentication, or key establishment.

However, these works are often restricted by a lack of realistic data about a system's communications and activities, as well as potential cyber-attacks to evaluate their proposed model [70]. This is while it is often extremely difficult to obtain such data from an actual environment due to security and privacy reasons. Therefore, real testbeds can be employed to obtain a much richer dataset with further details compared to those obtained from simulations and emulations. Obliviously, the better the dataset, the more verifiable results are expected [69].

Further motivations can also exist to construct a testbed for security research in IIoT. For example, there may be cases where discovering knowledge on a particular subject is not feasible through simulations, or they could be used for a much better applicability demonstration of research findings, or they could be Leveraged to educate people such as students, researchers, stakeholders.

Hence, it seems essential to develop real IIoT testbeds that can be used for credibility measurement of proposed security solutions, analyzing IIoT attack landscapes, and extracting threat intelligence, before they are used on a real IIoT infrastructure [69, 70].

Although real testbeds may require considerable financial, time, and human resources to set up and maintain, they provide many advantages. For example, academics can employ such testbeds to (1) understand state-of-the-art industrial protocols and devices, (2) perform experiments on various security attacks leveraging physical processes, and (3) develop novel security solutions, while testing them with actual attacks [69].

However, it should be noted that hybrid models of IIoT testbeds can also be helpful in some scenarios. By hybrid testbeds we mean that several components of the testbed can be implemented physically, while other components are emulated or simulated depending on the requirements of the system and the application of security research. Clearly, cyber-physical testbeds with real devices and processes will require more effort, time, and money to reconfigure and adapt to different settings, while fully simulated testbeds are more adaptable to changes [69]. Hence, a tradeoff is needed with respect to all these issues and the requirements of the application.

In this regard, many funding agencies have supported deploying real testbeds to proceed researches in the venue of IIoT security [70, 73–75]. Since the implementation of such testbeds often present significant investments and efforts, several studies have discussed the IIoT testbed implementation process as a guide through generic, accurate, inexpensive, and easy deployment of IIoT testbeds [69, 70].

6 Conclusion

In this chapter, we discussed the necessity of security concepts in industrial IoT environments as different cyber-attacks may happen on different surfaces. We explained that efficient mechanisms are required to assure security at different abstraction layers. For example, devices should be authenticated before any data transmission happens. Also, key establishment protocols are needed for message encryption. Beside all these employed techniques to provide a secure IIoT environment, efficient mechanisms are also required to detect any anomalies in the IIoT environment, such as intrusion detection systems.

In this respect, we reviewed the existing works in the literature which have developed novel authentication techniques, key establishment techniques, and intrusion detection techniques. Each of these techniques ae used to address security in a different abstraction layer of the IIoT system.

In this chapter, we also discussed the necessity of developing real IIoT testbeds for a proper evaluation of novel security mechanisms before they are actually used on a real-world infrastructure.

References

1. S. Yousefi, F. Derakhshan, and H. Karimipour, "Applications of big data analytics and machine learning in the internet of things," in *Handbook of Big Data Privacy*: Springer, 2020, pp. 77–108.
2. H. HaddadPajouh, A. Dehghantanha, R. M. Parizi, M. Aledhari, and H. Karimipour, "A survey on internet of things security: Requirements, challenges, and solutions," *Internet of Things,* p. 100129, 2019.
3. S. M. Tahsien, H. Karimipour, and P. Spachos, "Machine learning based solutions for security of Internet of Things (IoT): A survey," *Journal of Network and Computer Applications,* vol. 161, p. 102630, 2020.
4. Z. Bakhshi, A. Balador, and J. Mustafa, "Industrial IoT security threats and concerns by considering Cisco and Microsoft IoT reference models," in *2018 IEEE Wireless Communications and Networking Conference Workshops (WCNCW)*, 2018: IEEE, pp. 173–178.
5. A. Al-Abassi, H. Karimipour, H. HaddadPajouh, A. Dehghantanha, and R. M. Parizi, "Industrial big data analytics: challenges and opportunities," in *Handbook of Big Data Privacy*: Springer, 2020, pp. 37–61.
6. A.-R. Sadeghi, C. Wachsmann, and M. Waidner, "Security and privacy challenges in industrial internet of things," in *2015 52nd ACM/EDAC/IEEE Design Automation Conference (DAC)*, 2015: IEEE, pp. 1–6.
7. Y. Zhang and X. Huang, "Security and Privacy Techniques for the Industrial Internet of Things," in *Security and Privacy Trends in the Industrial Internet of Things*: Springer, 2019, pp. 245–268.
8. H. Haddadpajouh, A. Mohtadi, A. Dehghantanaha, H. Karimipour, X. Lin, and K.-K. R. Choo, "A Multi-Kernel and Meta-heuristic Feature Selection Approach for IoT Malware Threat Hunting in the Edge Layer," *IEEE Internet of Things Journal*, 2020.
9. J. Sakhnini, H. Karimipour, A. Dehghantanha, R. M. Parizi, and G. Srivastava, "Security aspects of Internet of Things aided smart grids: A bibliometric survey," *Internet of things,* p. 100111, 2019.

10. M. Begli, F. Derakhshan, and H. Karimipour, “A layered intrusion detection system for critical infrastructure using machine learning,” in *2019 IEEE 7th International Conference on Smart Energy Grid Engineering (SEGE)*, 2019: IEEE, pp. 120–124.
11. A. L. Buczak and E. Guven, “A survey of data mining and machine learning methods for cyber security intrusion detection,” *IEEE Communications surveys & tutorials,* vol. 18, no. 2, pp. 1153–1176, 2015.
12. S. Mohammadi, H. Mirvaziri, M. Ghazizadeh-Ahsaee, and H. Karimipour, “Cyber intrusion detection by combined feature selection algorithm,” *Journal of information security and applications,* vol. 44, pp. 80–88, 2019.
13. R. Samrin and D. Vasumathi, “Review on anomaly based network intrusion detection system,” in *2017 International Conference on Electrical, Electronics, Communication, Computer, and Optimization Techniques (ICEECCOT)*, 2017: IEEE, pp. 141–147.
14. A. Al-Abassi, J. Sakhnini, and H. Karimipour, “Unsupervised Stacked Autoencoders for Anomaly Detection on Smart Cyber-physical Grids,” in *2020 IEEE International Conference on Systems, Man, and Cybernetics (SMC)*, 2020: IEEE, pp. 3123–3129.
15. M. F. Elrawy, A. I. Awad, and H. F. Hamed, “Intrusion detection systems for IoT-based smart environments: a survey,” *Journal of Cloud Computing,* vol. 7, no. 1, p. 21, 2018.
16. S. Naseer *et al.*, “Enhanced network anomaly detection based on deep neural networks,” *IEEE Access,* vol. 6, pp. 48231–48246, 2018.
17. Y. Li *et al.*, “Robust detection for network intrusion of industrial IoT based on multi-CNN fusion,” *Measurement,* vol. 154, p. 107450, 2020.
18. M. Tavallaee, E. Bagheri, W. Lu, and A. A. Ghorbani, “A detailed analysis of the KDD CUP 99 data set,” in *2009 IEEE symposium on computational intelligence for security and defense applications*, 2009: IEEE, pp. 1–6.
19. B. Li, Y. Wu, J. Song, R. Lu, T. Li, and L. Zhao, “DeepFed: Federated deep learning for intrusion detection in industrial cyber-physical systems,” *IEEE Transactions on Industrial Informatics,* 2020.
20. A. Al-Abassi, H. Karimipour, A. Dehghantanha, and R. M. Parizi, “An ensemble deep learning-based cyber-attack detection in industrial control system,” *IEEE Access,* vol. 8, pp. 83965–83973, 2020.
21. S. Singh, H. Karimipour, H. HaddadPajouh, and A. Dehghantanha, “Artificial intelligence and security of industrial control systems,” *Handbook of Big Data Privacy*, pp. 121–164, 2020.
22. Y. Hu, A. Yang, H. Li, Y. Sun, and L. Sun, “A survey of intrusion detection on industrial control systems,” *International Journal of Distributed Sensor Networks,* vol. 14, no. 8, p. 1550147718794615, 2018.
23. S. Yousefi, F. Derakhshan, H. S. Aghdasi, and H. Karimipour, “An energy-efficient artificial bee colony-based clustering in the internet of things,” *Computers & Electrical Engineering,* vol. 86, p. 106733, 2020.
24. K. Renuka, S. Kumari, D. Zhao, and L. Li, “Design of a secure password-based authentication scheme for M2M networks in IoT enabled cyber-physical systems,” *IEEE Access,* vol. 7, pp. 51014–51027, 2019.
25. A. Esfahani *et al.*, “A lightweight authentication mechanism for M2M communications in industrial IoT environment,” *IEEE Internet of Things Journal,* vol. 6, no. 1, pp. 288–296, 2017.
26. D. Wang, H. Cheng, D. He, and P. Wang, “On the challenges in designing identity-based privacy-preserving authentication schemes for mobile devices,” *IEEE Systems Journal,* vol. 12, no. 1, pp. 916–925, 2016.
27. J.-Y. Lee, W.-C. Lin, and Y.-H. Huang, “A lightweight authentication protocol for internet of things,” in *2014 International Symposium on Next-Generation Electronics (ISNE)*, 2014: IEEE, pp. 1–2.
28. X. Yao, X. Han, X. Du, and X. Zhou, “A lightweight multicast authentication mechanism for small scale IoT applications,” *IEEE Sensors Journal,* vol. 13, no. 10, pp. 3693–3701, 2013.

29. Y. Qiu and M. Ma, "A mutual authentication and key establishment scheme for m2m communication in 6lowpan networks," *IEEE transactions on industrial informatics,* vol. 12, no. 6, pp. 2074–2085, 2016.
30. J. L. Hernandez-Ramos, M. P. Pawlowski, A. J. Jara, A. F. Skarmeta, and L. Ladid, "Toward a lightweight authentication and authorization framework for smart objects," *IEEE Journal on Selected Areas in Communications,* vol. 33, no. 4, pp. 690–702, 2015.
31. W.-L. Chin, Y.-H. Lin, and H.-H. Chen, "A framework of machine-to-machine authentication in smart grid: a two-layer approach," *IEEE Communications Magazine,* vol. 54, no. 12, pp. 102–107, 2016.
32. M.-C. Chuang and J.-F. Lee, "TEAM: Trust-extended authentication mechanism for vehicular ad hoc networks," *IEEE systems journal,* vol. 8, no. 3, pp. 749–758, 2013.
33. F. Zhu, W. Wu, Y. Zhang, and X. Chen, "Privacy-preserving authentication for general directed graphs in industrial IoT," *Information Sciences,* vol. 502, pp. 218–228, 2019.
34. M. T. Goodrich, R. Tamassia, and N. Triandopoulos, "Efficient authenticated data structures for graph connectivity and geometric search problems," *Algorithmica,* vol. 60, no. 3, pp. 505–552, 2011.
35. A. Kundu and E. Bertino, "Structural signatures for tree data structures," *Proceedings of the VLDB Endowment,* vol. 1, no. 1, pp. 138–150, 2008.
36. C. Brzuska *et al.*, "Redactable signatures for tree-structured data: definitions and constructions," in *International Conference on Applied Cryptography and Network Security*, 2010: Springer, pp. 87–104.
37. A. Kundu and E. Bertino, "How to authenticate graphs without leaking," in *Proceedings of the 13th International Conference on Extending Database Technology*, 2010, pp. 609–620.
38. A. Kundu and E. Bertino, "Privacy-preserving authentication of trees and graphs," *International journal of information security,* vol. 12, no. 6, pp. 467–494, 2013.
39. M. Shen *et al.*, "Blockchain-assisted secure device authentication for cross-domain industrial IoT," *IEEE Journal on Selected Areas in Communications,* vol. 38, no. 5, pp. 942–954, 2020.
40. M. F. Moghadam, A. Mohajerzdeh, H. Karimipour, H. Chitsaz, R. Karimi, and B. Molavi, "A privacy protection key agreement protocol based on ECC for smart grid," in *Handbook of Big Data Privacy*: Springer, 2020, pp. 63–76.
41. B. Lai, S. Kim, and I. Verbauwhede, "Scalable session key construction protocol for wireless sensor networks," in *IEEE Workshop on Large Scale RealTime and Embedded Systems (LARTES)*, 2002, vol. 7: Citeseer.
42. S. Zhu, S. Setia, and S. Jajodia, "LEAP+ Efficient security mechanisms for large-scale distributed sensor networks," *ACM Transactions on Sensor Networks (TOSN),* vol. 2, no. 4, pp. 500–528, 2006.
43. H. Chan, A. Perrig, and D. Song, "Random key predistribution schemes for sensor networks," in *2003 Symposium on Security and Privacy, 2003.* IEEE, pp. 197–213.
44. L. Eschenauer and V. D. Gligor, "A key-management scheme for distributed sensor networks," in *Proceedings of the 9th ACM conference on Computer and communications security*, 2002, pp. 41–47.
45. D. Liu, P. Ning, and R. Li, "Establishing pairwise keys in distributed sensor networks," *ACM Transactions on Information and System Security (TISSEC),* vol. 8, no. 1, pp. 41–77, 2005.
46. W. Du, J. Deng, Y. S. Han, P. K. Varshney, J. Katz, and A. Khalili, "A pairwise key predistribution scheme for wireless sensor networks," *ACM Transactions on Information and System Security (TISSEC),* vol. 8, no. 2, pp. 228–258, 2005.
47. W. Du, J. Deng, Y. S. Han, and P. K. Varshney, "A key predistribution scheme for sensor networks using deployment knowledge," *IEEE Transactions on dependable and secure computing,* vol. 3, no. 1, pp. 62–77, 2006.
48. R. Blom, "An optimal class of symmetric key generation systems," in *Workshop on the Theory and Application of of Cryptographic Techniques*, 1984: Springer, pp. 335–338.
49. L. Xu and Y. Zhang, "Matrix-based pairwise key establishment for wireless mesh networks," *Future Generation Computer Systems,* vol. 30, pp. 140–145, 2014.

50. Y. Zhang, L. Xu, X. Huang, and J. Li, 'Matrix-based key pre-distribution schemes in WMNs using pre and post deployment knowledge," *International Journal of Ad Hoc and Ubiquitous Computing,* vol. 20, no. 4, pp. 262–273, 2015.
51. J. Katz, R. Ostrovsky, and M. Yung, "Efficient password-authenticated key exchange using human-memorable passwords," in *International Conference on the Theory and Applications of Cryptographic Techniques*, 2001: Springer, pp. 475–494.
52. S. Jiang and G. Gong, "Password based key exchange with mutual authentication," in *International Workshop on Selected Areas in Cryptography*, 2004: Springer, pp. 267–279.
53. J. Katz and V. Vaikuntanathan, "Round-optimal password-based authenticated key exchange," in *Theory of Cryptography Conference*, 2011: Springer, pp. 293–310.
54. M. Abdalla, J.-M. Bohli, M. I. G. Vasco, and R. Steinwandt, "(Password) authenticated key establishment: from 2-party to group," in *Theory of Cryptography Conference*, 2007: Springer, pp. 499–514.
55. M. Abdalla, C. Chevalier, L. Granboulan, and D. Pointcheval, "Contributory password-authenticated group key exchange with join capability," in *Cryptographers' Track at the RSA Conference*, 2011: Springer, pp. 142–160.
56. Q. Tang and K.-K. R. Choo, "Secure password-based authenticated group key agreement for data-sharing peer-to-peer networks," in *International Conference on Applied Cryptography and Network Security*, 2006: Springer, pp. 162–177.
57. J. M. Hamamreh, H. M. Furqan, and H. Arslan, "Classifications and applications of physical layer security techniques for confidentiality: A comprehensive survey," *IEEE Communications Surveys & Tutorials,* vol. 21, no. 2, pp. 1773–1828, 2018.
58. Y. Peng, P. Wang, W. Xiang, and Y. Li, "Secret key generation based on estimated channel state information for TDD-OFDM systems over fading channels," *IEEE Transactions on Wireless Communications,* vol. 16, no. 8, pp. 5176–5186, 2017.
59. S. N. Premnath *et al.*, "Secret key extraction from wireless signal strength in real environments," *IEEE Transactions on mobile Computing,* vol. 12, no. 5, pp. 917–930, 2012.
60. X. Zhu, F. Xu, E. Novak, C. C. Tan, Q. Li, and G. Chen, "Using wireless link dynamics to extract a secret key in vehicular scenarios," *IEEE Transactions on Mobile Computing,* vol. 16, no. 7, pp. 2065–2078, 2016.
61. S. Jana, S. N. Premnath, M. Clark, S. K. Kasera, N. Patwari, and S. V. Krishnamurthy, "On the effectiveness of secret key extraction from wireless signal strength in real environments," in *Proceedings of the 15th annual international conference on Mobile computing and networking*, 2009, pp. 321–332.
62. Y. Zhang, Y. Xiang, T. Wang, W. Wu, and J. Shen, "An over-the-air key establishment protocol using keyless cryptography," *Future Generation Computer Systems,* vol. 79, pp. 284–294, 2018.
63. R. Di Pietro and G. Oligeri, "COKE crypto-less over-the-air key establishment," *IEEE transactions on information forensics and security,* vol. 8, no. 1, pp. 163–173, 2012.
64. Y. Zhang, Y. Xiang, W. Wu, and A. Alelaiwi, "A variant of password authenticated key exchange protocol," *Future Generation Computer Systems,* vol. 78, pp. 699–711, 2018.
65. Y. Zhang, Y. Xiang, and X. Huang, "Password-authenticated group key exchange: A cross-layer design," *ACM Transactions on Internet Technology (TOIT),* vol. 16, no. 4, pp. 1–20, 2016.
66. Y. Zhang, Y. Xiang, X. Huang, X. Chen, and A. Alelaiwi, "A matrix-based cross-layer key establishment protocol for smart homes," *Information Sciences,* vol. 429, pp. 390–405, 2018.
67. Y. Zhang, X. Huang, X. Chen, L. Y. Zhang, J. Zhang, and Y. Xiang, "A hybrid key agreement scheme for smart homes using the merkle puzzle," *IEEE Internet of Things Journal,* vol. 7, no. 2, pp. 1061–1071, 2019.
68. Y. Zhang, Y. Xiang, and X. Huang, "A cross-layer key establishment model for wireless devices in cyber-physical systems," in *Proceedings of the 3rd ACM Workshop on Cyber-Physical System Security*, 2017, pp. 43–53.
69. N. O. Tippenhauer, "Design and Realization of Testbeds for Security Research in the Industrial Internet of Things," in *Security and Privacy Trends in the Industrial Internet of Things*: Springer, 2019, pp. 287–310.

70. A.-H. Muna and E. Sitnikova, "Developing a Security Testbed for Industrial Internet of Things," *IEEE Internet of Things Journal*, 2020.
71. A. C. Panchal, V. M. Khadse, and P. N. Mahalle, "Security issues in IIoT: A comprehensive survey of attacks on IIoT and its countermeasures," in *2018 IEEE Global Conference on Wireless Computing and Networking (GCWCN)*, 2018: IEEE, pp. 124–130.
72. X. Yu and H. Guo, "A Survey on IIoT Security," in *2019 IEEE VTS Asia Pacific Wireless Communications Symposium (APWCS)*, 2019: IEEE, pp. 1–5.
73. A. P. Mathur and N. O. Tippenhauer, "SWaT: a water treatment testbed for research and training on ICS security," in *2016 International Workshop on Cyber-physical Systems for Smart Water Networks (CySWater)*, 2016: IEEE, pp. 31–36.
74. A. Siddiqi, N. O. Tippenhauer, D. Mashima, and B. Chen, "On practical threat scenario testing in an electric power ICS testbed," in *Proceedings of the 4th ACM Workshop on Cyber-Physical System Security*, 2018, pp. 15–21.
75. C. M. Ahmed, V. R. Palleti, and A. P. Mathur, "WADI: a water distribution testbed for research in the design of secure cyber physical systems," in *Proceedings of the 3rd International Workshop on Cyber-Physical Systems for Smart Water Networks*, 2017, pp. 25–28.

Blockchain Applications in the Industrial Internet of Things

Samad Najjar-Ghabel, Shamim Yousefi, and Hadis Karimipour

1 Introduction

Over the past few years, the fifth generation of cellular mobile communication is becoming one of the leading automation and intelligent data exchange contexts [1, 2]. It has been motivated by various communication and traffic-based factors, such as the Internet of Things (IoT), to use millions of smart sensors, embedded machines, and everyday physical devices in industrial applications [3, 4]. The Industrial Internet of Things (IIoT) is a new technology paradigm in this realm [5]. It is defined as a typical cyber-physical system for connecting physical devices, communication protocols, and internet infrastructure to automate industrial processes and ensure intelligence, performance, and safety [6–8]. To satisfy the requirements of IIoT-based systems, security represents the most significant weak point besides the energy-efficiency and real-time processing [9–11]. Since IIoT consists of resource-constrained devices and low-bandwidth channels, platform security, privacy management, identity control, and industrial rights checking must be considered throughout massive data exchange [12, 13].

To deal with privacy and security threats on IIoT, blockchain is one of the critical technologies to guarantee transaction safety, facilitate data collection, and improve

S. Najjar-Ghabel
Faculty of Electrical and Computer Engineering, University of Tabriz, Tabriz, Iran
e-mail: S.Najjar@tabrizu.ac.ir

S. Yousefi (✉)
Faculty of Electrical and Computer Engineering, University of Tabriz, Tabriz, Iran
e-mail: sh.yousefi@tabrizu.ac.ir

H. Karimipour
Department of Electrical and Software Engineering, University of Calgary, Calgary, AB, Canada
e-mail: hadis.karimipour@ucalgary.ca

H. Karimipour, F. Derakhshan (eds.), *AI-Enabled Threat Detection and Security Analysis for Industrial IoT*, https://doi.org/10.1007/978-3-030-76613-9_4

the storage process [14, 15]. Blockchain (underpins the crypto-currency Bitcoin) is a fully distributed database, formed by a set of blocks, eliminating third-party verification [14, 16, 17]. Capabilities like transparency, operational flexibility, immutability, and data encryption enable blockchain to manage big data generated by IIoT efficiently [18].

Our studies indicate that several review papers on blockchain applications in IIoT have been published [19–22]. However, we aim to prepare an up to date literature review for covering all major issues on IIoT by exploiting blockchain capabilities. The fundamental contributions of this chapter are summarized as follows:

- The literature on blockchain applications in the industrial internet of things is reviewed until the most recent articles.
- The IIoT systems, architecture, applications, and challenges are presented.
- Blockchain technology and its advantages/disadvantages to address various IIoT applications are comprehended.
- A state-of-the-art review has been provided on blockchain-based smart city, manufacturing, healthcare, energy management, supply chain/logistics, agriculture, smart homes, autonomous vehicles, and multimedia right management on the industrial Internet of things.
- Finally, possible challenges on the blockchain applications in IIoT and some recommendations or research direction are presented.

The rest of this chapter is organized as follows: Sect. 2 presents an overview of IIoT, consisting of the architecture, applications, and challenges in these systems. Section 3 describes blockchain technology and its combination with the industrial Internet of things. Blockchain applications in IIoT and their challenges have been presented in Sect. 4. Section 5 shows an analysis of reviewed literature on blockchain-based applications in the industrial Internet of things. The challenges of blockchain technology in IIoT and some recommendations have been explained in Sect. 6. Finally, the chapter is concluded in Sect. 7.

2 Industrial Internet of Things

The concepts of the industrial internet of things and Industry 4.0 (the fourth industrial revolution) have a lot in common but cannot be interchangeably placed. In 2011, Germany presented a term called "Industry 4.0 for 2020" at the Hanover event, which is now globally visible and universally accepted for utilizing the Internet-based technologies to enhance production systems' efficiency and offer intelligent services in industrial environments [23]. Industry 4.0 employs some technological enablers to introduce automated, decentralized, and dynamic production process, including the Internet (the essential technology of Industry 4.0 that a majority of the other enablers are dependent on it), big data, industrial internet of things, blockchain, artificial intelligence, robotic, edge computing, open-source software, cloud computing, and human-machine interaction [24]. An overview of Industry 4.0 enablers is illustrated in Fig. 1.

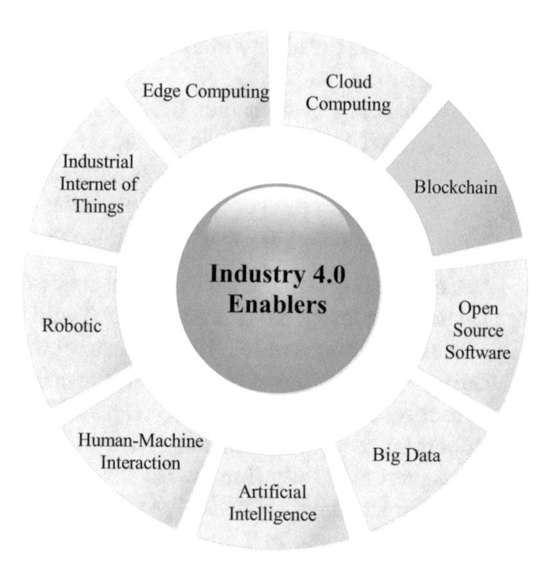

Fig. 1 Industry 4.0 enablers

Industry 4.0 aims to augment workers' capability and build a cooperative industrial environment for collaborative human-machine manufacturing infrastructures. Accordingly, it must have three central paradigms to support the human operator, confirming its centrality facing with different tasks. The paradigms are as follows [25]:

1. A smart product manages the resources and plans the manufacturing process during its lifecycle.
2. A smart machine (cyber-physical system) maps the conventional manufacturing processes to decentral, self-organizing, and flexible production lines.
3. An augmented operator adds the capability and adaptability of a human operator in smart systems.

In summary, the processes based on Industry 4.0 interconnect the industrial internet of things with the Cyber-Physical Systems (C.P.S.s[1]) to automate and dynamite industrial infrastructure and innovative services [26]. The industrial internet of things refers to the systems, machines, and users providing smart industrial operations, using intelligent data analytics for transformational business results [27]. Indeed, IIoT is defined as an IoT-based trend of a combination of current manufacturing process automation with wireless sensor networks [28–31], Internet infrastructure, and communication protocols to enhance efficiency, intelligence, productivity, quality of products, and safety in the industrial domain. To address these features, it must possess four design principles [21, 32], including:

[1] A Cyber-Physical System (CPS) is a combination of physical and software components in which its mechanisms are monitored or controlled by computer algorithms.

1. *Distribution* is the ability to independently decisions making and technical function execution.
2. *Interoperability* is defined as the possibility of connection and communication between IIoT equipment with humans.
3. *Technical assistance* consists of essential data collection and information visualization to support human-users' capabilities.
4. *Transparency* refers to the need to create virtual copies of the real world.

As a concluding remark, IIoT is a subset of IoT which monopolized by industrial applications. Industry 4.0 is where the IoT and C.P.S.s meet, originating from the IIoT. The intersections of IoT, IIoT, Industry 4.0, and C.P.S.s are illustrated in Fig. 2.

The primary concept of IIoT was introduced in the programmable logic controller project by Dick Morley in 1968 (General Motors used the programmable logic controllers in automatic transmission manufacturing) [33]. However, the current IIoT was presented following the development of cloud technology in 2002, which provides data storage for historical processes, and the emergence of the O.P.C. Unified Architecture[2] protocol in 2006, which allows for secure communications between system equipment without human intervention [34].

Nowadays, IIoT offers an incredible innovation in the technology-based industry, which tries to communicate between machines and humans, providing smart services. On the other, relying on the recent advances in multiple hardware and computer algorithms, IIoT components are equipped with sensing, recognizing, processing, computing, and communicating capabilities [35]. According to these abilities, it is a futuristic idea for industrial environments and how business is done. Thus, the vision of IIoT applications is couched in various domains, including smart city [36–38], manufacturing [39–41], healthcare 4.0 [42], energy management [43–45], supply chain/logistics [46], agriculture 4.0 [10], smart homes [47], autonomous

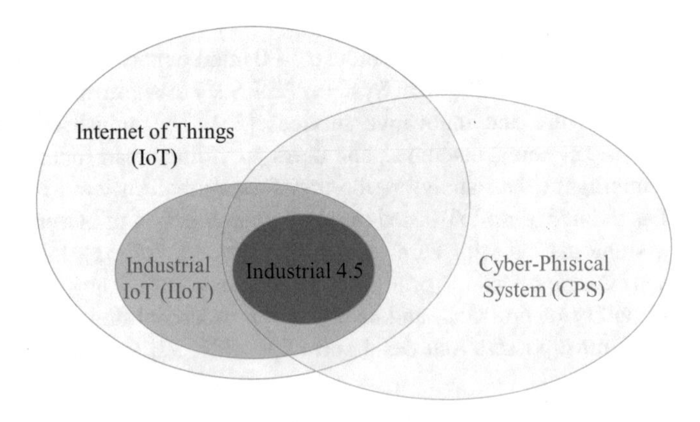

Fig. 2 Intersections of IoT, IIoT, Industry 4.0, and C.P.S.s [10]

[2] OPC Unified Architecture (OPC UA) is a machine to machine, service-oriented, and platform-independent architecture for industrial automation.

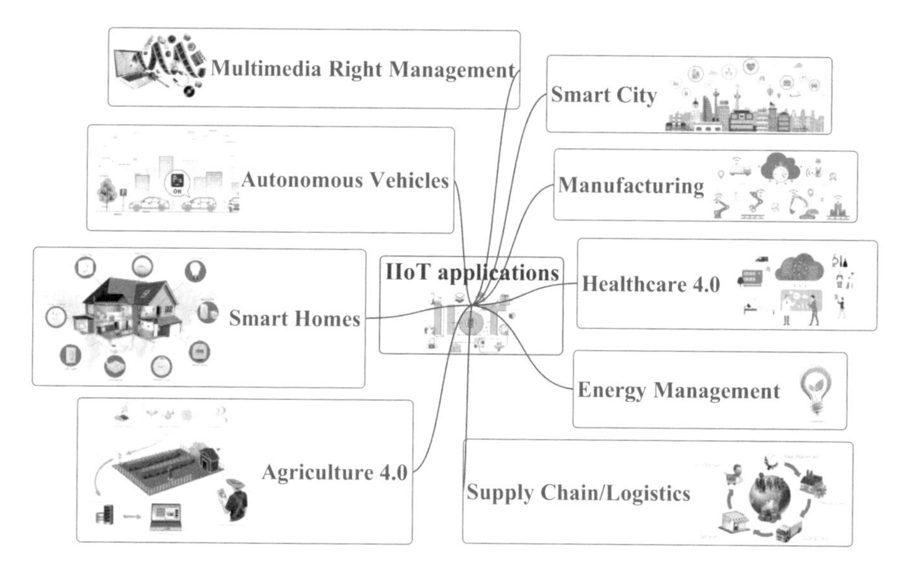

Fig. 3 IIoT applications

vehicles [48], and multimedia right management [49]. An overview of the IIoT applications is shown in Fig. 3.

Versatile applications emphasize that everything was dreaming about the smart industry in the past years, relying on IIoT services advancements. The IIoT market is estimated to catch 77.3 billion U.S. dollars in 2020 and 110.6 billion U.S. dollars by 2025 [50]. Asia Pacific Accreditation Cooperation (APAC) has been the central hub for smart manufacturing based on the growing population, rising attention to global information technology, and increasing industrial investments. In other words, APAC has become an international focal point for significant business expansions, and companies are expected to drive the growth of the IIoT market relying on the development of big data and cloud computing in this domain [50]. Thus, research and analysis on the industrial internet of things, its architecture, and challenges are located at the center of computer-since researchers' attention in the last years. The remainder of Sect. 2 is organized as follows: Sect. 2.1 presents IIoT architecture in summary and Sect. 2.2 explains the significant challenges in the field of IIoT systems.

2.1 IIoT Architecture

The first challenge that industry 4.0 faces while launching IIoT is to select an appropriate architecture [7]. Since IIoT consists of components connecting in a network-based system for data transmission, choosing the right architecture plays a critical role in various services' efficiency [51]. Given the nature of IIoT equipment and its extensive use of the software, the system architecture could be organized into a layered digital technology model according to business requirements. However, the standard architecture of IIoT is composed of three layers: edge, platform, and cloud [33, 52–54].

Edge Layer (Sensing/Perception Layer) supports the interaction between smart equipment for data collection, such as humidity, pressure, vibration, light, and chemistry. Indeed, it consists of millions of heterogeneous, resource-constrained, tiny, and inexpensive sensors, devices, controllers, and actuators interconnected by a local network to a gateway. They are responsible for the data collection process, sense the target factors from the monitoring environment, and transmit information to the gateway by wired and wireless networking protocols for the different functions. Energy consumption is a significant challenge in the edge layer that could be solved using wireless recharging technologies for demanding events and filtering sensed data for reducing transmission costs.

Platform Layer (Gateway Layer) consists of high-processing components to perform the routing protocols, facilitate communications, establish links with the cloud layer, drives action, and manage the lower equipment. Indeed, gateways preprocess the smart sensors/devices' data before sending it to the cloud layer, then summarize and locally analyze it to reduce transmission costs in the cellular networks. It should be noted that there is no constraint on communications among the components of the platform layer.

Cloud Layer (Datacenter/Control Layer) focuses on data transmission from the smart sensors/devices to the cloud servers. It fetches the users' input data and decides how to allocate the edge layer equipment to the gateways, based on the application requirements. In others, the cloud layer performs large-scale back-end event analysis to generate business value. The primary functions of an IIoT datacenter layer are connectivity and data routing/storage, event computing, sensor/device management, and application integration.

Figure 4 illustrates the layers of standard IIoT architecture.

2.2 IIoT Challenges

IIoT tries to employ smart devices/sensors to connect various manufacturing processes, provide energy-efficient mechanisms, and offer high-performance services [55]. However, it faces different challenges, including big data management, energy efficiency, trust, security, privacy, real-time performance, and interoperability of heterogeneous devices, wireless protocols, and specified operating systems [41]. Figure 5 illustrates the significant challenges of IIoT-based systems.

Big Data Management is a challenging task to provide real-time, robust, and flexible processing, decision making, storage, transmission, and availability on IIoT. The vast adaptation of heterogeneous sensors, actuators, industrial devices, edge servers, local gateway, and cloud servers generates an increased amount of high-velocity data. Efficient management methods are required to handle the

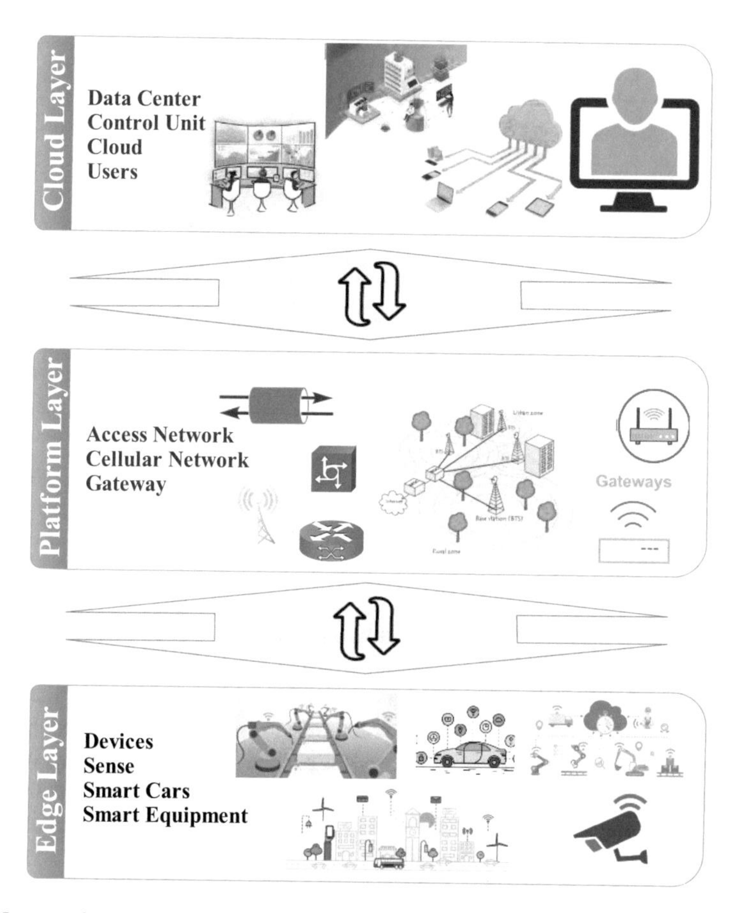

Fig. 4 Layers of standard IIoT architecture

massive amount of data, provide exceptional services with high-speed data processing, enhance reliability and security of data storage, and get full benefits from IIoT [56]. Besides, adaptive database management systems should be established to save big data, analyze critical information, and offer real-time industrial automation, such as performance prediction and anomaly detection [57]. To sum up, big data management technologies should provide efficient data processing for supporting the manufacturing product lifecycle to reach full business advantages.

Energy Efficiency is a fundamental challenge that influences system lifetime, and as a result, reliable and uninterrupted processing/communication on IIoT. Despite significant developments in hardware and software technologies, many IIoT applications are dependents on low-power sensors/devices to run for years on batteries. Furthermore, the sensor-based IIoT devices are often employed in inaccessible spots of industrial environments for a long time without the possibility of recharging or replacing the batteries [35, 58]. Indeed, power harvesting

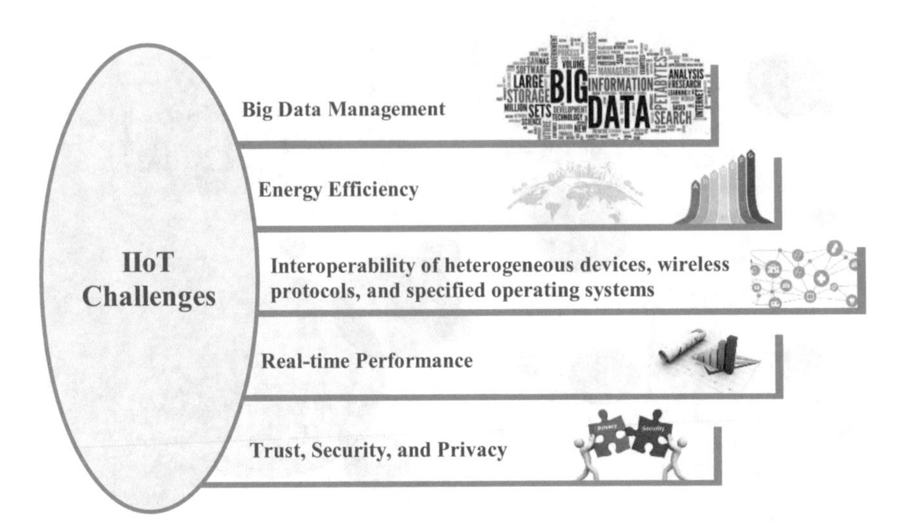

Fig. 5 IIoT challenges

creates a demand for energy-efficient processing algorithms, communication protocols, and radio transceivers.

Trust, Security, and Privacy are other significant concerns in almost all IIoT applications. IIoT is a resource-constrained communication system, which strongly relies on low-bandwidth links and light-weight devices. Thus, existing protection techniques, such as privacy assurance, security protocols, encryption methods, and traditional cryptography approaches do not satisfy its security requirements [59]. Besides, user acceptance and trust affect the success of any complex system like IIoT. Most recent research works highlighted privacy as a significant challenge in these systems, linked with the users' trust so that low privacy will discourage the industry from adopting IIoT [60]. To address security challenges, IIoT infrastructure should have some properties, including [61]:

- IIoT devices should be protected against potential physical attacks, like unauthorized updates or passive security thefts, while the authorized operators update the device's firmware.
- The communication links among equipment need to be secured to guarantees IIoT integrity and confidentiality.
- The storage of sensors/devices should be resistant to malicious manipulations using encryption mechanisms.
- IIoT services need to be available within the regular operation in the physical damage period to satisfy the robustness requirements.
- IIoT infrastructure requires safe authorization and identification techniques to prevent unauthorized access to network resources.

Real-time performance is the essential requirement of vital IIoT applications to guarantee the quality of services. Sensor-based systems are often deployed in noisy industrial areas to satisfy security-critical applications, timely data collection, reliability requirements, and real-time control decisions. In such environments, providing real-time performance while maintaining scalability could handle the unexpected disturbances [62]. Furthermore, distributed resource management should be considered in the IIoT community, focusing on real-time end-to-end constraints to ensure bounded response-time and deal with concurrent troubles [63].

The interoperability of heterogeneous devices, wireless protocols, and specified operating systems should be prioritized in IIoT applications to keep them operational. Based on the rapid growth of IIoT technology, a large number of heterogeneous sensors/devices and communication links are deployed densely in large-scale monitoring environments. It leads to an unprecedented number of interferers. Memory-limited and intelligent IIoT devices could likely eliminate interference or minimize it. However, it becomes an essential issue that IIoT equipment can reduce maximum external interference to ensure standard coexistence [64]. On the other, communications based on wireless protocols in IIoT are critical for efficient data aggregation from the monitoring environment. Thus, IIoT communication protocols should provide the required bandwidth to connect many heterogeneous sensors/devices for exchanging data with low latency, high reliability, and standard security requirements [65–67].

Furthermore, standard operating systems must be designed for IIoT systems to guarantee the requirements of different applications, including memory restrictions, real-time performance, power limitation, security/privacy protocols support, bandwidth consumption, heterogeneous devices management, and interoperability [9]. The summary of major challenges on IIoT systems is illustrated in Fig. 5.

3 Blockchain

Satoshi Nakamoto introduced the term blockchain 1.0 as the next disruptive technology, highlighting its advantages in offering transaction security in 2008 when he described the whitepaper Bitcoin [68]. The blockchain is a decentralized and immutable ledger that can record financial transactions between two users without trusted intermediary authentication. Indeed, it provides a decentralized Peer-to-Peer (P2P) cryptographic hashes-based network to creates an efficient transaction process and reduces its cost below 1% [69, 70]. In 2009, Satoshi Nakamoto created Bitcoin with the Genesis block. It has led to chaos in financial systems and made an opportunity for many people to make money. In the following, big companies, such as Google, focused on blockchain projects to combine intelligent contracts with digital currency for optimizing a wide range of financial applications. It has developed to blockchain 2.0 to support the bitcoin-based trading market, currency exchanges, and decentralized solutions for IoT/IIoT applications [18, 71, 72].

The investments in blockchain projects have started from 93millions U.S. dollars to 550millions U.S. dollars from 2013, and it is anticipated that the blockchain market grows to 2.3 billion U.S. dollars by 2021 [19]. Although people in business still regard Bitcoin and blockchain as the same concepts, it should be noted that Bitcoin is originated from the blockchain. It works at the heart of the decentralized industrial applications relying on cryptographically secured IIoT [73]. The remainder of Sect. 3 is organized as follows: Sect. 3.1 presents the blockchain structure and Sect. 3.2 explains the blockchain usage in IIoT-based systems.

3.1 Blockchain Structure

An essential objective of blockchain is establishing trust, which is achieved by constructing the current block based on the resulting hash from the previous one [74–77]. Indeed, the method known as the Proof-of-Work (P.O.W.) organizes a set of transactions in some blocks exploiting a Merkle tree, and the hash of the root block is added to the tree awarding for the performed task on the system [18, 78–82]. Blockchain technology is different from traditional distributed systems based on specified features [19]:

- *Permission-less* There is no requirement of permission in blockchain systems, i.e., each entity with specific properties can use the network.
- *Trust-less* There is no need for a digital certification to perform transactions in blockchain systems, i.e., the users communicating, cooperating, and collaborating in the network are unknown to each other.
- *Resistant to censorship* Since blockchain is a control-less system, any entity can transact or interact with others with no requirement of any modification or censor.

To address the trust requirements, a standard blockchain consists of three significant parts [83]:

- *The block* is known as a set of modified-less bills so that any entity can access the information that you record into it. It should be noted that period, size, and triggering events of blocks can vary based on the applications.
- *The network* is a group of blocks (nodes) over digital interconnections to share transaction information.
- *The chain* is a function to link all the blocks in a blockchain system.

The blockchain structure, block components, and relationship between blocks are illustrated in Fig. 6. It shows that each block is divided into two sectors, including the block header and its body. The block header consists of the previous block hash, a timestamp for determining data writing time, Nonce for keeping the hash of the current block more extensive than the hash of the next one, and the Merkle root for verifying data integrity. The block body involves the transaction number and its information details [84]. Once a new transaction happens, its information is broadcast over the system. All miner blocks (responsible for validating hash) receive the information and validate the transaction signature to verify it. Then, the transaction is packed into timestamp blocks and broadcast over the system again. The system blocks prove valid transactions and refer to the chain's previous block by employing

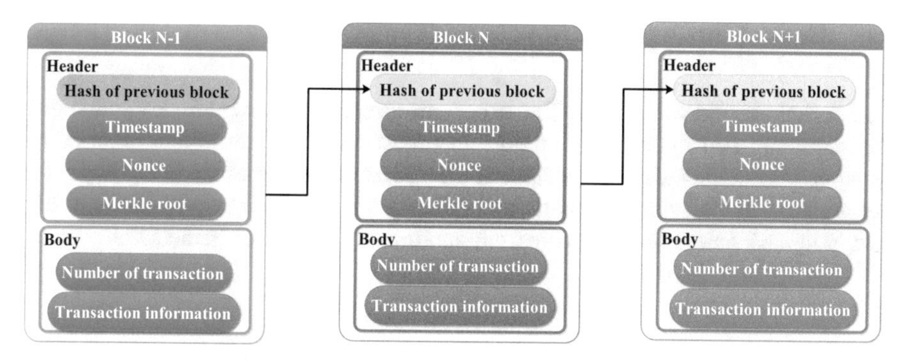

Fig. 6 The blockchain structure

a specified hash algorithm or discard invalid transactions [85]. During the validation process, the blockchain offers a trusting framework for data/information transmission or trading applications.

3.2 *Blockchain Usage in IIoT*

In the smart world, IIoT plays a vital role in enhancing people's life quality by digitizing various information, applications, services, and data storage technologies. However, it acts as a black box that fails to guarantee security, privacy, and data transparency. To deal with security challenges in IIoT, Blockchain is a viable technology for providing trusted, open, and auditable sharing solutions to revolutionize processes in industrial environments [86]. It leads to any data transmission to be traceable [87, 88]. The main advantages of IIoT and blockchain integration are as follows [89]:

- *Security* Blockchain technology provides secure communications among IIoT devices based on smart contracts, where information exchanges are assumed as transactions. Moreover, the update history is traced easily in industrial environments to allow secure device updating.
- *Identity* Employing blockchain systems offer a trusted identification and authorization process for IIoT devices to trace information origin.
- *Scalability* Decentralized blockchain-based IIoT improves fault tolerance and controlling data collection and processing in large-scale environments, where many users corporate.
- *Reliability* Users can verify the accountability and authenticity of transactions with certainty using blockchain technology in IIoT applications.
- *Autonomy* In blockchain-based networks, IIoT devices can interact with each other without the intermediary unit's control. It leads to provide device-agnostic industrial applications.

4 Blockchain Applications in IIoT

Blockchain applications for IIoT have become an emerging research field, attracting the academy and industry's attention. In this section, different blockchain-based applications in the industrial internet of things have been presented in eight groups: smart city, manufacturing, healthcare 4.0, energy management, agriculture 4.0, smart homes, autonomous vehicles, and multimedia right management.

4.1 Smart City

The primary objective of IIoT is to offer autonomous services for supporting the requirements of Smart Cities. It aims to connect technology, the economy, government, citizens, and society to address the necessities such as water, quality of services, transportation, healthcare, operational costs of public administration, and energy [90]. In others, increasing the population migrating to urban areas and the associated urbanization process lead to the scarcity of resources. Thus, optimal utilization of available vital resources is a fundamental challenge, required leveraging technologies such as IIoT [91].

Blockchain technology has been applied to IIoT-based smart city applications to provide a distributed, secure, and efficient platform for supporting people's life. For example, B2EExpand is a gaming company that has invested in Steam to introduce cross-game video games based on an Ethereum blockchain. Ripple has provided a global payment solution to address the financial challenges by communicating between financial institutions such as banks, payment providers, and digital asset exchanges. To promote healthcare services in smart cities, the MedRec Company has combined IIoT and blockchain for secure, time-optimal, and cost-efficient data aggregation from facilities, patients, and medical providers. Guts have used blockchain concepts to offer a fraud-preventing ticketing ecosystem. Finally, Warranteer has utilized blockchain-based networks to simplify access to product data [19]. The overall scheme of blockchain-based smart cities using IIoT is shown in Fig. 7.

Parking systems are one of the essential services in IIoT-based smart cities. It aims to develop a traffic management system for minimizing costs. In this regard, Pham et al. [92] have presented an efficient cloud-based parking system in smart cities. It tries to decrease the number of instances in which drivers cannot find a parking spot. Although the system reduced the average waiting time of drivers for parking, it suffers from security challenges in real-world applications. To address these challenges, Lazaroiu et al. [93] have proposed a distributed blockchain-based smart parking system model. The model consists of two entities: entity A demonstrates the visitor who paid the system authenticator's parking fees (entity B). The online blockchain contains the transaction data so that a block is appended to the blockchain when most of the devices verify its authenticity. Finally, the smart contract triggers guarantee the wallet transferring from A to B. The results of analyzing

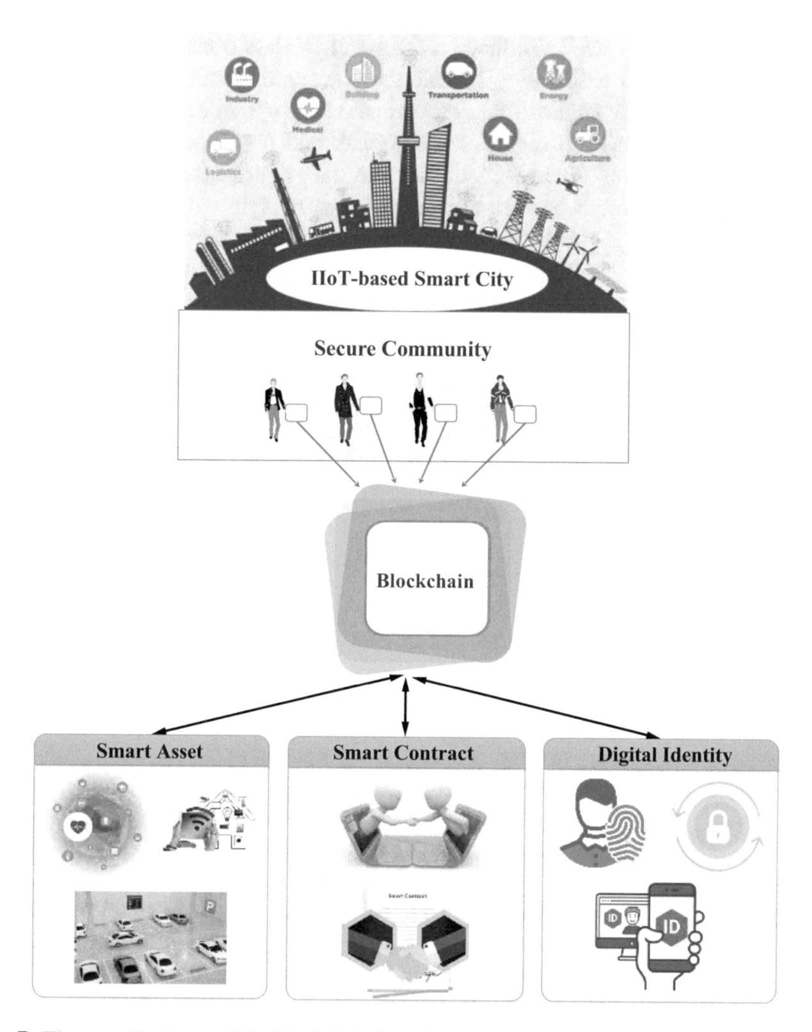

Fig. 7 The overall scheme of the blockchain-based smart city using IIoT

illustrate that the paper offers a replicable model to build a parking system in smart cities based on IIoT and blockchain approaches to meet security needs along with energy efficiency.

The amount of data generated by IIoT devices and sensors is becoming unpredictable based on the increasing number and type of network equipment and software technologies. The classic client/server-based communication models are centralized and store data at a central server. It poses some problems for automated communication among system equipment. Fan et al. [94] have offered a blockchain-based solution to deal with communication challenges on large-scale data management systems in smart cities.

Zhang et al. [36] have introduced a light-weight data transmission mechanism based on blockchain technology to guarantee accurate data aggregation in IIoT, focusing on smart city applications. The process employs a decentralized ledger on multiple edge gateways to record, synchronize, and maintain the information. It deals with consistency challenges during the data aggregation using a two-path routing strategy. The mechanism also uses a light-weight data block structure to improve traditional blockchain technology and save resources. The simulation results show that the combination of IIoT and blockchain technology could reduce the hop count of data aggregation, energy consumption, and transmission delay in smart city applications. Furthermore, it enhances data accuracy, network reliability, and system security.

The Kasperky Labs report demonstrated that IIoT devices, such as data kiosks and autonomous machines have many security gaps that are vulnerable to external malicious attacks [95]. To enhance the privacy and security of IIoT, Biswas [96] has introduced a secure blockchain-based model that enables smart devices to communicate with each other in a smart city framework. It aims to deal with various security bags, including accountability, authenticity, integrity, confidentiality, and availability threats. Blockchain applications in IIoT are mature in slight domains, such as tracking information, and could reduce the production cost and computational complexity of processes. However, most applications, especially in the smart city field, still stay in the experimental step. Thus, there is a lack of practical international standards for blockchain applications in IIoT.

4.2 Manufacturing

To address after-sales services, repair and maintenance departments of manufacturers publish technical manuals for almost all products. Distribution and update of such records contain tons of paperwork. However, technical manuals are accessible to the users on the manufacturing framework using the blockchain technology without worrying about deliberate changes or losing updated versions. In others, blockchain could significantly improve the smart manufacturing industry's performance by exploiting data-sharing ledgers to guarantee security and privacy needs. Wan et al. [97] have introduced a secure blockchain-based IIoT architecture for enormous modern factories' data interactions. It first reorganizes traditional IIoT architecture to form a smart multicenter model for data aggregation and optimizes results. Then, an automated production platform is provided to discuss the specific implementation. The experimental results indicate that the blockchain-based IIoT model guarantees system privacy, making a light-weight and secure solution.

Huang et al. [98] have presented a credit-based consensus system for IIoT-based factories. The system uses a directed acyclic graph-based blockchain on power-constrained IIoT devices/sensors to adjust the proof of work difficulty based on users' behavior. It also employs a symmetric cryptography-based access control model to manage the data authority process flexibly in the blockchain system. The simulation results show that using the asynchronous consensus model in smart factories can enhance system throughput. Zhang et al. [99] have provided a blockchain-based IIoT to address the security requirements of traceable configuring intelligent manufacturing systems to deal with manufacturing disturbances and limited-capacity security models. It involves hardware infrastructures, software-defined components, resource-efficient encryption methods, and consensus-oriented transaction logic to support the autonomous manufacturing process. The application examples justify that the practical fault-tolerant IIoT offers secure, stable, traceable, and decentralized manufacturing environments with reasonable throughput and latency.

Blockchain technology is being exploited in other IIoT-based manufacturing applications. Industrial Blockchain of Things (IBoT) is defined as an integrated framework for industrial data exchange in a distributed network [100]. It aims to prevent credential data tampering and employ consensus methods for cybersecurity in industrial environments. Researchers in [101] have simulated a decentralized blockchain-based IIoT on Ethereum to provide a tolerable automotive manufacturing ecosystem. Finally, Angrish et al. [102] have exploited blockchain technology to introduce a distributed mechanism to handle manufacturing data generated by various entities. In this mechanism, called FabRec, data are available to all IIoT devices/sensors in a peer-to-peer network. However, fiduciary equipment is used to guarantee transparency and data protection through a verifiable audit trail. Although IBoT-based models improve security criteria in manufacturing applications, managing the chain of multiple blockchains used by different manufacturers is a considerable challenge to consider. Researchers need to find methods to standardize other blockchain implementations in smart organizations for more interoperability.

4.3 Healthcare 4.0

Healthcare 4.0 is an expression of Industry 4.0, which combines IIoT and cyber-physical systems to virtually address real health applications' needs [103]. It employs data-driven digital health technologies, including mobile health, smart health, wireless health, online health, telehealth, eHealth, and digital medicine, to provide more informed decisions [104]. Blockchain technology plays a vital role in enhancing the performance of smart health systems. For example, it can efficiently share critical patients' data to improve health services delivery by reducing mismatched data and errors. Besides, blockchain-enabled health systems can enhance interoperability by providing automatic access to the patient's records. It also allows for clinical drug tracking by investigating drug side effects without the risk of modification. Finally, blockchain-based IIoT can monitor and manage diseases using smart pills and wearable devices in healthcare applications [105]. An illustration of blockchain-based IIoT for healthcare 4.0 is shown in Fig. 8.

To protect medical data, Xia et al. [106] have provided a blockchain-based data-sharing system in a trust-less big area called MeDShare. The system monitors users

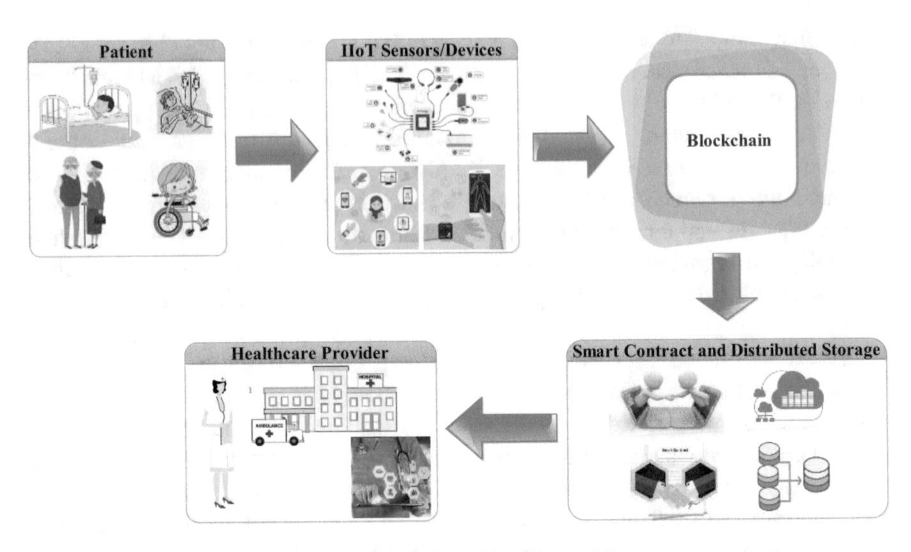

Fig. 8 The overall scheme of the blockchain-based healthcare 4.0

who access data for identifying malicious attacks to guarantee data provenance, auditing, and management for big medical data in cloud platforms. MeDShare also records data transmission among users in a tamper-proof manner, with side-blocks and smart contracts. Side-blocks include contract reports and manage the block fetching process to enhance the accuracy of the logs. The performance analysis of MeDShare demonstrates that the system achieves data provenance while the data sharing with minimal risk to users' privacy. Jiang et al. [107] have implemented and evaluated a minimal-viable-product blockchain-based plan to exchange healthcare data generated by medical institutions and individuals in IIoT. It first analyzes the requirements of healthcare data from various users using two loosely-coupled blockchain mechanisms. Then, the IIoT-based system combines off-chain storage and on-chain verification to address privacy challenges. Finally, the plan employs two fairness-based packing methods to enhance the IIoT throughput and fairness. Other similar approaches combine blockchain technology, machine learning algorithms, and group-based secret sharing to provide privacy-preserving IIoT in medical applications [57, 108].

Wang [109] has presented a medical data management model using blockchain technology over healthcare 4.0. The model communicates between industrial IoT, cloud storage, and blockchain in medical applications to improve eHealth services. It provides secure monitoring of patients' vital signs from remote hospitals using a blockchain-based data-sharing system so that only authorized users can access the sensors/devices. The data management model also uses a web-based frontend and representational state transfer application programming interfaces to offer product-centric services based on blockchain technology. The simulation results show that the overall throughput of health systems can be improved using this model because of its low latency.

To ensure privacy and accuracy in healthcare data sharing, Wu and Tsai [110] have proposed two blockchain-based methods in IIoT environments. The first method employs a public-key cryptography algorithm to allocate a public-private key to the users. The second method uses an elliptical curve cryptography-based key-pairing process to guarantee data privacy and distribute keys efficiently. In another approach, Zhang et al. [111] have claimed that saving the addresses of IIoT devices/sensors in the blockchain instead of raw health data leads to a secure system for pervasive social networks. The design illustrates a potential mechanism for employing blockchain technology for healthcare applications by allowing data access only to authorized IIoT devices/sensors in a secure manner.

To ensure quality control of medical products, Bocek et al. [112] have presented a start-up that employs blockchain-based IIoT devices/sensors in the pharma supply chain. The model exploits smart contracts to assess the temperature and humidity levels during the drug transportation process for reducing bureaucracy and costs. MedRec is another blockchain-based architecture for efficient storing of medical data that enables the patient and health providers to develop a longitudinal record containing a lifetime of healthcare cases [113]. Finally, Medicalchain is a distributed blockchain-based platform used in the U.K. for managing healthcare providers' access to health data [114]. Indeed, the platform enables users to record data securely and transparently on a distributed ledger.

Although blockchain is on the top of healthcare 4.0 technologies, patients using IIoT-based healthcare systems have a right to delete their data whenever they want. However, this regulation is in contradiction to blockchain principles. Medical multimedia data storage with high quality is another major challenge in the scalable implementation of blockchain-based IIoT in healthcare applications. Since hashing is not a reversible function, only hash storage in health systems is not useful for retrieving primary data.

4.4 Energy Management

Over the past decades, the energy management process in industrial applications has been undergoing specified transformations. The energy industry is becoming more complicated to handle with the emergence of smart technologies and IIoT devices. Blockchain is a newfound technology in IIoT-based energy management systems that accelerate the recent transformation by enabling smart contracts between IIoT devices/sensors, optimizing transaction costs, and operating the grid efficiently [115–117]. In this context, Li et al. [118] have designed a credit-based secure payment model to support energy trading IIoT. It first establishes a hybrid model of blockchain technology and energy management systems to minimize IIoT costs. A credit-based payment model is then proposed to support a massive flood of energy trading payments and reduce the transaction confirmation delays. Finally, the model utilizes an effective Stackelberg game-based pricing strategy for loans to enhance credit bank performance. Security analysis on a real dataset shows that the blockchain-based energy management model is low-cost and secure in IIoT-based peer-to-peer networks dealing with a trusted intermediary. Kang et al. [119] have introduced another localized peer-to-peer energy trading scheme among plug-in hybrid vehicles to balance energy demand. It exploits blockchain technology to guarantee transaction security, energy pricing, and social welfare in energy trading. The Texas real map's numerical results demonstrate that the blockchain-based peer-to-peer energy trading scheme improves transaction security and social welfare while protecting users' privacy.

To manage energy resources in smart cities, Khattak et al. [120] have presented an intelligent open-source blockchain called Hyperledger. It provides a decentralized ledger and secure system to manage smart contracts. Another article has introduced an energy-efficient and privacy-preserving data aggregation mechanism based on blockchain in smart communities [121]. It divides devices into some groups; each group has a particular blockchain for data recording. Pseudonyms are exploited to hide users' identities and support their privacy. It also uses the bloom filter for fast authentication. The simulation results show that the mechanism addresses security and energy-efficiency challenges. Finally, Aitzhan et al. [122] have focused on transaction security in distributed smart energy trading systems. It uses multi-signatures, blockchain technology, and anonymous encryption to communicate energy prices and securely enforce trading transactions anonymously. Security analysis of case studies indicates the superiority of blockchain-based mechanisms in IIoT-based energy management systems.

Some commercial IIoT projects have been implemented based on blockchain technology for energy management along with academic approaches. PowerLedger is a blockchain-based energy trading platform, enabling users of renewable sources to offer their extra energy at a pre-determined price over a decentralized system [123]. Bankymoon is another blockchain-enabled energy management project, targetting smart schools to address affordable energy supply [124]. It not only benefits the energy users but also the suppliers get on-time payment. Despite the privilege that blockchain technology offers in industrial energy management applications, current solutions' scalability is a significant challenge. Since 0.13% of energy consumption worldwide is originated from Bitcoin transactions [121], the energy needed for transactions in the energy sector (which is safely multiplied 1000 times) will be prohibitively high.

4.5 Agriculture 4.0

Agriculture 4.0 refers to industrial agriculture, which exploits smart IIoT devices/sensors and big data to drive food supply efficiencies in the face of population growth and climate change [125]. With the replacement of Agriculture 4.0 to the traditional agriculture schemes, farmers can:

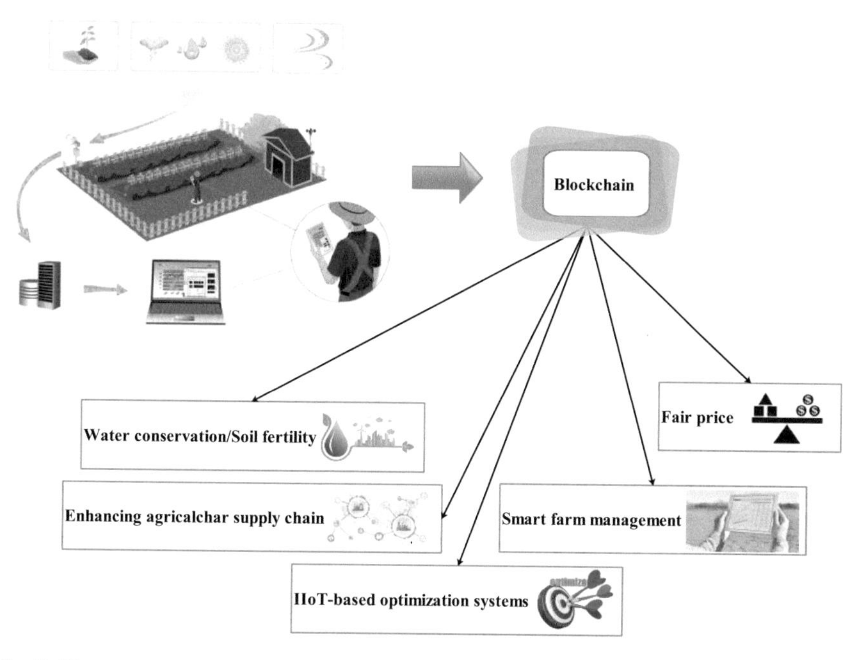

Fig. 9 The overall scheme of the blockchain-based agriculture 4.0

- Ensure product security and minimize dependency on imports.
- Address net exporters of products and smart mechanisms.
- Support productivity and innovation-based economy [126].

However, the performance of agricultural 4.0 depends on some external factors such as moisture content, climate, soil type, quality of reap, and supply chains. Blockchain technology could make these factors increasingly transparent and lead to a controllable product journey from the farm to the supermarket. Besides, blockchain-based IIoT permits local producers to access big data as more affluent farms to making the agriculture industry a more transparent and secure area [127]. The overall scheme of a blockchain-based agriculture 4.0 is depicted in Fig. 9.

To tackle the food safety challenges, Lin et al. [128] have combined trustworthy blockchain technology and the low-power wide-area industrial internet of things to monitor products' lifetime from the agriculture fields to the consumers' home, including raw material process, product transportation, storing, and distribution. The trusted ecological food traceability system consists of all components of a smart, self-organized, and open agricultural ecosystem to record data, verify products/users, and minimize the human intervention to the platform. Authors believe that their blockchain model for agriculture 4.0 will help users to guarantee food safety status. Hua et al. [129] have presented a decentralized, trust, and reliable agricultural product tracing system to address food safety. It uses the advantages of blockchain technology to solve the trust challenges in the product supply chain. System analyzing results show that exploiting blockchain in agricultural data management systems widens the IIoT application domain and supports a reliable community among agriculture entities. Caro et al. [130] have provided another decentralized agricultural supply chain traceability system, which can rely on Ethereum or the Hyperledger Sawtooth blockchain implementations. It integrates IIoT devices to produce valuable data along the supply chain and then stores data directly in the underlying blockchain to ensure the traceability system's transparency and audibility. Comparing the classical food traceability scenario and the smart one indicates the superiority of the blockchain-based IIoT platforms in terms of latency, CPU load, and network usage.

Awan et al. [131] have introduced a secure blockchain-based routing method for distributed agricultural 4.0 applications to use the communication links effectively. It exploits intelligent collaboration within heterogeneous IIoT-based sensors to find the best route from the source devices to the sink. The comparison results confirm that the method removes redundant data, blocks architecture attacks, and reduces the energy consumption of sensors to improve IIoT lifetime in agriculture applications. To predict raw material needs, farmers' payment transactions, and distribution costs, Putri et al. [132] have proposed a Hyperledger Blockchain-based IIoT system for supply chain and logistics in the agricultural domain. Hyperledger Blockchain provides high data security for fast data exchange among farmers and consumers over a trusted distribution system. It also guarantees data transparency in Semarang District Agriculture Service.

iGrow is a commercial center that offers an agricultural supply chain for organic food [133]. Its team first identifies stable demand, prices, and growing characteristics of crops to find suitable farmers. Then, iGrow provides an opportunity for urban people to control their investment growth. It aims to exploit the agricultural blockchain-based platform to connect farmers with the real-time market for healthy food distribution. Avenews-GT is another commercial platform that provides a secure and transparent ecosystem for agricultural product trading [134]. Indeed, blockchain record interactions and transparent data sharing among the partners improve the performance of Avenews-GT in agriculture 4.0 applications.

Blockchain technology enables data traceability in IIoT-based agriculture applications. It could be a concern to convince small-scale users to share their data over open and transparent systems. Furthermore, paying for the blockchain subscription and IIoT enablement leads to staggering costs for the agricultural industry.

4.6 Smart Homes

A smart home refers to a ubiquitous computing-based building where appliances and operations are automatically managed from anywhere using IIoT devices/sensors to improve inhabitants' quality of life [135]. Blockchain technology provides a secure platform that all devices within the home request information from the others to satisfy home requirements. In such platforms, owners have a list of IIoT devices that can communicate with other internal ones and protect device communications with a shared key [136, 137]. In other words, data can be stored locally in IIoT

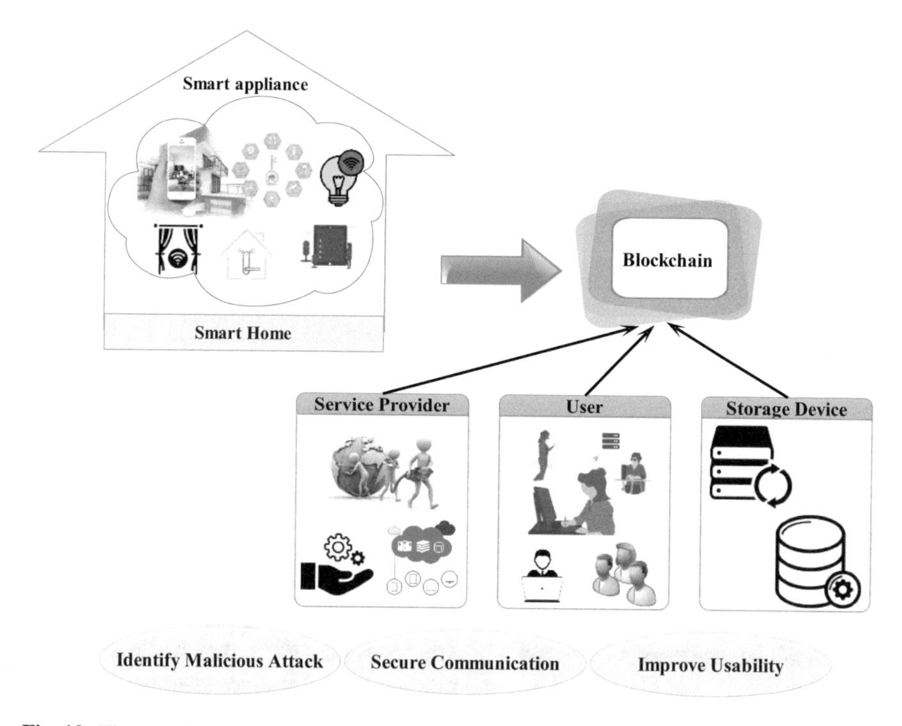

Fig. 10 The overall scheme of the blockchain-based smart homes

devices and authenticated using a shared key without the need for cloud storage in blockchain-based smart homes. An illustration of blockchain-based IIoT for smart homes is shown in Fig. 10.

The first essential part of any smart home is the IIoT-based door lock system to prevent unauthorized entrance. In smart door lock mechanisms, the inhabitants' information is stored in a central server to allow only the authorized users to access the interiors. However, the information managed by such systems could be attacked by malicious who tries to gain unauthorized access to the house. To tackle this challenge, Han et al. [138] have designed a blockchain-based smart door clock system to guarantees non-repudiation, authentication, and data integrity in smart home applications. It employs passive infrared, ultrasonic, and motion sensors for immediate indoor/outdoor intrusion detection [139]. The system also exploits the immutable nature of the blockchain to prevent unauthenticated users access to the house. Although the smart door clock system provides intrusion detection and block mining of the transactions, the IIoT networks' latency in real-world applications can breach the intrusion detection process. To ensure data integrity in the real environment, Nadiya et al. [140] have simulated a manipulating door clock system combining IIoT and smart contracts on the Ethereum blockchain. Analyzing the security level test using an avalanche effect method shows that the proposed system can be considered a secure approach for real-world smart homes. The presented solution in [141] has exploited Infura API to connect Ethereum blockchain and IIoT infrastructure for reducing security costs in smart homes. The results of model evaluation using different Ethereum test networks, including Ganache, Ropsten, and Rinkeby, demonstrate low smart lock costs and high security, privacy, and convenience. Other articles have presented practical authentication systems for smart home applications exploiting message authentication, IIoT, blockchain, and group signature [142–144]. The systems aim to authenticate users' access history to the gateways for guaranteeing security needs, including traceability, anonymity, and confidentiality.

Yang et al. [145] have proposed a personalized, secure, and versatile smart home system exploiting inherent authentication and permission control features of blockchain technology to support complex real-life IIoT scenarios. It uses a blockchain-based encryption approach on devices/sensors to ensure content privacy and verify authenticated users. The system also simulates a hierarchical deterministic fundamental model for multi-level controls and a distributed artificial intelligence-based component for enhancing security and customization of IIoT-based smart homes. To minimize smart homes' electricity consumption and the total energy cost in modern communities, Afzal et al. [136] have formulated a game-based distributed demand-side management system. The users living in smart homes play a game with the best strategy to reduce the energy consumption of appliances exploiting a private industrial internet of things. Since the presented game plan is distributed on a blockchain, it provides users secure communication, autonomous monitoring of smart appliances, and billing of energy consumption via intelligent contracts. Comparison results demonstrate that the game-based energy management model minimizes the total electricity costs in smart communities.

Finally, Zhang et al. [146] have introduced a privacy-protection, confident, and unforgeable blockchain-based smart meter mechanism for distributed smart home applications. It uses elliptic curve point multiplication in the encryption algorithm to minimize the computing and communication costs. Although blockchain technology enhances people's living standards in IIoT-based smart homes, some challenges such as high energy consumption of devices/sensors, unreasonable data transmission delay, and overhead should be resolved in this area. Furthermore, miners are a type of blockchain entity that centrally handless the incoming/outgoing transactions in smart home applications; they have a critical role in supporting secure authentication and local storage management [147]. A miner cannot assign tasks when the attack happens on it, so unauthorized users can steal vital data. Thus, focusing on miners' security is one of the essential issues in blockchain-based smart home applications.

4.7 Autonomous Vehicles

Recent developments in identification, communication, and computation technologies have led to smart transportation systems' growth. Autonomous vehicles are used in smart transportation systems to provide intelligent, safe, and convenient services [148]. However, such systems are generally centralized, unreliable, and temporarily down in external malicious attacks. To tackle such challenges, Yuan et al. [149] have introduced a blockchain-based autonomous ecosystem for smart transportation systems. It exploits the parallel transportation management mechanism to support secure, trusted, and distributed real-world transportation systems.

To control peer-to-peer transactions between smart vehicles, Liu et al. [150] have presented a blockchain-based power trading model in vehicle-to-grid networks. It realizes the data equivalence and power trading transparency based on the blockchain and smart contract technology. The model also uses a dynamic pricing-based reverse auction method to guarantee the transaction matching. Simulations verify the feasibility of the peer-to-peer transaction model. Deshpande et al. [151] have exploited the blockchain capabilities to provide a secure framework for smart vehicle applications. Based on the electronic control unit needs, the framework is presented on the NXP IMX6Q, Multos M5-P19, MultiChain testbed to show its low memory, storage, and processing overhead with high immutability and security. To deal with the unwanted decisions and privacy attacks imposed on the central pricing systems in IIoT-based smart transportations, Huckle et al. [152] have proposed an automatic payment model that allows smart vehicles to synchronize automatically with users' services and pay the costs of external services such as fuel payment. Analyzing the results of distributed IIoT architecture using blockchain demonstrates the opportunity for offering securely shared economy distributed applications in the autonomous vehicle domain. Another research in [153] has studied the potential of IIoT and blockchain technology in smart logistics and transportation systems to address vulnerability under security attacks.

Finally, Lin et al. [154] have provided a peer-to-peer computing resource trading model to handle dynamic demands on the internet of vehicles. It utilizes a consortium blockchain method to support transaction security and privacy protection in smart transportation systems. The model also constructs a two-step trading game to encourage smart vehicles to join the system and improve buyers' and sellers' utilities. Security analysis proves the efficiency of resource trading strategy in collaboration between the buyer and smart vehicle systems. However, access control to install/uninstall smart vehicles' batteries affects the cybersecurity procedure in IIoT-based transportation systems. Thus, secure verification of legitimate access by users involved in the battery installation/replacements is a significant concern in autonomous vehicle applications.

4.8 Multimedia Right Management

Nowadays, the IIoT -based delivery medium becomes the expected standard for multimedia transmission instead of past platforms like compact discs. Although current digital rights management systems offer reasonable availability, costs, and performance, their centralized architecture leads to security and privacy concerns [21, 155]. Blockchain technology addresses the sharing and reusing challenges in

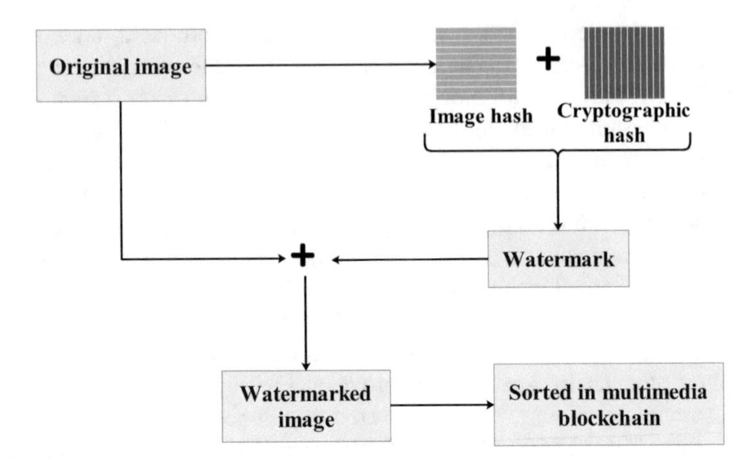

Fig. 11 The general overview of blockchain-based multimedia right management [155]

multimedia right management over IIoT-based systems. The general overview of blockchain-based multimedia right management is shown in Fig. 11.

In this aspect, Ghaffar et al. [156] have presented a decentralized research datasets management method based on peer-to-peer networks, a limited number of computing resources, and blockchain technology. It uses blockchain capabilities to maintain transaction records in a decentralized public digital ledger. Moreover, a smart contract for digital right management is provided for data sharing over IIoT-based networks. Finally, a data-sharing case study on the Ethereum contract platform is proposed to achieve access. To address multi-device data redundancy and lack of credit guarantee in IIoT, Si et al. [157] have proposed a light-weight blockchain-based data-sharing security framework. It exploits a combination of Byzantine fault tolerant-based data and transaction blockchain for secure data storage. The framework also uses a high-trust dynamic game mechanism for device cooperation to deal with local malicious attacks. Simulation results demonstrate that the framework is efficient, safe, and feasible to verify secure storage system information.

Finally, Bhowmik and Feng [158] have introduced a watermarking-based multimedia blockchain framework to retrieve the transaction trails and the modification histories. The watermark data involves two parts, including a cryptographic hash to preserve transaction histories and a picture hash to store retrievable original media content. The cryptographic hash is passed to a distributed ledger when the watermark is extracted for retrieving the historical transaction trail. The picture hash is employed to recognize the tampered regions. The simulations prove the efficiency of the trusted multimedia right management mechanism. However, current approaches have not discussed data storage and its watermarks in detail. Besides, unreasonable latency in vital applications and the lack of incentive plans are other issues that should be considered in blockchain-based multimedia right management in IIoT.

5 Analysis

This chapter has used Google Scholar, Web of Science, Science Direct, and Scopus as the search engines to investigate several keywords about blockchain-based IIoT applications in recent journals, books, and conference articles. Reports, editorial notes, commentaries, book reviews, and written materials in non-English languages are neglected in this chapter-book. We highlight the literature, which exploits blockchain technology to improve IIoT services in various applications, including smart city, manufacturing, healthcare 4.0, energy management, agriculture 4.0, smart homes, autonomous vehicles, and multimedia right management from 2013 to 2020.

The dispersion of blockchain-based applications in the industrial internet of things is shown in Fig. 12 by publication years (from 2013 to 2020). The numerical results demonstrate that recently, special attention has been paid to blockchain technology on IIoT to improve the quality of humans' everyday life, security/privacy of services, and decrease resource consumption. In particular, due to its distribution and data disclosure, blockchain technology has been introduced as a decentralized approach to support security requirements in IIoT-based applications. It tracks peer-to-peer transactions and saves billions of IIoT devices based on decentralized credits in modern systems to offer insecure data management solutions in centralized organizations.

Figure 13 illustrates statistical results on different blockchain-based applications in IIoT from 2013 to 2020. As shown in this figure, the reviewed applications are

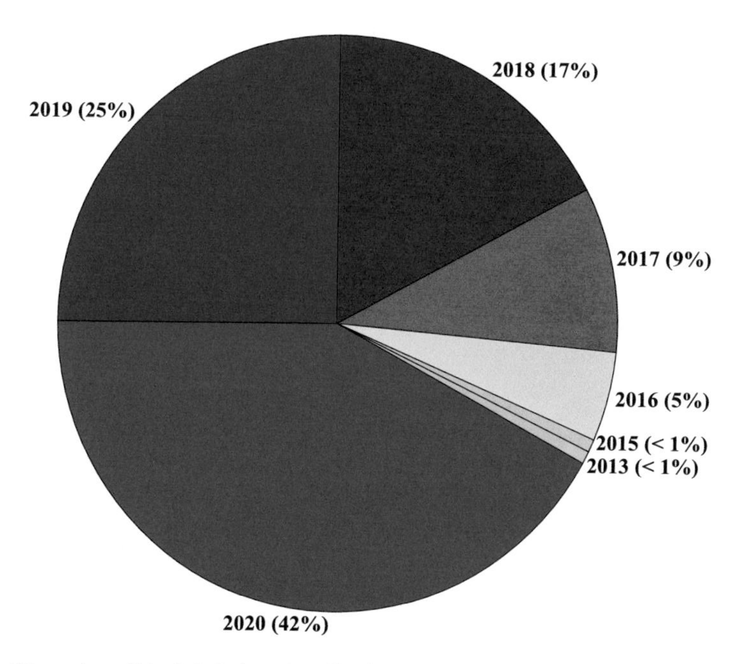

Fig. 12 Dispersion of blockchain-based applications in IIoT from 2013 to 2020

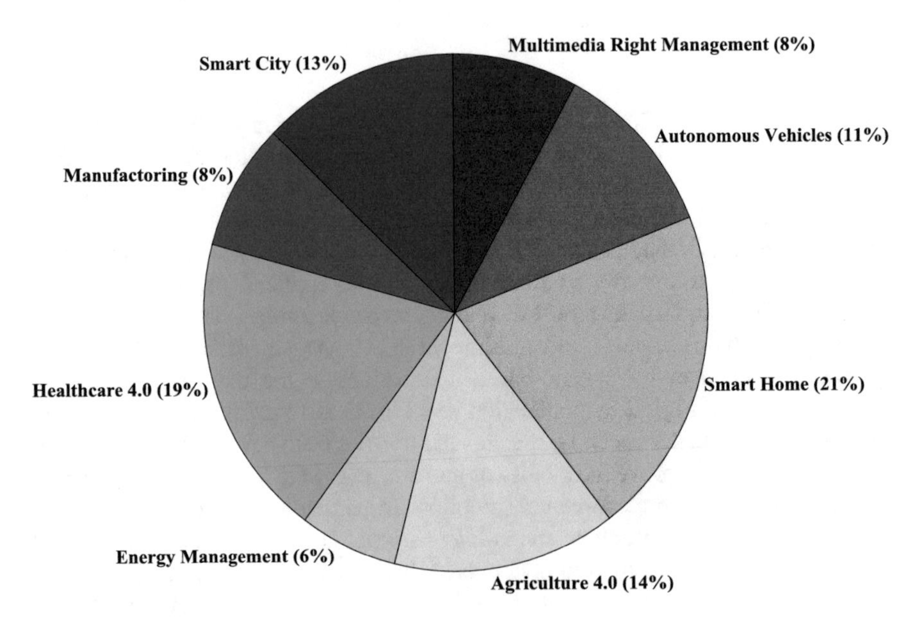

Fig. 13 Diversity of blockchain-based applications in IIoT from 2013 to 2020

classified into eight categories, including smart city, manufacturing, healthcare 4.0, energy management, agriculture 4.0, smart homes, autonomous vehicles, and multimedia right management. The results show that authors have focused on smart cities and healthcare 4.0 to enhance the quality of secure IIoT-based services compared to other blockchain applications. Furthermore, our analysis demonstrates that little work has been done to improve security in manufacturing and multimedia right management. Finally, it can be concluded that the smart city and healthcare 4.0 are more critical than other IIoT-based applications in the blockchain-based scientific and industrial domain. In contrast, there are more weaknesses to address in the manufacturing and multimedia right management fields.

6 Challenges and Open Issues

The heterogeneous nature of IIoT has brought together some technical challenges in blockchain-based approaches. In others, the heterogeneity in IIoT resources and massive data flows need powerful computing, storing, and networking services on the Internet. Exploiting distributed edge-cloud systems can address IIoT challenges. The edge-cloud systems, coupled with blockchain technology, provide immutable traces of IIoT resources from devices/sensors and applications without depending on the centralized controller entities. However, distributed blockchain-based systems face some limitations. In this section, we sum up the concerns about blockchain-based applications in the industrial internet of things and introduce some research directions to enhance smart society's performance.

Connectivity Secure connection of millions of devices/sensors is the first challenge in future IIoT. It affects the structure of classic communication mechanisms and underlying technologies [72]. Thus, blockchain-based IIoT applications require some paradigms to authorize, authenticate, and connect different entities in a smart system. Besides, IIoT devices/sensors should be upgraded to be compatible with the high-speed 5G connectivity.

Scalability Current blockchain technologies face another significant concern in the wide acceptance and deployment of secure environments, called scalability. Since each battery-powered and resource-restricted device/sensor must maintain a blockchain copy, it faces enormous storage overhead. This limitation is compounded in IIoT scenarios with millions of users and the growing amount of generated data. Although light-weight platforms like Ethereum have been introduced to address scalability challenges in blockchain-based IIoT, exploring more scalable approaches is a milestone to support the resource performance of IIoT devices/sensors [70].

Computational constraints Blockchain technology has been introduced to upgrade traditional client-server systems. However, its data should be stored in IIoT devices/sensors, which usually have computational limitations and low storage capacity. So handling low-power devices/sensors is a critical hurdle in adopting blockchain technology [19].

Security and privacy IIoT systems are prone to blockchain-based security vulnerabilities (message hijack and smart-contract program attacks) along with common safety risks (eavesdropping and replay attacks). Besides, privacy leakage is another primary concern in blockchain-based IIoT applications due to stored transaction data on the blocks [157]. Therefore, the systems' security and privacy requirements combined with IIoT and blockchain need to be explored further.

Industry Standards Since blockchain is a distributed technology without a third-party authority, country regulations and industry standards must be executed in such systems [94]. With the development of smart IIoT platforms, the need for online execution of industry standards for blockchain should be increased even more [39]. It also requires to provide international policies for trust and data security in blockchain systems.

7 Conclusion

In this chapter, we provide a review of blockchain technology applications in the industrial internet of things. First, the significant challenges behind IIoT and the efficiency of exploiting blockchain technology to tackle them were discussed. Then, the blockchain-based approaches in the industrial internet of things were classified into eight categories: smart city, manufacturing, healthcare 4.0, energy management, agriculture 4.0, smart homes, autonomous vehicles, and multimedia right management. Practically, using blockchain technology capabilities could improve the quality of real-world services, but despite all progress in the IIoT domain, connectivity, scalability, computational constraints, security, and industry standards are significant challenges in blockchain-based applications. Thus, it is necessary to address the system connectivity, scalability, computational limitations of devices/sensors, security and privacy factors, and international industry standards for the next-generation applications of blockchain technology in the industrial internet of things.

References

1. A. H. Sodhro *et al.*, "Towards 5G-Enabled Self Adaptive Green and Reliable Communication in Intelligent Transportation System," *IEEE Transactions on Intelligent Transportation Systems*, pp. 1–9, 2020.
2. A. Goudarzi, M. M. Honari, and R. Mirzavand, "Resonant Cavity Antennas for 5G Communication Systems: A Review," *Electronics*, vol. 9, no. 7, p. 1080, Jul. 2020.
3. Saqlain, Piao, Shim, and Lee, "Framework of an IoT-based Industrial Data Management for Smart Manufacturing," *Journal of Sensor and Actuator Networks*, vol. 8, no. 2, p. 25, Apr. 2019.
4. R. W. L. Coutinho and A. Boukerche, "Modeling and Analysis of a Shared Edge Caching System for Connected Cars and Industrial IoT-Based Applications," *IEEE Transactions on Industrial Informatics*, vol. 16, no. 3, pp. 2003–2012, Mar. 2020.
5. X. Wang, L. T. Yang, Y. Wang, L. Ren, and M. J. Deen, "ADTT: A Highly-Efficient Distributed Tensor-Train Decomposition Method for IIoT Big Data," *IEEE Transactions on Industrial Informatics*, pp. 1–1, 2020.
6. A. N. Jahromi, J. Sakhnini, H. Karimpour, and A. Dehghantanha, "A deep unsupervised representation learning approach for effective cyber-physical attack detection and identification on highly imbalanced data," in *Proceedings of the 29th Annual International Conference on Computer Science and Software Engineering*, 2019, pp. 14–23.
7. H. Boyes, B. Hallaq, J. Cunningham, and T. Watson, "The industrial internet of things (IIoT): An analysis framework," *Computers in Industry*, vol. 101, pp. 1–12, Oct. 2018.
8. A. Gilchrist, *Industry 4.0*. Berkeley, CA: Apress, 2016.
9. W. Z. Khan, M. H. Rehman, H. M. Zangoti, M. K. Afzal, N. Armi, and K. Salah, "Industrial internet of things: Recent advances, enabling technologies and open challenges," *Computers & Electrical Engineering*, vol. 81, p. 106522, Jan. 2020.
10. E. Sisinni, A. Saifullah, S. Han, U. Jennehag, and M. Gidlund, "Industrial Internet of Things: Challenges, Opportunities, and Directions," *IEEE Transactions on Industrial Informatics*, vol. 14, no. 11, pp. 4724–4734, Nov. 2018.
11. K. T. Park *et al.*, "Cyber Physical Energy System for Saving Energy of the Dyeing Process with Industrial Internet of Things and Manufacturing Big Data," *International Journal of Precision Engineering and Manufacturing-Green Technology*, vol. 7, no. 1, pp. 219–238, Jan. 2020.
12. H. Hui, C. Zhou, S. Xu, and F. Lin, "A novel secure data transmission scheme in industrial internet of things," *China Communications*, vol. 17, no. 1, pp. 73–88, Jan. 2020.
13. S. M. Tahsien, H. Karimipour, and P. Spachos, "Machine learning based solutions for security of Internet of Things (IoT): A survey," *Journal of Network and Computer Applications*, vol. 161, p. 102630, Jul. 2020.
14. A. Yazdinejad, R. M. Parizi, A. Dehghantanha, H. Karimipour, G. Srivastava, and M. Aledhari, "Enabling Drones in the Internet of Things with Decentralized Blockchain-based Security," *IEEE Internet of Things Journal*, pp. 1–1, 2020.

15. J. Chi *et al.*, "A secure and efficient data sharing scheme based on blockchain in industrial Internet of Things," *Journal of Network and Computer Applications*, vol. 167, p. 102710, Oct. 2020.
16. O. Novo, "Blockchain Meets IoT: An Architecture for Scalable Access Management in IoT," *IEEE Internet of Things Journal*, vol. 5, no. 2, pp. 1184–1195, Apr. 2018.
17. A. Panarello, N. Tapas, G. Merlino, F. Longo, and A. Puliafito, "Blockchain and IoT Integration: A Systematic Survey," *Sensors*, vol. 18, no. 8, p. 2575, Aug. 2018.
18. F. Jameel, U. Javaid, W. U. Khan, M. N. Aman, H. Pervaiz, and R. Jäntti, "Reinforcement Learning in Blockchain-Enabled IIoT Networks: A Survey of Recent Advances and Open Challenges," *Sustainability*, vol. 12, no. 12, p. 5161, Jun. 2020.
19. Q. Wang, X. Zhu, Y. Ni, L. Gu, and H. Zhu, "Blockchain for the IoT and industrial IoT: A review," *Internet of Things*, vol. 10, p. 100081, Jun. 2020.
20. N. Siegfried, T. Rosenthal, and A. Benlian, "Blockchain and the Industrial Internet of Things," *Journal of Enterprise Information Management*, vol. ahead-of-p, no. ahead-of-print, Jan. 2020.
21. T. Alladi, V. Chamola, R. M. Parizi, and K.-K. R. Choo, "Blockchain Applications for Industry 4.0 and Industrial IoT: A Review," *IEEE Access*, vol. 7, pp. 176935–176951, 2019.
22. S. Zhao, S. Li, and Y. Yao, "Blockchain Enabled Industrial Internet of Things Technology," *IEEE Transactions on Computational Social Systems*, vol. 6, no. 6, pp. 1442–1453, Dec. 2019.
23. E. Oztemel and S. Gursev, "Literature review of Industry 4.0 and related technologies," *Journal of Intelligent Manufacturing*, vol. 31, no. 1, pp. 127–182, Jan. 2020.
24. G. Büchi, M. Cugno, and R. Castagnoli, "Smart factory performance and Industry 4.0," *Technological Forecasting and Social Change*, vol. 150, p. 119790, Jan. 2020.
25. U. Bodkhe *et al.*, "Blockchain for Industry 4.0: A Comprehensive Review," *IEEE Access*, vol. 8, pp. 79764–79800, 2020.
26. J. Sakhnini, H. Karimipour, A. Dehghantanha, R. M. Parizi, and G. Srivastava, "Security aspects of Internet of Things aided smart grids: A bibliometric survey," *Internet of Things*, p. 100111, Sep. 2019.
27. Y. Liao, E. de Freitas Rocha Loures, and F. Deschamps, "Industrial Internet of Things: A Systematic Literature Review and Insights," *IEEE Internet of Things Journal*, vol. 5, no. 6, pp. 4515–4525, Dec. 2018.
28. H. S. Aghdasi and S. Yousefi, "Enhancing lifetime of visual sensor networks with a preprocessing-based multi-face detection method," *Wireless Networks*, vol. 24, no. 6, pp. 1939–1951, Aug. 2018.
29. S. Najjar-Ghabel and S. Yousefi, "Enhancing Performance of Face Detection in Visual Sensor Networks with a Dynamic-based Approach," *Wireless Personal Communications*, vol. 97, no. 4, pp. 6151–6166, Dec. 2017.
30. S. Najjar-Ghabel, L. Farzinvash, and S. N. Razavi, "Mobile sink-based data gathering in wireless sensor networks with obstacles using artificial intelligence algorithms," *Ad Hoc Networks*, vol. 106, p. 102243, Sep. 2020.
31. L. Farzinvash, S. Najjar-Ghabel, and T. Javadzadeh, "A distributed and energy-efficient approach for collecting emergency data in wireless sensor networks with mobile sinks," *AEU—International Journal of Electronics and Communications*, vol. 108, pp. 79–86, Aug. 2019.
32. F. Derakhshan and S. Yousefi, "A review on the applications of multiagent systems in wireless sensor networks," *International Journal of Distributed Sensor Networks*, vol. 15, no. 5, p. 155014771985076, May 2019.
33. S. Munirathinam, "Industry 4.0: Industrial Internet of Things (IIOT)," 2020, pp. 129–164.
34. S. Schneider, "THE INDUSTRIAL INTERNET OF THINGS (IIoT)," in *Internet of Things and Data Analytics Handbook*, Hoboken, NJ, USA: John Wiley & Sons, Inc., 2016, pp. 41–81.
35. S. Yousefi, F. Derakhshan, and H. Karimipour, "Applications of Big Data Analytics and Machine Learning in the Internet of Things," in *Handbook of Big Data Privacy*, Cham: Springer International Publishing, 2020, pp. 77–108.

36. W. Zhang, Z. Wu, G. Han, Y. Feng, and L. Shu, "LDC: A light-weight dada consensus algorithm based on the blockchain for the industrial Internet of Things for smart city applications," *Future Generation Computer Systems*, vol. 108, pp. 574–582, Jul. 2020.
37. K. N. Qureshi, S. S. Rana, A. Ahmed, and G. Jeon, "A novel and secure attacks detection framework for smart cities industrial internet of things," *Sustainable Cities and Society*, vol. 61, p. 102343, Oct. 2020.
38. S. M. H. Fard, H. Karimipour, A. Dehghantanha, A. N. Jahromi, and G. Srivastava, "Ensemble sparse representation-based cyber threat hunting for security of smart cities," *Computers & Electrical Engineering*, vol. 88, p. 106825, Dec. 2020.
39. K. A. Abuhasel and M. A. Khan, "A Secure Industrial Internet of Things (IIoT) Framework for Resource Management in Smart Manufacturing," *IEEE Access*, vol. 8, pp. 117354–117364, 2020.
40. X. Xu, M. Han, S. M. Nagarajan, and P. Anandhan, "Industrial Internet of Things for smart manufacturing applications using hierarchical trustful resource assignment," *Computer Communications*, vol. 160, pp. 423–430, Jul. 2020.
41. A. Al-Abassi, H. Karimipour, A. Dehghantanha, and R. M. Parizi, "An Ensemble Deep Learning-Based Cyber-Attack Detection in Industrial Control System," *IEEE Access*, vol. 8, pp. 83965–83973, 2020.
42. F. Al-Turjman and S. Alturjman, "Context-Sensitive Access in Industrial Internet of Things (IIoT) Healthcare Applications," *IEEE Transactions on Industrial Informatics*, vol. 14, no. 6, pp. 2736–2744, Jun. 2018.
43. E. Koç, "How Can Industrial Internet of Things (IIoT) Improve Enterprise Productivity?," 2020, pp. 112–133.
44. I. Mugarza, A. Amurrio, E. Azketa, and E. Jacob, "Dynamic Software Updates to Enhance Security and Privacy in High Availability Energy Management Applications in Smart Cities," *IEEE Access*, vol. 7, pp. 42269–42279, 2019.
45. S. Yousefi, F. Derakhshan, H. S. Aghdasi, and H. Karimipour, "An energy-efficient artificial bee colony-based clustering in the internet of things," *Computers & Electrical Engineering*, vol. 86, p. 106733, Sep. 2020.
46. S. K. Kaya, "Industrial Internet of Things," 2020, pp. 134–155.
47. J. Y. Cho *et al.*, "Significant power enhancement method of magneto-piezoelectric energy harvester through directional optimization of magnetization for autonomous IIoT platform," *Applied Energy*, vol. 254, p. 113710, Nov. 2019.
48. M. Liu, F. R. Yu, Y. Teng, V. C. M. Leung, and M. Song, "Performance Optimization for Blockchain-Enabled Industrial Internet of Things (IIoT) Systems: A Deep Reinforcement Learning Approach," *IEEE Transactions on Industrial Informatics*, vol. 15, no. 6, pp. 3559–3570, Jun. 2019.
49. J. BUTSCHAN, S. HEIDENREICH, B. WEBER, and T. KRAEMER, "TACKLING HURDLES TO DIGITAL TRANSFORMATION—THE ROLE OF COMPETENCIES FOR SUCCESSFUL INDUSTRIAL INTERNET OF THINGS (IIoT) IMPLEMENTATION," *International Journal of Innovation Management*, vol. 23, no. 4, p. 1950036, May 2019.
50. https://www.marketsandmarkets.com/Market-Reports/industrial-internet-of-things-market-129733727.html, "Industrial IoT (IIoT) Market," 2020.
51. M. Younan, E. H. Houssein, M. Elhoseny, and A. A. Ali, "Challenges and recommended technologies for the industrial internet of things: A comprehensive review," *Measurement*, vol. 151, p. 107198, Feb. 2020.
52. K. Wang, Y. Wang, Y. Sun, S. Guo, and J. Wu, "Green Industrial Internet of Things Architecture: An Energy-Efficient Perspective," *IEEE Communications Magazine*, vol. 54, no. 12, pp. 48–54, Dec. 2016.
53. C. Vijayakumaran, B. Muthusenthil, and B. Manickavasagam, "A reliable next generation cyber security architecture for industrial internet of things environment," *International Journal of Electrical and Computer Engineering (IJECE)*, vol. 10, no. 1, p. 387, Feb. 2020.

54. H. HaddadPajouh, A. Dehghantanha, R. M. Parizi, M. Aledhari, and H. Karimipour, "A survey on internet of things security: Requirements, challenges, and solutions," *Internet of Things*, p. 100129, Nov. 2019.
55. S. Najjar-Ghabel, L. Farzinvash, and S. N. Razavi, "HPDMS: high-performance data harvesting in wireless sensor networks with mobile sinks," *The Journal of Supercomputing*, vol. 76, no. 4, pp. 2748–2776, Apr. 2020.
56. Y. E. Oktian, S.-G. Lee, and B.-G. Lee, "Blockchain-Based Continued Integrity Service for IoT Big Data Management: A Comprehensive Design," *Electronics*, vol. 9, no. 9, p. 1434, Sep. 2020.
57. L. Faramondi, G. Oliva, R. Setola, and L. Vollero, "IIoT in the Hospital Scenario: Hospital 4.0, Blockchain and Robust Data Management," 2019, pp. 271–285.
58. D. Jiang, Y. Wang, Z. Lv, W. Wang, and H. Wang, "An Energy-Efficient Networking Approach in Cloud Services for IIoT Networks," *IEEE Journal on Selected Areas in Communications*, vol. 38, no. 5, pp. 928–941, May 2020.
59. J. Sengupta, S. Ruj, and S. Das Bit, "A Comprehensive Survey on Attacks, Security Issues and Blockchain Solutions for IoT and IIoT," *Journal of Network and Computer Applications*, vol. 149, p. 102481, Jan. 2020.
60. Z. Guan, X. Lu, N. Wang, J. Wu, X. Du, and M. Guizani, "Towards secure and efficient energy trading in IIoT-enabled energy internet: A blockchain approach," *Future Generation Computer Systems*, vol. 110, pp. 686–695, Sep. 2020.
61. H. Al-Aqrabi, A. P. Johnson, R. Hill, P. Lane, and T. Alsboui, "Hardware-Intrinsic Multi-Layer Security: A New Frontier for 5G Enabled IIoT," *Sensors*, vol. 20, no. 7, p. 1963, Mar. 2020.
62. R. Zhohov, D. Minovski, P. Johansson, and K. Andersson, "Real-time Performance Evaluation of LTE for IIoT," in *2018 IEEE 43rd Conference on Local Computer Networks (LCN)*, 2018, pp. 623–631.
63. C. Xia, X. Jin, C. Xu, Y. Wang, and P. Zeng, "Real-time scheduling under heterogeneous routing for industrial Internet of Things," *Computers & Electrical Engineering*, vol. 86, p. 106740, Sep. 2020.
64. H. Wu, X. Lyu, and H. Tian, "Online Optimization of Wireless Powered Mobile-Edge Computing for Heterogeneous Industrial Internet of Things," *IEEE Internet of Things Journal*, vol. 6, no. 6, pp. 9880–9892, Dec. 2019.
65. S. Yousefi, F. Derakhshan, and A. Bokani, "Mobile Agents for Route Planning in Internet of Things Using Markov Decision Process," in *2018 IEEE International Conference on Smart Energy Grid Engineering (SEGE)*, 2018, pp. 303–307.
66. S. Yousefi, F. Derakhshan, H. Karimipour, and H. S. Aghdasi, "An efficient route planning model for mobile agents on the internet of things using Markov decision process," *Ad Hoc Networks*, vol. 98, p. 102053, Mar. 2020.
67. S. Najjar-Ghabel, S. Yousefi, and L. Farzinvash, "Reliable data gathering in the Internet of Things using artificial bee colony," *TURKISH JOURNAL OF ELECTRICAL ENGINEERING & COMPUTER SCIENCES*, vol. 26, no. 4, pp. 1710–1723, Jul. 2018.
68. G. Ateniese, B. Magri, D. Venturi, and E. Andrade, "Redactable Blockchain—or—Rewriting History in Bitcoin and Friends," in *2017 IEEE European Symposium on Security and Privacy (EuroS&P)*, 2017, pp. 111–126.
69. S. Singh and N. Singh, "Blockchain: Future of financial and cyber security," in *2016 2nd International Conference on Contemporary Computing and Informatics (IC3I)*, 2016, pp. 463–467.
70. H.-N. Dai, Z. Zheng, and Y. Zhang, "Blockchain for Internet of Things: A Survey," *IEEE Internet of Things Journal*, vol. 6, no. 5, pp. 8076–8094, Oct. 2019.
71. X. Wang *et al.*, "Survey on blockchain for Internet of Things," *Computer Communications*, vol. 136, pp. 10–29, Feb. 2019.

72. Y. Xiao, N. Zhang, W. Lou, and Y. T. Hou, “A Survey of Distributed Consensus Protocols for Blockchain Networks,” *IEEE Communications Surveys & Tutorials*, vol. 22, no. 2, pp. 1432–1465, 2020.
73. Q. Feng, D. He, S. Zeadally, M. K. Khan, and N. Kumar, “A survey on privacy protection in blockchain system,” *Journal of Network and Computer Applications*, vol. 126, pp. 45–58, Jan. 2019.
74. G. Ayoade, V. Karande, L. Khan, and K. Hamlen, “Decentralized IoT Data Management Using BlockChain and Trusted Execution Environment,” in *2018 IEEE International Conference on Information Reuse and Integration (IRI)*, 2018, pp. 15–22.
75. A. Yazdinejad, R. M. Parizi, A. Dehghantanha, Q. Zhang, and K.-K. R. Choo, "An energy-efficient SDN controller architecture for IoT networks with blockchain-based security," IEEE Transactions on Services Computing, vol. 13, no. 4, pp. 625-638, 2020.
76. A. Yazdinejad, G. Srivastava, R. M. Parizi, A. Dehghantanha, H. Karimipour, and S. R. Karizno, “SLPoW: Secure and Low Latency Proof of Work Protocol for Blockchain in Green IoT Networks,” in 2020 IEEE 91st Vehicular Technology Conference (VTC2020-Spring), 2020, pp. 1–5: IEEE.
77. A. Yazdinejadna, R. M. Parizi, A. Dehghantanha, and M. S. Khan, “A kangaroo-based intrusion detection system on software-defined networks,” Computer Networks, vol. 184, p. 107688, 2021.
78. A. Yazdinejad, R. M. Parizi, A. Dehghantanha, and K.-K. R. Choo, “Blockchain-enabled authentication handover with efficient privacy protection in SDN-based 5G networks,” IEEE Transactions on Network Science and Engineering, 2019.
79. A. Yazdinejad, G. Srivastava, R. M. Parizi, A. Dehghantanha, K.-K. R. Choo, and M. Aledhari, “Decentralized authentication of distributed patients in hospital networks using blockchain,” IEEE journal of biomedical and health informatics, vol. 24, no. 8, pp. 2146-2156, 2020.
80. A. Yazdinejad, R. M. Parizi, G. Srivastava, A. Dehghantanha, and K.-K. R. Choo, “Energy efficient decentralized authentication in internet of underwater things using blockchain,” in 2019 IEEE Globecom Workshops (GC Wkshps), 2019, pp. 1–6: IEEE
81. A. Yazdinejad, R. M. Parizi, A. Dehghantanha, and K.-K. R. Choo, “P4-to-blockchain: A secure blockchain-enabled packet parser for software defined networking,” Computers & Security, vol. 88, p. 101629, 2020.
82. B. Mo, K. Su, S. Wei, C. Liu, and J. Guo, “A Solution for Internet of Things based on Blockchain Technology,” in *2018 IEEE International Conference on Service Operations and Logistics, and Informatics (SOLI)*, 2018, pp. 112–117.
83. K. Kataoka, S. Gangwar, and P. Podili, “Trust list: Internet-wide and distributed IoT traffic management using blockchain and SDN,” in *2018 IEEE 4th World Forum on Internet of Things (WF-IoT)*, 2018, pp. 296–301.
84. V. Dedeoglu *et al.*, “Blockchain Technologies for IoT,” 2020, pp. 55–89.
85. S. Ferretti and G. D’Angelo, “On the Ethereum blockchain structure: A complex networks theory perspective,” *Concurrency and Computation: Practice and Experience*, vol. 32, no. 12, Jun. 2020.
86. W. Chen *et al.*, “Cooperative and Distributed Computation Offloading for Blockchain-Empowered Industrial Internet of Things,” *IEEE Internet of Things Journal*, vol. 6, no. 5, pp. 8433–8446, Oct. 2019.
87. P. W. Khan and Y. Byun, “A Blockchain-Based Secure Image Encryption Scheme for the Industrial Internet of Things,” *Entropy*, vol. 22, no. 2, p. 175, Feb. 2020.
88. H. Yao, T. Mai, J. Wang, Z. Ji, C. Jiang, and Y. Qian, “Resource Trading in Blockchain-Based Industrial Internet of Things,” *IEEE Transactions on Industrial Informatics*, vol. 15, no. 6, pp. 3602–3609, Jun. 2019.
89. K. Zhang, Y. Zhu, S. Maharjan, and Y. Zhang, “Edge Intelligence and Blockchain Empowered 5G Beyond for the Industrial Internet of Things,” *IEEE Network*, vol. 33, no. 5, pp. 12–19, Sep. 2019.

90. M. H. ur Rehman, I. Yaqoob, K. Salah, M. Imran, P. P. Jayaraman, and C. Perera, "The role of big data analytics in industrial Internet of Things," *Future Generation Computer Systems*, vol. 99, pp. 247–259, Oct. 2019.
91. H. Al-Aqrabi, A. P. Johnson, R. Hill, P. Lane, and L. Liu, "A Multi-layer Security Model for 5G-Enabled Industrial Internet of Things," 2019, pp. 279–292.
92. T. N. Pham, M.-F. Tsai, D. B. Nguyen, C.-R. Dow, and D.-J. Deng, "A Cloud-Based Smart-Parking System Based on Internet-of-Things Technologies," *IEEE Access*, vol. 3, pp. 1581–1591, 2015.
93. C. Lazaroiu and M. Roscia, "Smart district through IoT and Blockchain," in *2017 IEEE 6th International Conference on Renewable Energy Research and Applications (ICRERA)*, 2017, pp. 454–461.
94. L. Fan, J. R. Gil-Garcia, D. Werthmuller, G. B. Burke, and X. Hong, "Investigating blockchain as a data management tool for IoT devices in smart city initiatives," in *Proceedings of the 19th Annual International Conference on Digital Government Research Governance in the Data Age—dgo '18*, 2018, pp. 1–2.
95. D. Ron and A. Shamir, "Quantitative Analysis of the Full Bitcoin Transaction Graph," 2013, pp. 6–24.
96. T. Robinson, "Bitcoin is not anonymous," *ResPublica*, 2020. [Online]. Available: http://www.respublica.org.uk/disraeli-room-post/2015/03/24/bitcoin-is-not-anonymous/.
97. J. Wan, J. Li, M. Imran, D. Li, and Fazal-e-Amin, "A Blockchain-Based Solution for Enhancing Security and Privacy in Smart Factory," *IEEE Transactions on Industrial Informatics*, vol. 15, no. 6, pp. 3652–3660, Jun. 2019.
98. J. Huang, L. Kong, G. Chen, M.-Y. Wu, X. Liu, and P. Zeng, "Towards Secure Industrial IoT: Blockchain System With Credit-Based Consensus Mechanism," *IEEE Transactions on Industrial Informatics*, vol. 15, no. 6, pp. 3680–3689, Jun. 2019.
99. C. Zhang, G. Zhou, H. Li, and Y. Cao, "Manufacturing Blockchain of Things for the Configuration of a Data-and Knowledge-Driven Digital Twin Manufacturing Cell," *IEEE Internet of Things Journal*, pp. 1–1, 2020.
100. Z. Zhang, L. Huang, R. Tang, T. Peng, L. Guo, and X. Xiang, "Industrial Blockchain of Things: A Solution for Trustless Industrial Data Sharing and Beyond," in *2020 IEEE 16th International Conference on Automation Science and Engineering (CASE)*, 2020, pp. 1187–1192.
101. P. K. Sharma, N. Kumar, and J. H. Park, "Blockchain-Based Distributed Framework for Automotive Industry in a Smart City," *IEEE Transactions on Industrial Informatics*, vol. 15, no. 7, pp. 4197–4205, Jul. 2019.
102. A. Angrish, B. Craver, M. Hasan, and B. Starly, "A Case Study for Blockchain in Manufacturing: 'FabRec': A Prototype for Peer-to-Peer Network of Manufacturing Nodes," *Procedia Manufacturing*, vol. 26, pp. 1180–1192, 2018.
103. G. Aceto, V. Persico, and A. Pescapé, "Industry 4.0 and Health: Internet of Things, Big Data, and Cloud Computing for Healthcare 4.0," *Journal of Industrial Information Integration*, vol. 18, p. 100129, Jun. 2020.
104. S. Tanwar, K. Parekh, and R. Evans, "Blockchain-based electronic healthcare record system for healthcare 4.0 applications," *Journal of Information Security and Applications*, vol. 50, p. 102407, Feb. 2020.
105. P. P. Jayaraman, A. R. M. Forkan, A. Morshed, P. D. Haghighi, and Y. Kang, "Healthcare 4.0: A review of frontiers in digital health," *WIREs Data Mining and Knowledge Discovery*, vol. 10, no. 2, Mar. 2020.
106. Q. Xia, E. B. Sifah, K. O. Asamoah, J. Gao, X. Du, and M. Guizani, "MeDShare: Trust-Less Medical Data Sharing Among Cloud Service Providers via Blockchain," *IEEE Access*, vol. 5, pp. 14757–14767, 2017.
107. S. Jiang, J. Cao, H. Wu, Y. Yang, M. Ma, and J. He, "BlocHIE: A BLOCkchain-Based Platform for Healthcare Information Exchange," in *2018 IEEE International Conference on Smart Computing (SMARTCOMP)*, 2018, pp. 49–56.

108. M. Amrollahi, S. Hadayeghparast, H. Karimipour, F. Derakhshan, and G. Srivastava, "Enhancing Network Security Via Machine Learning: Opportunities and Challenges," in *Handbook of Big Data Privacy*, Cham: Springer International Publishing, 2020, pp. 165–189.
109. D. H. Wang, "IoT based Clinical Sensor Data Management and Transfer using Blockchain Technology," *Journal of ISMAC*, vol. 2, no. 3, pp. 154–159, Jul. 2020.
110. H.-T. Wu and C.-W. Tsai, "Toward Blockchains for Health-Care Systems: Applying the Bilinear Pairing Technology to Ensure Privacy Protection and Accuracy in Data Sharing," *IEEE Consumer Electronics Magazine*, vol. 7, no. 4, pp. 65–71, Jul. 2018.
111. J. Zhang, N. Xue, and X. Huang, "A Secure System For Pervasive Social Network-Based Healthcare," *IEEE Access*, vol. 4, pp. 9239–9250, 2016.
112. T. Bocek, B. B. Rodrigues, T. Strasser, and B. Stiller, "Blockchains everywhere—a use-case of blockchains in the pharma supply-chain," in *2017 IFIP/IEEE Symposium on Integrated Network and Service Management (IM)*, 2017, pp. 772–777.
113. A. Lipman, "MedRec," *Robert Wood Johnson Foundation*, 2020. [Online]. Available: https://medrec.media.mit.edu/.
114. A. Albeyatti and M. Tayeb, "MedicalChain," *Medicalchain*, 2020. [Online]. Available: https://medicalchain.com/en/.
115. S. Maitra, V. P. Yanambaka, D. Puthal, A. Abdelgawad, and K. Yelamarthi, "Integration of Internet of Things and blockchain toward portability and low-energy consumption," *Transactions on Emerging Telecommunications Technologies*, Aug. 2020.
116. K. Gai, Y. Wu, L. Zhu, Z. Zhang, and M. Qiu, "Differential Privacy-Based Blockchain for Industrial Internet-of-Things," *IEEE Transactions on Industrial Informatics*, vol. 16, no. 6, pp. 4156–4165, Jun. 2020.
117. F. Darbandi, A. Jafari, H. Karimipour, A. Dehghantanha, F. Derakhshan, and K.-K. Raymond Choo, "Real-time stability assessment in smart cyber-physical grids: a deep learning approach," *IET Smart Grid*, vol. 3, no. 4, pp. 454–461, Aug. 2020.
118. Z. Li, J. Kang, R. Yu, D. Ye, Q. Deng, and Y. Zhang, "Consortium Blockchain for Secure Energy Trading in Industrial Internet of Things," *IEEE Transactions on Industrial Informatics*, pp. 1–1, 2017.
119. J. Kang, R. Yu, X. Huang, S. Maharjan, Y. Zhang, and E. Hossain, "Enabling Localized Peer-to-Peer Electricity Trading Among Plug-in Hybrid Electric Vehicles Using Consortium Blockchains," *IEEE Transactions on Industrial Informatics*, vol. 13, no. 6, pp. 3154–3164, Dec. 2017.
120. H. A. Khattak, K. Tehreem, A. Almogren, Z. Ameer, I. U. Din, and M. Adnan, "Dynamic pricing in industrial internet of things: Blockchain application for energy management in smart cities," *Journal of Information Security and Applications*, vol. 55, p. 102615, Dec. 2020.
121. Z. Guan *et al.*, "Privacy-Preserving and Efficient Aggregation Based on Blockchain for Power Grid Communications in Smart Communities," *IEEE Communications Magazine*, vol. 56, no. 7, pp. 82–88, Jul. 2018.
122. N. Z. Aitzhan and D. Svetinovic, "Security and Privacy in Decentralized Energy Trading Through Multi-Signatures, Blockchain and Anonymous Messaging Streams," *IEEE Transactions on Dependable and Secure Computing*, vol. 15, no. 5, pp. 840–852, Sep. 2018.
123. "PowerLedger," *powerledger*, 2020. [Online]. Available: https://powerledger.io/whitepaper/.
124. A. Goranovic, M. Meisel, L. Fotiadis, S. Wilker, A. Treytl, and T. Sauter, "Blockchain applications in microgrids an overview of current projects and concepts," in *IECON 2017—43rd Annual Conference of the IEEE Industrial Electronics Society*, 2017, pp. 6153–6158.
125. Z. Zhai, J. F. Martínez, V. Beltran, and N. L. Martínez, "Decision support systems for agriculture 4.0: Survey and challenges," *Computers and Electronics in Agriculture*, vol. 170, p. 105256, Mar. 2020.
126. S. K. Roy and D. De, "Genetic Algorithm based Internet of Precision Agricultural Things (IopaT) for Agriculture 4.0," *Internet of Things*, p. 100201, Mar. 2020.

127. Y. Liu, X. Ma, L. Shu, G. P. Hancke, and A. M. Abu-Mahfouz, "From Industry 4.0 to Agriculture 4.0: Current Status, Enabling Technologies, and Research Challenges," *IEEE Transactions on Industrial Informatics*, pp. 1–1, 2020.
128. J. Lin, Z. Shen, A. Zhang, and Y. Chai, "Blockchain and IoT based Food Traceability for Smart Agriculture," in *Proceedings of the 3rd International Conference on Crowd Science and Engineering—ICCSE'18*, 2018, pp. 1–6.
129. J. Hua, X. Wang, M. Kang, H. Wang, and F.-Y. Wang, "Blockchain Based Provenance for Agricultural Products: A Distributed Platform with Duplicated and Shared Bookkeeping," in *2018 IEEE Intelligent Vehicles Symposium (IV)*, 2018, pp. 97–101.
130. M. P. Caro, M. S. Ali, M. Vecchio, and R. Giaffreda, "Blockchain-based traceability in Agri-Food supply chain management: A practical implementation," in *2018 IoT Vertical and Topical Summit on Agriculture—Tuscany (IOT Tuscany)*, 2018, pp. 1–4.
131. S. H. Awan *et al.*, "BlockChain with IoT, an Emergent Routing Scheme for Smart Agriculture," *International Journal of Advanced Computer Science and Applications*, vol. 11, no. 4, 2020.
132. A. N. Putri, M. Hariadi, and A. D. Wibawa, "Smart Agriculture Using Supply Chain Management Based On Hyperledger Blockchain," *IOP Conference Series: Earth and Environmental Science*, vol. 466, no. 1, 2020.
133. "iGrow," 2019. [Online]. Available: https://www.igrowchain.com.
134. avenews team, "Avenews-GT," 2020. [Online]. Available: https://www.avenews-gt.com/about.
135. T. K. L. Hui, R. S. Sherratt, and D. D. Sánchez, "Major requirements for building Smart Homes in Smart Cities based on Internet of Things technologies," *Future Generation Computer Systems*, vol. 76, pp. 358–369, Nov. 2017.
136. M. Afzal, Q. Huang, W. Amin, K. Umer, A. Raza, and M. Naeem, "Blockchain Enabled Distributed Demand Side Management in Community Energy System With Smart Homes," *IEEE Access*, vol. 8, pp. 37428–37439, 2020.
137. P. K. Singh, R. Singh, S. K. Nandi, and S. Nandi, "Managing Smart Home Appliances with Proof of Authority and Blockchain," 2019, pp. 221–232.
138. D. Han, H. Kim, and J. Jang, "Blockchain based smart door lock system," in *2017 International Conference on Information and Communication Technology Convergence (ICTC)*, 2017, pp. 1165–1167.
139. M. Begli, F. Derakhshan, and H. Karimipour, "A Layered Intrusion Detection System for Critical Infrastructure Using Machine Learning," in *2019 IEEE 7th International Conference on Smart Energy Grid Engineering (SEGE)*, 2019, pp. 120–124.
140. U. Nadiya, M. I. Rizqyawan, and O. Mahnedra, "Blockchain-based Secure Data Storage for Door Lock System," in *2019 4th International Conference on Information Technology, Information Systems and Electrical Engineering (ICITISEE)*, 2019, pp. 140–144.
141. L. de Camargo Silva, M. Samaniego, and R. Deters, "IoT and Blockchain for Smart Locks," in *2019 IEEE 10th Annual Information Technology, Electronics and Mobile Communication Conference (IEMCON)*, 2019, pp. 0262–0269.
142. C. Lin, D. He, N. Kumar, X. Huang, P. Vijayakumar, and K.-K. R. Choo, "HomeChain: A Blockchain-Based Secure Mutual Authentication System for Smart Homes," *IEEE Internet of Things Journal*, vol. 7, no. 2, pp. 818–829, Feb. 2020.
143. Y. Lee, S. Rathore, J. H. Park, and J. H. Park, "A blockchain-based smart home gateway architecture for preventing data forgery," *Human-centric Computing and Information Sciences*, vol. 10, no. 1, p. 9, Dec. 2020.
144. D. Minoli, "Positioning of blockchain mechanisms in IOT-powered smart home systems: A gateway-based approach," *Internet of Things*, vol. 10, p. 100147, Jun. 2020.
145. L. Yang, X.-Y. Liu, and W. Gong, "Secure Smart Home Systems: A Blockchain Perspective," in *IEEE INFOCOM 2020—IEEE Conference on Computer Communications Workshops (INFOCOM WKSHPS)*, 2020, pp. 1003–1008.

146. S. Zhang, J. Rong, and B. Wang, "A privacy protection scheme of smart meter for decentralized smart home environment based on consortium blockchain," *International Journal of Electrical Power & Energy Systems*, vol. 121, p. 106140, Oct. 2020.
147. R. Yang, X. Chang, J. Mišić, and V. B. Mišić, "Assessing blockchain selfish mining in an imperfect network: Honest and selfish miner views," *Computers & Security*, vol. 97, p. 101956, Oct. 2020.
148. A. Miglani, N. Kumar, V. Chamola, and S. Zeadally, "Blockchain for Internet of Energy management: Review, solutions, and challenges," *Computer Communications*, vol. 151, pp. 395–418, Feb. 2020.
149. Y. Yuan and F.-Y. Wang, "Towards blockchain-based intelligent transportation systems," in *2016 IEEE 19th International Conference on Intelligent Transportation Systems (ITSC)*, 2016, pp. 2663–2668.
150. H. Liu, Y. Zhang, S. Zheng, and Y. Li, "Electric Vehicle Power Trading Mechanism Based on Blockchain and Smart Contract in V2G Network," *IEEE Access*, vol. 7, pp. 160546–160558, 2019.
151. V. Deshpande, L. George, and H. Badis, "SaFe: A Blockchain and Secure Element Based Framework for Safeguarding Smart Vehicles," in *2019 12th IFIP Wireless and Mobile Networking Conference (WMNC)*, 2019, pp. 181–188.
152. S. Huckle, R. Bhattacharya, M. White, and N. Beloff, "Internet of Things, Blockchain and Shared Economy Applications," *Procedia Computer Science*, vol. 98, pp. 461–466, 2016.
153. M. Humayun, N. Jhanjhi, B. Hamid, and G. Ahmed, "Emerging Smart Logistics and Transportation Using IoT and Blockchain," *IEEE Internet of Things Magazine*, vol. 3, no. 2, pp. 58–62, Jun. 2020.
154. X. Lin, J. Wu, S. Mumtaz, S. Garg, J. Li, and M. Guizani, "Blockchain-based On-Demand Computing Resource Trading in IoV-Assisted Smart City," *IEEE Transactions on Emerging Topics in Computing*, pp. 1–1, 2020.
155. I. Mistry, S. Tanwar, S. Tyagi, and N. Kumar, "Blockchain for 5G-enabled IoT for industrial automation: A systematic review, solutions, and challenges," *Mechanical Systems and Signal Processing*, vol. 135, p. 106382, Jan. 2020.
156. A. Ghaffar *et al.*, "Smart Contracts for Research Lab Sharing Scholars Data Rights Management over the Ethereum Blockchain Network," 2020, pp. 70–81.
157. H. Si, C. Sun, Y. Li, H. Qiao, and L. Shi, "IoT information sharing security mechanism based on blockchain technology," *Future Generation Computer Systems*, vol. 101, pp. 1028–1040, Dec. 2019.
158. D. Bhowmik and T. Feng, "The multimedia blockchain: A distributed and tamper-proof media transaction framework," in *2017 22nd International Conference on Digital Signal Processing (DSP)*, 2017, pp. 1–5.

Application of Deep Learning on IoT-Enabled Smart Grid Monitoring

Ibrahim Al-Omari, Shahrzad Hadayeghparast, and Hadis Karimipour

1 Introduction

The electrical power utilities have been a vital industry of the modern businesses and societies due its importance in enhancing people's life. It went through a great development to make this system reliable and efficient. Moreover, the conventional power system (PS) has moved toward the smart grid (SG) during the last two decades as a reason of the increasing of the integrated distributed energy recourses (DERs), micro-grid, aggregated demand response, and the costumers' participation in generating electric power [1]. That has improved the reliability and offered an affective chance to transform the distribution system to be active and controllable system (see Fig. 1) [2]. Along beside that development, new challenges have been posed in controlling, protecting and monitoring the power system specifically in the distribution systems [3]. That has increased the need for providers to improve the traditional controlling system in order to mitigate the gap between the technological advances in the SG and the conventional power system [4]. The energy management system (EMS), which is the responsible of monitoring and optimizing the PS, has different departments such as controlling, load forecasting and optimizing. The backbone of the EMS is the state estimation (SE) in which it plays a key role to deal with the massive network, less predictable load profiles and passive variables. The accuracy and reliability of the SE have a significant impact on the operational conditions of the SG in which it moderates losses under drastic changes in the SG behaviors by formulating the time-varying nature of the system's model [5–7].

The concept of SE in power system was established in 1970 [8]. The contribution of [8] is to obtain the best state estimates of the power system variables by providing

I. Al-Omari · S. Hadayeghparast (✉) · H. Karimipour
Department of Electrical and Software Engineering, University of Calgary, Calgary, AB, Canada
e-mail: ialomari@uoguelph.ca; shadayeg@uoguelph.ca; hadis.karimipour@ucalgary.ca

H. Karimipour, F. Derakhshan (eds.), *AI-Enabled Threat Detection and Security Analysis for Industrial IoT*, https://doi.org/10.1007/978-3-030-76613-9_5

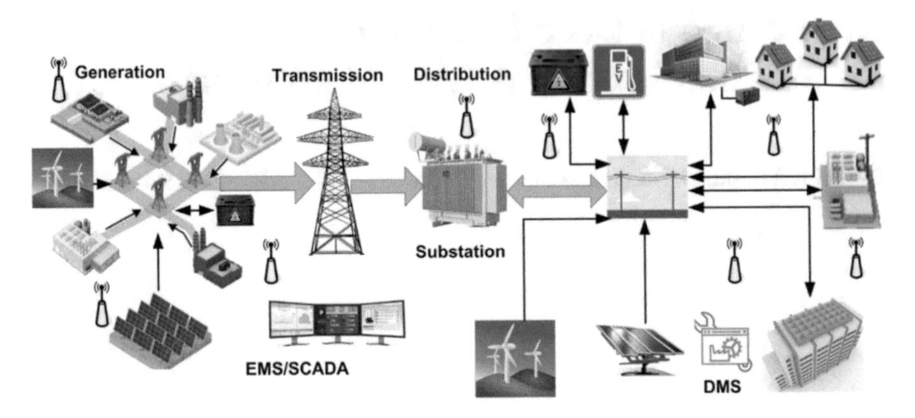

Fig. 1 The smart grid and active DS [2]

a methodology for processing the measurements from the electric power network. Since then, the SE in transmission system (TS) has been used widely and successfully. On the other hand, the distribution system (DS) at that time was considered as easy to predict and manage due to the unidirectional power flow in which the design and operation of the DS was being a passive system. However, the DS has become a bidirectional communication infrastructure based on the concept of SG which makes the SG more complicated. That requires to develop the modern EMS; in particular, the smart grid state estimation (SGSE) which plays a vital rule in the EMS to improve the system efficiency and real time operational conditions. A general figure of the SGSE and a description of the SGSE general requirements is shown in Fig. 2 [5]. Although the state estimation is well developed and widely used in the TS, the distribution system or smart grid SG is still the subject of active research. A comparison between the transmission system and smart grid factors will give an overall figure to overcome the main challenges in SGSE. Firstly, the TS is considered observable, unlike the highly uncertainties and unknown nodes in SG which means the number of metering devices is very small compare to such a widespread size of the SG networks [9]. The second factor is that SG is extremely unbalanced and has a magnificent level of complexity which leads to having a difficulty for designing the proper SGSE. The communication system is the third factor where the accuracy and the rate of data exchange are limited in the SG duo to the constraints [10]. The fourth factor is the integration of DERs which is another considerable issue in SGSE that increases the uncertainties in the SG. Lastly, The network configuration of the SG creates another degree of complexity duo to the lack of the stored topologies and its changeable configurations [11]. Moreover, the lack of the observed nodes and the availability of measurements in the SG is the main issue up to date, which makes the task of SE more complicated. All these factors drive challenges in designing the suitable state estimator and disable the implementation of the TS state estimators' methods in SGSE [12].

The main contribution of this review paper is to investigate the scope of SGSE and survey the most related papers to the learning-based SGSE. Moreover, it has

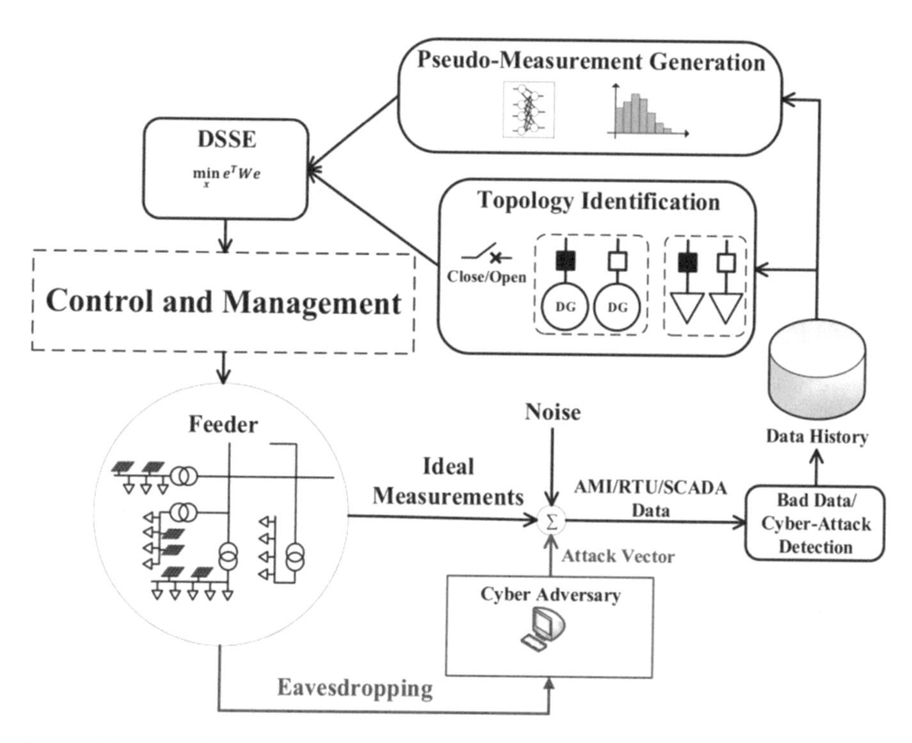

Fig. 2 SGSE line diagram in smart grid [5]

gone through different applications in the SGSE beside the Learning-based methods to provide a comprehensive survey for SGSE and to summarize the studies which are mainly have focused on learning methods SGSE.

The rest of the paper is organized as follows: Section 2 explains the smart grid state estimations. Sections 3 and 4 present fundamental concepts and formulation of SE for active distribution networks, respectively. Section 5 presents various applications in SGSE mainly the conventional approaches and the filter-based models. Section 6 defines the main focus of this paper the learning-based applications in SGSE. Simulation results for learning-based SE methods are presented in Sect. 7. Finally, this paper is concluded in Sect. 8.

2 Smart Grid State Estimation

The SE duty in SG has three main actions starting in processing the received data, filtering the measurements from noise and detect gross errors. Any state estimator in the SG has to follow certain stages to provide an accurate and reliable state estimates for the operational conditions in SG. These stages are described in five steps as shown in Fig. 3 and detailed in the following steps [5, 6, 12]:

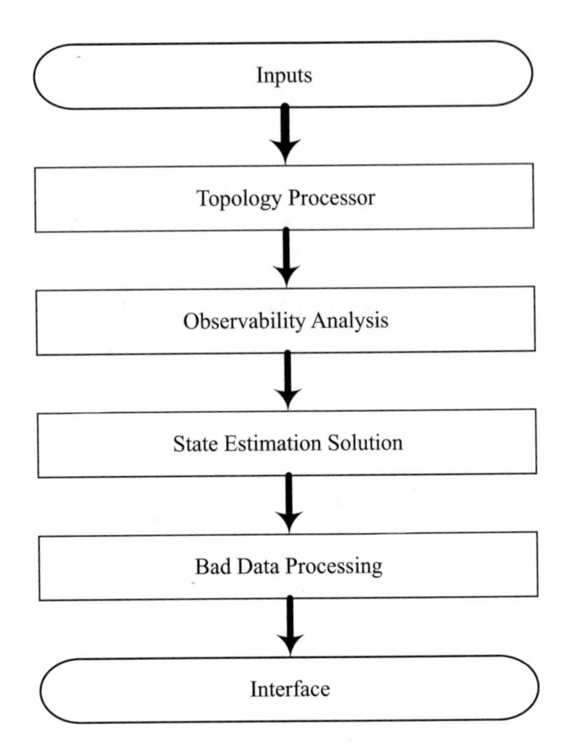

Fig. 3 State estimation block diagram

1. Topology processor: this stage is to aggregate the physical networks' status data in order to formulate the certain on-line configuration of the system.
2. Observability analysis: it identifies the provided set of measurement either it is observed all the required data for the next step or not, which these data are the network configuration and the measurements of all the buses.
3. The state estimation solution: this step is to apply the optimal calculation for the state estimates in the SG.
4. Bad data processing: it distinguishes the gross error as well as the injected bad data in order to eliminate them properly.
5. Interface: it characterizes the structural errors and assure that there is enough measurement redundancy by estimating different systemic parameters.

The SGSE can be categorized to three main forms: static state estimation (SSE), dynamic state estimation (DSE) and forecast-aided state estimation (FASE) in which each form has its own procedure based on the inputs and outputs intervals time. The SSE in SG is simply assuming the system is steady state in which it applies a conventional computation method for detection, statistics and filtering in real time and long term estimation e.g. load forecasting for days or even weeks [13].

The DSE is mainly for estimating the states of the system's dynamics in real time or stamp time a head. It is the preferable state estimation form in the EMS since it deals with the real time measurements and network topologies, which these topologies are greatly changeable in the SG [14]. The DSE facilitates the monitoring system in EMS effectively in which its estimation procedures are recursively updated the system's state estimates based on the previous and current state estimates in order to detect the changes in the network configuration and to estimate the next time state vector. The DSE is superior in supervising the operational conditions in the SG and then to act based on the SE results against emergent physical inconstancy or bad data injections [7]. Along beside the SSE and DSE is the FASE which is technically works like the DSE. However, the DSE is greatly concerned about the transient stability studies in the power system whereas the FSAE is considered as a combinational method of SSE and DSE. FASE is a new terminology for dynamic-static state estimation [15]. All these forms of state estimation in EMS is considering for planning and analysis purposes in which SSE is significantly used for planning and DSE is used in analysis and real time functionality.

In the recent years, the employing of the measurement devices (e.g. Phasor Measurement Unit (PMU), Advanced Metering Infrastructure (AMI)) have improved the state estimation methods in TS; however, it infeasible to be implemented in the whole SG because of its highly economic costs. The PMU has been applied in different SGSE methods with a limited number of devices [15]. The PMU is considered as a unique device in the power system in which it has the ability to provide the phase angle of the electrical waveforms with sampling rate compare to the conventional supervisory control and data acquisition (SCADA) as well as providing the root mean square (RMS) [16].

3 Fundamental Concepts for State Estimation Concepts in Active Distribution System

Distribution system SE (DSSE) is the process of obtaining system state variables utilizing a limited number of measurements deployed at specific locations of the power grid [5]. Various types of measurements can be used to determine the state of the system, but in order to do that, it is necessary to find the relationship between the state variables and measurements. This relationship, which is nonlinear in general, can be expressed by Eq. (1) [17]:

$$z = \begin{bmatrix} z_1 \\ z_2 \\ \vdots \\ z_m \end{bmatrix} = \begin{bmatrix} h_1(x_1,\ldots,x_n) \\ h_2(x_1,\ldots,x_n) \\ \vdots \\ h_n(x_1,\ldots,x_n) \end{bmatrix} + \begin{bmatrix} w_1 \\ w_2 \\ \vdots \\ w_m \end{bmatrix} = \boldsymbol{h}(\boldsymbol{x}) + \boldsymbol{w} \tag{1}$$

where **x** is the vector of state variables and **z** is the vector of the measurements. The number of state variables and the number of measurements are represented by n and m, respectively. The vector of nonlinear functions that related state variables to measurements is denoted by **h**(**x**). The noise contribution to the measurement vector is represented by $\mathbf{w} \sim \mathcal{N}(\mathbf{0}, \boldsymbol{R})$. The measurement covariance matrix is

$\mathbf{R} = diag\left\{\sigma_1^2, \sigma_2^2, \ldots, \sigma_m^2\right\}$ assuming that the noise in the measurements is uncorrelated.

The measurements vector denoted by **z** includes measurements of all types that is [2]:

- Real-time Synchronized measurements from PMUs like Voltage and current measurements along with phase angles.
- Real-time non-synchronized measurements like bus power injections, current, and voltage magnitudes, line power flows.
- Pseudo-measurements acquired based on statistical load profiles.

Based on the choice of state variables, there are two main formulations of SE for distribution systems: branch-current-based DSSE (BC-DSSE) and Node-voltage-based DSSE (NV-DSSE) [5]. Complex node-voltages are defined as state variables in the NV-DSSE. The state variables are represented in rectangular coordinates containing the real and imaginary parts of node-voltages such as $x = \left[V_1^r, V_2^r, \ldots, V_N^r, V_1^x, V_2^x, \ldots, V_N^x\right]$, or in polar coordinates such as $x = [\theta_2, \theta_3, \ldots, \theta_N, V_1, V_2, \ldots, V_N]$, where the number of system buses is represented by N [2].

Bus-1 is normally considered as reference bus and its angle is considered zero (i.e. $\theta_1 = 0$) when the polar formulation is adopted. Therefore, this phase angle 'θ_1' is excluded from the state-vector and voltage angles of all other nodes are measured with respect to this angle. However, 'θ_1' may be included as one of the state variable if the PMU measurements are deployed, since the reference is not needed [2, 18].

In the BC-DSSE method, complex branch currents are considered as state variables. The rectangular coordinates of state variables are utilized in this formulation. When there is no PMU available in the system, the state vector only formed from branch currents (real and imaginary components), i.e. $x = \left[I_1^r, I_2^r, \ldots, I_{N_b}^r, I_1^x, I_2^x, \ldots, I_{N_b}^x\right]$, where the number of branches is denoted by N_b [2].

4 State Estimation Problem in Active Distribution Systems

Consider a multi-phase distribution feeder consisting of N buses and L lines that can be modeled as a graph $\mathcal{G} := (\mathcal{N}, \mathcal{L})$, where $\mathcal{N} := \{1, 2, \ldots, N\}$ includes all the buses, and $\mathcal{L} \in \mathcal{N} \times \mathcal{N}$ express the lines in the network. Voltage at all the phases of bus n is denoted by $\mathbf{v}_n = [v_{n,a}, v_{n,b}, v_{n,c}]^T$. Next, define $\mathbf{v} := \left[\mathbf{v}_1^T, \mathbf{v}_2^T, \ldots, \mathbf{v}_n^T\right]^T$ which collects the voltages at all the buses $\boldsymbol{n} \in \mathcal{N}$. For each line $(\boldsymbol{l}, \boldsymbol{m}) \in \mathcal{L}$, let $\boldsymbol{Z}_{lm} = \boldsymbol{Y}_{lm}^{-1}$ represent the phase impedance matrix in the $\boldsymbol{\pi}$-equivalent model [19].

The objective of DSSE problem (NV-DSSE is investigated in this section) is to obtain the system state vector $\mathbf{v} \in C^{3N}$ from real-time measurements, and pseudo measurements. Pseudo-measurements which relate to forecasted loads and renewable generation are used as surrogates, since there is limited number of real-time measurements. Naturally, the noise level corresponding to the real-time measured quantities is lower than the measurement noise level of the pseudo-measurements [19].

Real-time measurements are provided by placing advanced metering infrastructure (AMI), supervisory control and data acquisition (SCADA), and μPMUs at some locations in the network. The measured quantities are modeled as [19]:

$$\tilde{z}_\ell = \tilde{h}_\ell(v) + w_\ell, \quad 1 \le \ell \le L_m \tag{2}$$

where the measurement synthesizing functions are represented by functions $\tilde{h}_\ell(\mathrm{v})$ and can be either quadratic or linear relationships. The measurement noise and the modeling inaccuracies are modeled by w_ℓ.

Additionally, load and renewable generation forecasting methods are adopted to obtain pseudo-measurements, which can be used to enhance the observability of the system. The forecasted quantities, which are usually are modeled in Eq. (3) [19]:

$$\breve{z}_\ell = \breve{h}_\ell(\mathrm{v}) + u_\ell, \quad 1 \le \ell \le L_s \tag{3}$$

where $\breve{z}_\ell$ account for forecasted quantities and u_ℓ represents the forecast error.

Define **z** to be a vector consisting of all the pseudo-measurements and real-time measurements, and let $\mathbf{h}(\mathbf{v}) : \mathbb{C}^{3N} \rightarrow \mathbb{R}^{L_m + L_s}$ be the mapping from the voltage (state) vector **v** to the measurements. The weighted least-squares (WLS) formulation of the DSSE problem can be modeled as follows [19]:

$$\min_{\mathbf{v}} J(\mathbf{v}) = \sum_{\ell}^{L_m} \tilde{w}_\ell \left(\tilde{z}_\ell - \tilde{h}_\ell(\mathbf{v})\right)^2 + \sum_{\ell}^{L_m} \breve{w}_\ell \left(\breve{z}_\ell - \breve{h}_\ell(\mathbf{v})\right)^2 = \\ \left(\mathbf{z} - \mathbf{h}(\mathbf{v})\right)^T \mathbf{W}\left(z - h(v)\right) \tag{4}$$

Due to the fact that the measurement mappings **h**(**v**) inside the squares in nonlinear, the optimization problem in (4) is non-convex. The measurement functions $\tilde{h}_\ell(\mathbf{v})$ and $\breve{h}_\ell(\mathbf{v})$ is comprised of [19]:

- *phasor measurements* obtained by μPMUs comprise the complex nodal voltages $\mathbf{v}_n$, and/or current flows. Consequently, the corresponding measurement synthesizing function has a linear relation with the state variable v. complex measurements are expressed as two real measurements. For example, the measurement synthesizing functions in Eq. (5) and (6) are used to the obtain real and imaginary part of the complex nodal voltage at bus n for phase φ [19]:

$$\Re\{v_{n,\varphi}\} = \frac{1}{2} \mathrm{e}_\varphi^T \left(\mathbf{v}_n + \overline{\mathbf{v}}_n\right) \tag{5}$$

$$\Im\{v_{n,\varphi}\} = \frac{1}{2j} \mathrm{e}_\varphi^T \left(\mathbf{v}_n - \overline{\mathbf{v}}_n\right) \tag{6}$$

where e_φ is the φ-th canonical basis in R^3. Similarly, Eqs. (7) and (8) present the real and imaginary parts of the measured complex branch current [19]:

$$\Re\{i_{lm,\varphi}\} = \frac{1}{2} e_{\varphi}^{T} \left(\mathbf{Y}_{lm} (\mathbf{v}_l - \mathbf{v}_m) + \overline{\mathbf{Y}}_{lm} (\overline{\mathbf{v}}_l - \overline{\mathbf{v}}_m) \right) \tag{7}$$

$$\Im\{i_{lm,\varphi}\} = \frac{1}{2j} e_{\varphi}^{T} \left(\mathbf{Y}_{lm} (\mathbf{v}_l - \mathbf{v}_m) - \overline{\mathbf{Y}}_{lm} (\overline{\mathbf{v}}_l - \overline{\mathbf{v}}_m) \right) \tag{8}$$

- *real-valued measurements* are real and reactive power flow and injection measurements $p_{lm,\varphi}$, $q_{lm,\varphi}$, $p_{n,\varphi}$, $q_{n,\varphi}$, and voltage magnitudes $|v_{n,\varphi}|$, current magnitudes $|i_{lm,\varphi}|$. They are usually measured by SCADA systems, AMI, or μPMUs. The real-valued measurements are nonlinearly related to the state variable v. The measured active and reactive power flows, current magnitude square, and the voltage magnitude square can be represented as quadratic functions of the state variable v. Consequently, all the real-valued measurements can be written as quadratic functions of the state variable v [19].
- *pseudo-measurements* are estimated by exploiting historical data and locational information and adopting load and renewable energy generation forecast methods. Since the accuracy of these measurements is often less than real-time measurements, low weights are considered for their corresponding terms in the WLS formulation. Quadratic functions can be used for formulating the mapping from the state variable to the forecasted load and renewable energy source injections [19].

Consequently, Eq. (9) can be used for any measurement synthesizing function $h_\ell(\mathbf{v})$ [19]:

$$h_\ell(\mathbf{v}) = \overline{\mathbf{v}}^T D_\ell \mathbf{v} + c_\ell^T \mathbf{v} + \overline{c}_\ell^T \overline{\mathbf{v}} \tag{9}$$

where D_ℓ is a Hermitian matrix. Consequently, $J(\mathbf{v})$ is very difficult to optimize because it is a fourth order function of the state variable.

5 Various State Estimation Methods Used in Smart Grid

This section will go through different methods being used for SGSE. Firstly, the SE formulation and weighted least square (WLS) is investigated. Then, some Kalman filter (KF)-based methods are introduced.

5.1 Conventional Approach

The most common and basic method in the power system state estimation is weighted least square WLS [20]. The state estimation problem can be formulated as a WLS optimization problem [12]. There are different proposed methods to overcome some of the weaknesses of WLS (e.g. sensitivity to bad data, computational cost). A least trimmed square (LTS) and least median square (LMS) has been describe in detail in [21]. These two methods has slightly improved the problem of the presence of outliers. Another algorithm (least absolute value LAV) was applied in [22] which has automated the bad rejection. In [23], an approach was proposed to increase the robustness by using generalized maximum-likelihood (GM). However, the WLS, which is widely used in TSSE, is an optimal solution in the SG since it has a high sensitivity to bad data. WLS is being a preferred reference for many proposed methods in PSSE to validate their applied algorithms.

5.2 Kalman Filter-Based Approaches

Kalman filter (KF) was established in the early 1960s to achieve the properties of any linear dynamic system. It is also known by linear quadratic estimation (LQE) in statistics and control theory. The theory of KF relies on iterated procedures in two stages, so called prediction stage and update stage in which it recursively minimizes the root mean square of the predicted measurements to obtain an optimal estimator. The prediction stage is to update the time for the received measurement and states based on the previous iteration. While the update stage estimates the real time system's states and detects the uncertainties to update and revise the measurement and the weighted average [24]. There are many authors have revised KF to accommodate the nonlinear systems e.g. linearizing the nonlinear functions; hence, KF cannot be optimal for SGSE since SG has high uncertainties and nonlinear system. Authors in [25–27] have applied WLS, extended Kalman filter (EKF) and unscented Kalman filter (UKF) on a single-machine infinite-bus power system. These approaches were compared one to another in different scenarios e.g. noisy conditions, presence of outliers. Based on their results, they claim that the WLS method has the lowest accuracy under noise or faulty conditions compare to the more robust EKF and UKF. Moreover, the UKF has the suitable results while EKF present an acceptable result with less computational time in comparison to UKF due to its characteristics based on the first order of mean and covariance approximation. Additionally, the result of UKF is advantageous in tracking the true states and it is expeditious to adapt the DSE. Another study in [28] has applied EKF and compared with WLS which presents a high performance for EKF in estimating the states with desirable cost of time and iterations unlike the WLS. Authors in [29] used EKF and graphic processing units to propose a robust parallel DSE. An attack detection methodology in [30] was built upon a parallel Kalman filtering algorithm for robust DSE. A developed iterated Kalman filter (IKF) was proposed in [16] to be applied for nonlinear system which estimates the states in a selected power system properly with appropriate performance and accuracy. In [31], a particle filter (PF) was proposed and validates its results with EKF in which the PF is used a recursive Bayesian algorithm. It is greatly robust in the faulty conditions and high uncertain data compared with the EKF, it costs a high computational time though. Most of these approaches are dependent on using measurement devices; specifically PMUs, which is not preferable in the SGSE since it is effectively expensive and it is an obstacle for these techniques. In addition, KF-based is considerably inefficient when the uncertainties and nonlinearities are getting higher where the SG normally has these specifications.

6 Learning Based Applications in SGSE

Learn-Based method has been widely applied in many applications in the power system in which it is based on Machine Learning (ML) and an artificial neural network (ANN) [32–35]. Most of the recent research in PSSE is using different ANN approaches, unlike the ML algorithms (e.g. classification, support regression or support vector machine algorithms) which has applied in a few studies related to the PSSE, hence even the ANN state estimators has a limited number of studies in SGSE. The models of NN system can be categorized to two main models Feed-forward neural network and Recurrent neural network where each on of these models has different proposed sub-models [36]. Moreover, there are learning methods "NN-based" applied in SGSE which are summarized in the three well known branches of machine learning (Supervised learning, unsupervised learning and reinforcement learning) [37]. Additionally, the supervised learning techniques in SGSE can be classified to three main studies: support vector machine, regression analysis and Bayesian theorem. The unsupervised learning technique has mainly applied the Kernel Principle Component Analysis (KPCA) in PSSE. The learning- based algorithms has shown its ability to improve different issues in the power system. In SGSE different approaches has been applied associated with different methods (e,g, WLS, Pseudo measurements) or measurement devices (e,g, Phasor measurement units (PMU), Advanced Infrastructure Metering (AIM)). Each method has its benefits and drawbacks to the SGSE. The following subsections will explain and discuses the recent studies in the learning-based SGSE, hence these applications are mostly all the available direct studies in SGSE based on a deep research through IEEE library and Google Scholar. Moreover, the following learning-based methods in SGSE has been applied either by using the ML methods or a combined of ML methods as well as a mixture of WLS or filter-base and learning-based. These techniques are described in details as follows: support vector approaches, Bayesian theorem approaches, regression analysis approaches and artificial neural network approaches.

6.1 Support Vector Machine Approaches

Support vector machine (SVM) has been successfully applied in many applications which is based on analyzing data by using regression and classification. The most

SVM used in SGSE is the Kernel Principle Component Analysis (KPCA) duo its ability to deal with nonlinear functions [38]. SVM analyzes two classes to formulate a hyperplane between them to distinguish the classes with minimum error in the maximum margin [39].

A dynamic state estimator was proposed in [40] by using KPCA and SVM. It uses the KPCA to extract the nonlinear relationship between the SCADA inputs and the feature extraction [41]. Then, it employs the extracted data as inputs to the SVM to train its model and predict the state estimates. The main aim of this method is to handle the problem of the requirement of high dimension inputs for learning algorithms from SCADA and the cost of training time by preprocessing module KPCA. It also uses the estimated state after the state estimation iteration to be input beside the raw data from SCADA as a closed loop method. The KPCA-SVM estimator's results claim a better performance compare to SVM results in term of time cost and accuracy. Although it reduced the computational time and has high ability to generalize the nonlinear system, it still not optimal estimator for a dynamic state estimation because this paper does not include the extraction time by KPCA and it mentioned that the extraction time raises after each iteration.

A least square support vector machine (LS-SVM) was proposed in [42] which is used conjugate gradient optimization algorithm to reduce the training time. It was compared with the WLS estimator and demonstrate robust LS-SVM estimator and fairly accurate against bad data and topological errors. It improves the power system state estimation slightly but it enhance significantly the sparsity in the optimization problem. Joint SE and cyber-attack detection based on feature grouping was proposed in [43], in which LS-SVM is used for classification.

6.2 Bayesian Theorem Approaches

A Bayesian network is based on the probabilities of statistics theory and presents a set of interactions in probabilistic graphical model [44, 45]. It can be applied in the SG since the SG is naturally satisfy the local Markov property. It can observe the state parameters from the measurement by using SM or by random variables designed as a hidden state module [46].

Since the distribution system has a highly changeable network configurations which greatly affects the results of the state estimation, an identifier of network configuration changes in distribution system state estimation is proposed in [47] by using recursive Bayesian algorithm. A model bank is designed to store different network configurations that obtained from parallel WLS estimators as well as from the original stored configurations whereas each model has its own real measurements, pseudo and virtual measurements. Then, it uses the Recursive Bayesian Probability to compute the error of each model by comparing the real measurements and the estimator's results to provide the correct configuration at a given time. Authors of this work claim that this method has a suitable accuracy and acceptable computational time.

A new research field has opened to drive new techniques to handle the problem of the limitation of approximation the quantities of real measurement and pseudo measurementt by Gaussian distribution [48]. This problem of the different parametric uncertainties can be formulated as a problem of statistical inference over parameter space [49]. A sparse Bayesian learning based harmonic state estimator was proposed in [50]. It is applied to locate the harmonic sources and estimate the voltages at those points accurately. It was implemented in small scale TS which does not show its efficiency either in high scale TS or in the DS. Another study in [51] has proposed DSSE based on Bayes theorem to reduce the measurement uncertainties for the pseudo measurements and the available measurement from the sensors in the SG. For this reason, A Bayesian state estimator is used in [51] for non-Gaussian pseudo measurement. This study focus on improving the input of SGSE by generating a high quality of pseudo measurement from historical load profiles beside the real measurement received from smart meters. Since the uncertainties in pseudo measurement is greatly high, authors in [50, 51] claim that the generated pseudo measurement by using Bayesian algorithm has improved the quality of the inputs of state estimator which leads to having a suitable accuracy on the state estimates. This study shows a promising results of a high quality pseudo measurement and a rich description of the uncertainties in such a system which is helpful tool to be used in dynamic and static SGSE and in control applications since the available state estimators need more accurate input measurements to perform properly.

Another study in [52] has proposed the Bayesian inference with deep neural network (DNN) for SGSE. It uses the historical data to learn the distribution of that data and then generate the prior distribution data with the association of real measurements. This generate stochastic data goes to the DNN to train it by using gradient descent algorithm and then the NNs calculate the minimum mean squared error (MMSE) to produce the accurate state estimates. Its results in the SG shows its robustness against bad data and high computational efficiency; however, the training and testing could cost expensively computational time for widespread SG since this study was for giving an architecture of applying Bayesian and DNN regardless the computational time.

A belief propagation-based state estimator was implemented in [53] which is based on factor graph rules. It does not need the step of observability analysis in its procedures. It specifies the prior distribution and the real measurements of all states values which is generated from the historical data from measurement devices like SCADA, AMI or PMUs. Then, it estimates the states as a problem of statistical inference by the involving of Bayesian rules. It shows its superiority to deal with pseudo measurements with less real measurements which could be applied in large scale SG because it relies on the proper initialization of the state estimator. The main disadvantage of this field of research is that it needs to improve the measurement devices placement.

6.3 Regression Analysis Approaches

Regression analysis is to estimate the relationship between variables by processing a set of statistics. It basically deals with the numeric or continuous output values based on the input variables. The regression analysis is described in details in [54].

A numerical analysis was proposed in [55] to demonstrate the ability of describing unreliable and inaccurate data and the relationship between them by using Fuzzy regression analysis. This method was implemented in distribution power system in [56]. Authors in [51, 56] suggest that this approach could by adapted to a practical algorithm due to its simplicity whereas it is considered a time consumption based on the results of [56].

A FASE using regression analysis was proposed in [57]. It considers the effect of adjacent buses which creates a nondiagonal transition matrix and update the transition matrix in certain time by using regression analysis. It is combined of learning based and filter-based algorithm by employing time-variant state-transition matrix in which it relies on the previous value and the historical data obtained from the regression method. This joint technique shows an improvement in the accuracy of the DSE using EKF. The same authors of [57] has applied the regression analysis FASE algorithm in a large scale power system in [58]. It shows high and accurate performance compare to the EKF by itself. In [59] the same method was implemented in the presence of the PMUs. The results in [59] claim an improvement of the conventional filter technique for detecting bad data and topological changes as well. All the three studies have not mentioned the computational time which is considered as a significant factor in the DSE.

6.4 Artificial Neural Network Approaches

The ANN is based on human neural network to simulate a mathematical model or computational model. It has a single layer or multiple layers connected by neurons. These neurons can be adjusted to obtain the desired target. The main properties of the NNs is the nonlinear mapping which makes it desirable for the smart grid applications. Moreover, it deals with: the stochastic variations via the increase of data properly, it expedites the online processing and classifications, and it includes a potential built in nonlinear modeling for data filtration [60].

A closed-loop state estimation tool to monitor and operate an actual MV network was applied in [61]. This method creates a dataset based on adaptive nonlinear autoregressive exogenous (NARX) in which it updates and retrains continuously the models' NN state estimation (NN-SE) by using the feedback of the state estimator. NARX applications for a passive system's control is described in detail in [62]. Although the applied network has a limited number of online measurements, it relies on the SCADA, recorded smart meter measurements and pseudo measurements. This approach mainly is to reduce the load forecasting error and improve the state estimates' accuracy. The result of this implementation show a robust state estimator because of the implementation of the load estimation and NARX as closed loop in which the state estimates present the result and goes back to the load estimation step to retrain and update. However, this method was assumed a known network topology and a correct network parameters which is not preferable in applying SGSE duo to the sudden changes in the SG. Author in [63] has applied the same method for the purposes of short term planning in the distribution system by using NARX to train the NN-SE. The results in [63] has demonstrated the same result in [61] which has a robust static state estimator against bad data and gross errors.

Another state estimator using neural networks was applied in [64]. This technique samples and solves the input load profiles to feed them to the SE-NN model by using quadratic based backward/forward sweep method (QBBFS) [59]. Then, the SE-NN model is trained by the created initialize network (output of QBBFS) and training parameter. After the SE-NN is learned, the SE-NN can simulate and reproduce the output fast and accurately with no need for the QBBFS method. This method has a quick response and accurate results compare to other power flow solutions; however, it does not involve any renewable energy resources and large-scale distribution system in the simulation which could be challenging for this method.

A closed loop robust distribution state estimator based on machine learning algorithm was proposed in [65]. This algorithm has presented three mechanisms to develop a robust state estimator against the measurement errors and the gross errors that influences by the pseudo measurements and the smart meters. It directly inters the measurement from LV and MV nodes to M-estimator and a machine learning algorithm is involved to learn and train itself by having smart meter measurements as well as the output of the M-estimator. That makes this method a closed loop information flow which guarantees a robust performance for the state estimates in comparison with the WLS and re-weighted least square (RWLS). This method basically based on statistic load profiles to train the machine learning algorithm and to identify the patterns which was implemented in the parallel distributed processing model (PDP). This method shows a promising results against the measurement errors and the configuration errors even though there was temporary failure in the smart metering communication system.

A state estimator designed in three stages ANN training stage, error modeling stage and the state estimation application stage was applied in [66]. The ANN mainly used to generate an accurate pseudo measurement and associated with real measurement that used as inputs to the state estimator. The pseudo measurement was generated by using real power flow measurements to train the ANN in order to have high quality pseudo measurements compare to the traditional methods. Then, the output is compared to the load profiles as output target to minimize the error. After that, it is applied to WLS estimator with the following inputs: real measurements, pseudo measurements [67] obtained from the proposed method, and the network topology and parameters. Results demonstrate that the ANN generated pseudo measurement works well with limited real measurements and produces acceptable state estimates' accuracy. Moreover, this approach can train the network configuration in the future to tackle the issue of topological changes in the smart grids. In term of computational time, the training and testing time could be challenging for large systems; however, after it is being trained and tested, it can improve the computational time of the state estimator dramatically.

An online state estimator using NN for a distribution network was proposed in [68] and established the results on geographic information system (GIS). The aim of this NN-SE is to estimate the voltage of buses which do not have any measurement devices (sensors) for different applied networks that has a limited number of sensors. The NN-SE learns from the power flow patterns which is provided by K-matrix combined with particle swarm optimization (PSO). It simulates 50 power flow patterns that obtained from K-matrix-PSO power flow, hence the inputs are loads, line and generation data. The NN-SE is learned from the 50 power flow patterns and then export the estimation data to a database and the GIS to display the detailed information of location and utility. This method was proposed to improve the operating performance since the online monitoring is required in the smart grids. The main drawback of this method is that it does not explain the proper locations of this sensors as well as the minimum number of the sensors because it would be economically unsuitable for smart grids.

As it is mentioned before the main issue in the SGSE is the observability. If the measurements are not observed at all the buses, the Jacobian transforms to a matrix composed of undetermined system and the power system might be unobservable which diverge the state estimation solution. In [68] it demonstrate that an ANN-SE can obtain accurate state estimates in such unobservable system. Authors has implemented multilayers ANN-SE solution by just having the measurements form load buses (re/active powers). This method uses a back propagation NN which can accurately map the relationship between the measured variable and other state variables like Voltages and its angles. The ANN-SE is used the same equation of defining the states in WLS in order to train itself.

The same Authors of [69] has implemented in [70] a technique that used ANNs and principal component analysis (PCA) for power system state estimation. PCA is a non-parametric statistical method that create a set of uncorrelated variables by converting observation of possibly correlated variables. It basically capture all the main changes in the system by identifying the significant variables in that system. After that, it uses these variable as inputs for training the ANNs which the ANN has the ability to deal with less number of measurement compare with conventional method WLS. To provide a faster response, it is applied a feed-forward straight operation and it can provide a dynamic response due to ANN can give a time trajectory. This method is considered as a reduced model which will reduce the running cost. The main advantage of this simulation is to examine the robustness and accuracy of the ANN-SE based on a few number of real measurements. In other word, it demonstrates that the scope of ANN-SE in smart grids has promising solutions in SGSE.

In [71] a WLS estimator based on PMU was implemented with inserting less number of PMUs at strategic buses that can connect all the other nodes. It was applied NN-SE by using recursive WLS state estimation as a reference for a power system to reduce the number of PMUs at that system. This algorithm eliminates the inputs with minimal or zero effects in order to solve the issue of computational time for a large power system. It shows that the WLS estimator is time consuming which presents the NN-SE as a suggested solution. The NN will be well trained with less real measurement from PMUs which reduces the training time as it is the main concern of NN. The results was compared with the WLS estimator where the NN was superior than the WLS in computational cost and accuracy, hence, the WLS needs overdetermined system to obtain a reasonable accuracy [72].

A multilayer perceptron neural network (MLPNN) state estimator was proposed in [73]. The input/output data for training the NN-SE was reached from the load flow simulation. The MLPNN was trained by using the Resilient Propagation algorithm (RProp). This approach is desirable for unchangeable topological systems which been assumed in this study. The reason for not changing the topology is that MLPNN is capable to solve the issue of approximation where it can detect the pattern and it is fed to the relevant learning data. This approach has an advantage of reducing the input measurements as the other NN-SE approaches compare to the traditional method WLS. The accuracy of the result was precision even with erroneous measurements. However, it is obvious this method cannot be acceptable to be applied in the SGSE since the distribution system is considered highly unbalanced.

Based on the applied learning-based methods that was applied in SGSE, the ANN is an attractive method for enhancing and compensating the weaknesses in the SGSE. The learning-based algorithms has presented promising improvements for SGSE. Based on the available studies in this area, it is still undergoing field of research which requires more implementations to obtain the desired state estimator.

False Data Injection (FDI) attacks have the ability to bypass the standard SE used in SGs; therefore, a cyber-attacks detection using supervised learning and heuristic feature selection was proposed in [74]. An attack detection model was proposed in [75] that leverages Deep Neural Network (DNN) and Decision Tree (DT) classifiers to detect cyber-attacks from the new representations. A deep unsupervised representation learning approach for effective cyber-physical attack detection and identification on highly imbalanced data was proposed in [76]. A deep and scalable unsupervised machine learning system was presented in [77] for cyber-attack detection in large-scale smart grids.

7 Simulation

The performance of learning-based SE methods is investigated in two case studies:

Case Study 1

Numerical test on the IEEE 33-bus distribution system is presented using DNN-based SE model [78]. Real load data and solar generation were collected from Pecan Street project [79] with sampling rate of 1 second. The load aggregates at each bus were scaled to meet the constraints of the test system. Two different types of measurements are considered: SCADA power flow measurements [78] and line current sensors [80]. The MATPOWER toolbox [81] was used to produce the measurements and state variables by solving the AC power flow equations. The dataset includes 604,800 samples (voltages and measurements for 1 week with a second-by-second reporting rate). The training set and test set consist of 80% and 20% of the dataset with 483,840 and 120,960 samples, respectively.

Case Study 2
Numerical test on the IEEE 118-bus benchmark system is performed leveraging CNN-based SE method. Real load data is based on the 2012 Global Energy Forecasting Competition (GEFC) [82]. Dataset consisting of measurements and state variables is taken from [83]. The dataset consists of 18,528 samples, 80% and 20% of which were considered as the training set and test set respectively.

7.1 Case Study 1

The performance of the DNN-based SE model is demonstrated by applying it to the IEEE 33-bus test system. The single line diagram of the feeder is illustrated in Fig. 4. The DNN model aims to obtain distribution system states with a limited number of measurements [78]. The architecture of the DNN model is shown in Fig. 5. The inputs and outputs of the DNN model are measurements and estimated state variables (bus voltage magnitudes and angles), respectively. The DNN model has three hidden layers, each having 128 neurons. The number of neurons in the input layer is equal to the number of measurements. Due to the fact that bus-1 is considered as reference, the output layer consists of 64 neurons (32 voltage magnitudes and 32 phase angle). Two different types of measurements are considered in this study:

- SCADA measurements consisting of Line power flows depicted in Fig. 6 $(V_1, P_{1-2}, P_{2-19}, P_{3-23}, P_{6-26,}Q_{1-2}, Q_{2-19}, Q_{3-23}, Q_{6-26,})$ [78].
- Line currents Sensors shown in Fig. 7 $(I_8, I_{13}, I_{20}, I_{24}, I_{29})$ [80].

According to the best of the author's knowledge, this is the first study that performs learning-based SE utilizing low-cost line current sensors. The recent advancements in developing noncontact line current sensors is the reason for choosing this

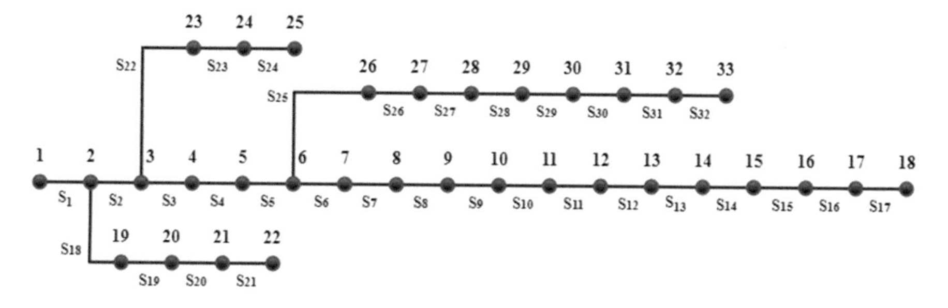

Fig. 4 The IEEE 33-bus test system [78]

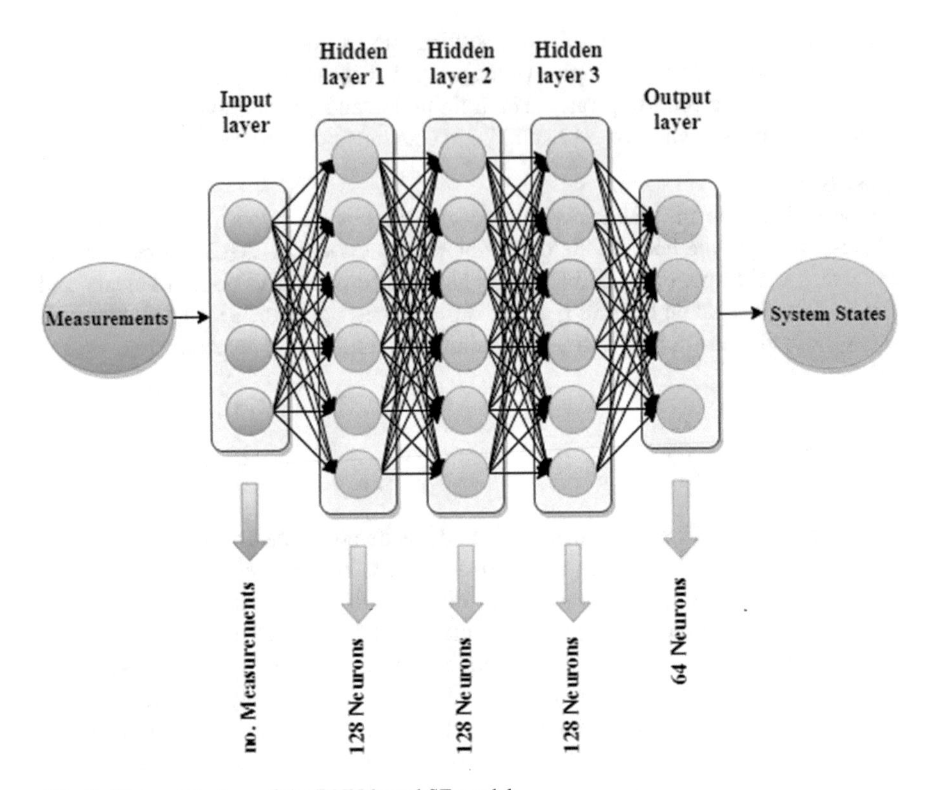

Fig. 5 The architecture of the DNN-based SE model

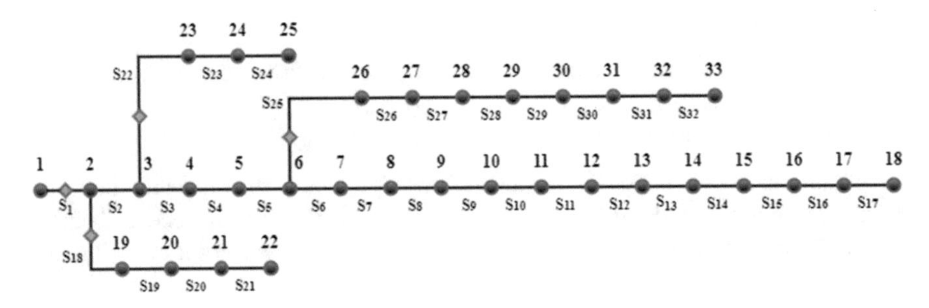

Fig. 6 SCADA measurements for line power flow are placed in branches shown with green diamonds

type of measurements. Two important features of this sensors are their easier installation and lower cost compared to the traditional utility sensors that measure voltage and power [80]. Recently the utility industry deployed line current sensors for installation on power distribution lines [80].

The performance of the DNN-based SE model is assessed in terms of normalized root mean-square error (RMSE) [83, 84]:

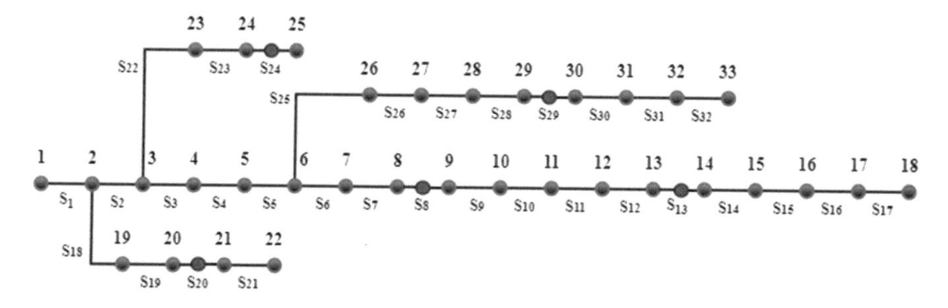

Fig. 7 The line current sensors shown with red circles are deployed on line segments of the IEEE 33-bus test system

$$RMSE = \frac{\hat{\mathbf{v}} - \mathbf{v}_2}{N} \tag{10}$$

where $\hat{\mathbf{v}}$ is the estimate obtained by DNN-based SE method and $\mathbf{v}$ is the true value.

The RMSE of the DNN model is determined by calculating the average RMSE over 20 independent runs. The average RMSE of the DNN-based SE model is 4.87×10^{-6} and 1.83×10^{-4} for the first type (SCADA power flow measurements) and the second type (line current sensors) of measurements, respectively. The results demonstrate the high accuracy of the DNN as a learning-based SE method. Also, the computation time takes a few milliseconds, which meets the real-time SE requirements.

The estimated bus voltages for the IEEE 33-bus test system using DNN-based SE at test instance 10,000 and 100,000 are shown in Figs. 8 and 9, respectively. According to the results, the DNN model using SCADA measurements achieves great accuracy.

7.2 *Case Study 2*

The performance of the CNN-based SE method is evaluated by applying it to the IEEE 118-bus benchmark system. In this study, CNN model is adopted for SE in power grids due to the excellent performance of CNN models in processing data with grid-like topology [85, 86]. The architecture of the proposed CNN-based SE model is illustrated in Fig. 10. Measurements, which consist of all voltage magnitudes, as well as all forwarding-end active (reactive) power flows are the input to the CNN model. The output of the CNN is passed through a dense fully connected layer with 236 neurons to produce state variables (118 voltage magnitudes and 118 phase angles).

The average RMSE of the CNN model is 2.65×10^{-4}, which demonstrates the high accuracy of the CNN-based SE method. The simulation results presented in Figs. 11 and 12 verify its great performance. Figure 11 shows the estimated bus

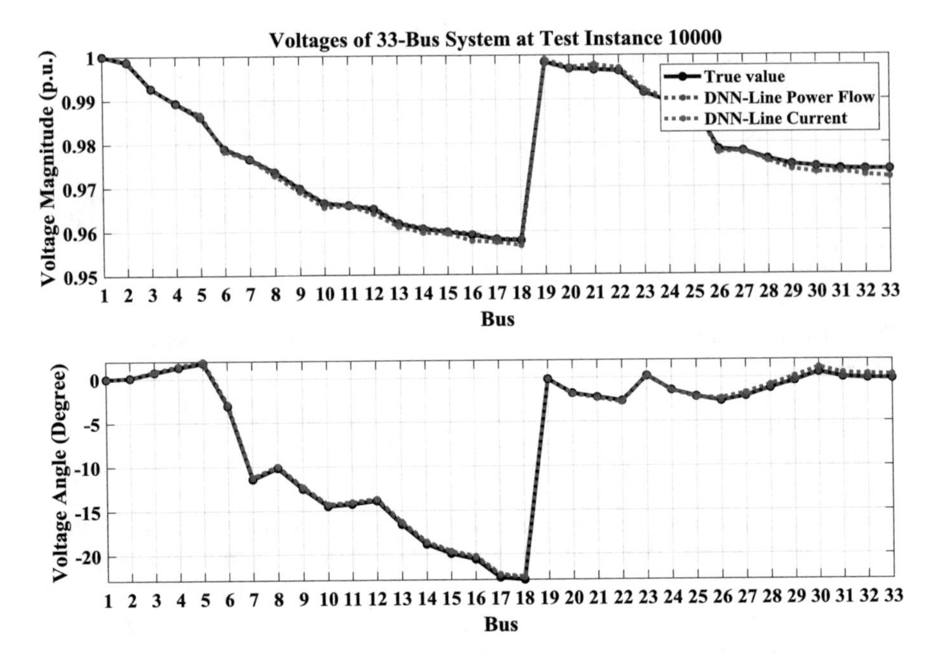

Fig. 8 Estimated voltage magnitudes and angles using DNN model for IEEE 33-bus system at test instance 10,000

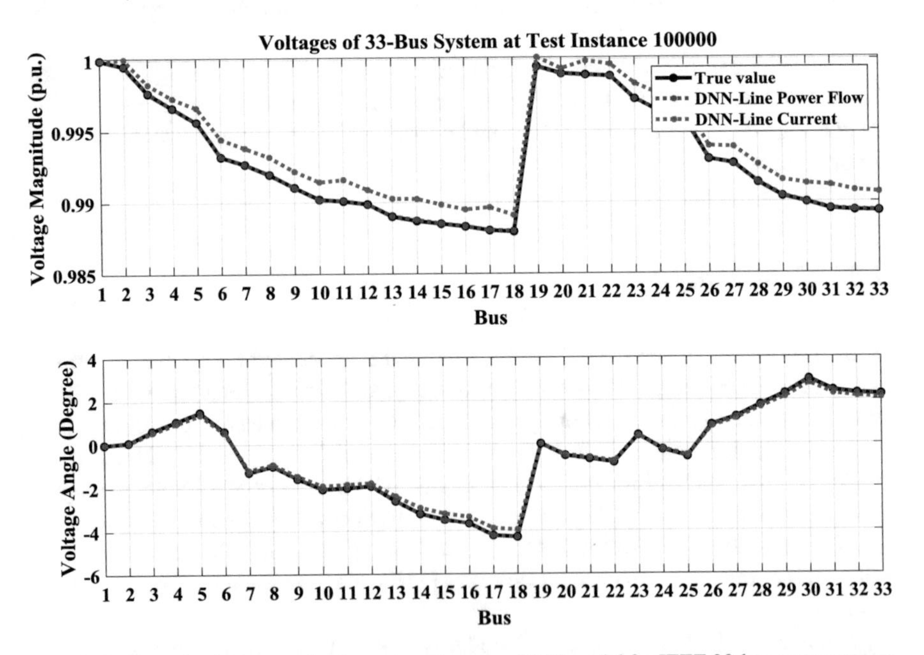

Fig. 9 Estimated voltage magnitudes and angles using DNN model for IEEE 33-bus system at test instance 100,000

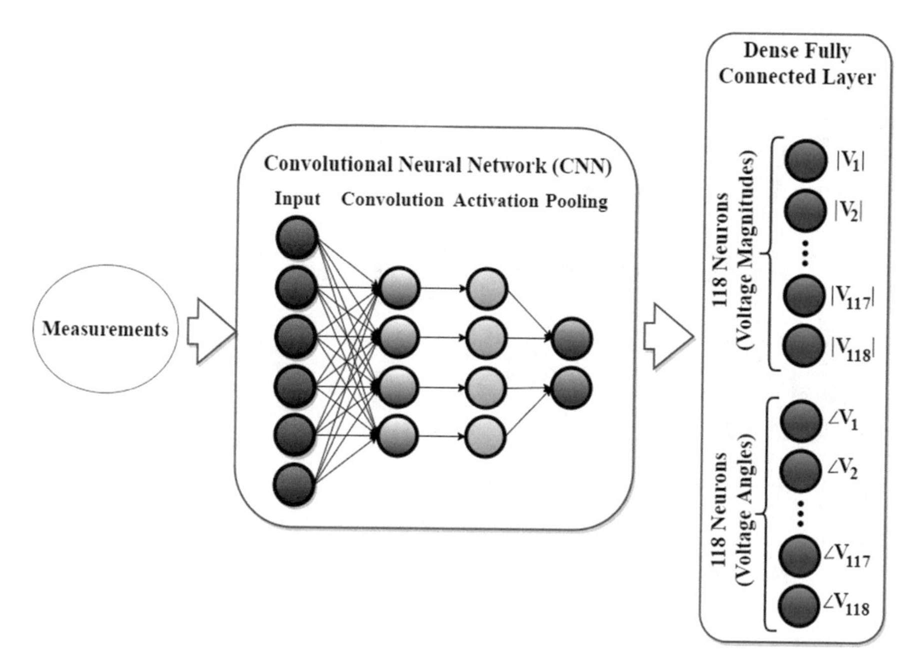

Fig. 10 The architecture of the CNN-based SE model

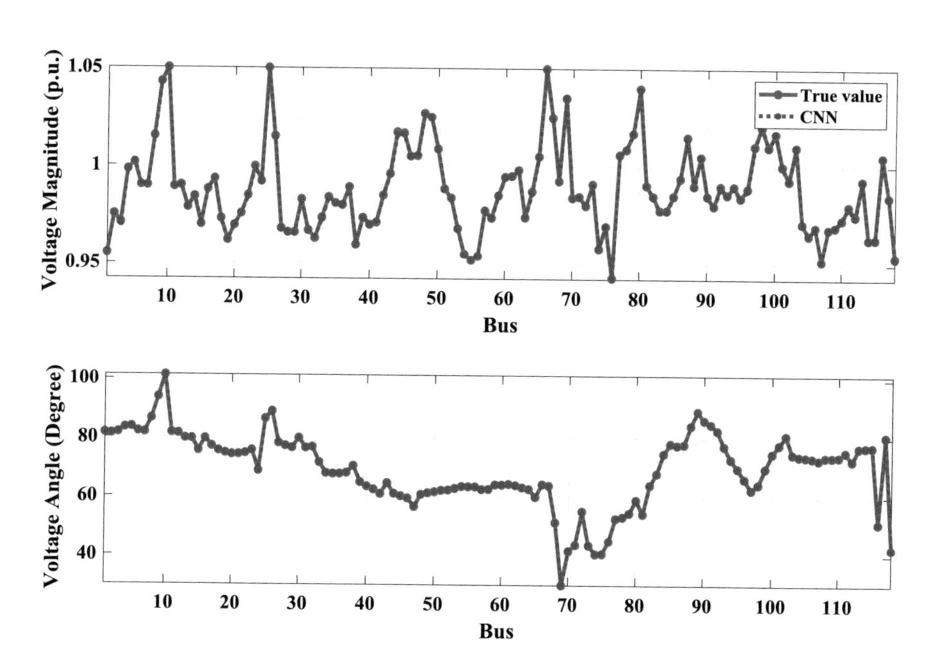

Fig. 11 Estimated voltage magnitudes and angles of the 118-bus system at test instance 1000

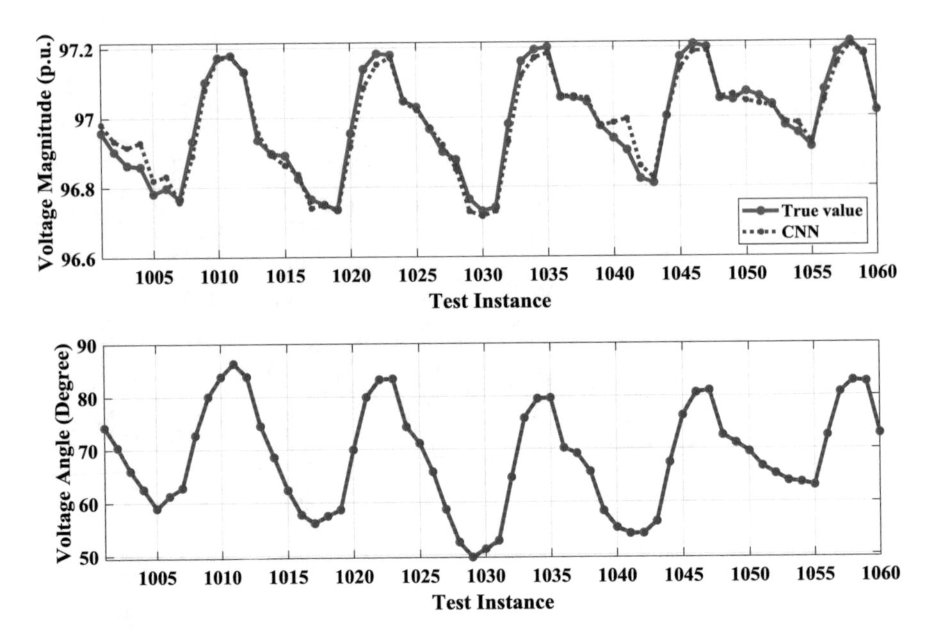

Fig. 12 Estimated voltage magnitudes and angles of bus 20 of the 118-bus system from instances 1000 to 1060

voltage magnitudes and angles of the 118-bus system at test instance 1000. Also, the estimated voltage magnitudes and phase angles of bus 20 from test instance 1000 to 1060 are illustrated in Fig. 12. It should be noted that the CNN-based SE is executed in a few milliseconds (around 2 ms).

According to the simulation results, learning-based methods provide real-time and accurate SE, which is critical for the implementation of various features envisioned by the smart grid concept.

8 Discussion

The aim of estimating the SG states is to provide a realistic and real-time knowledge about the entire system e.g. real measurement, network configuration. The recent studies have focused on different issues in the SE procedures like observability, topologies, detecting bad and gross error. To evaluate any estimator, the four criteria should be met: (1) Observability: allows the operator to have the entire system observed. (2) Reliability: having the ability to detect, identify and correct the bad or missed data. (3) Quality: to have the highest accuracy. (4) Robustness: during any sudden changes, it should be met all foregoing requisites. Based on these criteria and the studies on the SE techniques, we can say that each study normally focusses on one or two of these criteria since the SG networks has many challenges to be entirely observed. In this literature, the main aim is to investigate the learning-based methods in SGSE. The improvement of SGSE could be scoped down to three branches: state estimation and load forecasting tool, improving the input of a state estimator mainly pseudo measurement, and the available of measurement devices. The applications of support vector machine and regression analysis have shown their superiority for creating statistical models which are helpful for improving the observability with less number of measurements compare to other approaches e.g. WLS, EKF. Another finding is that the implementation of load forecasting and state estimation together in which the SE can use the load forecasting to improve the input in each iteration as well as the load forecast method improves itself by using the results of SE. That could be a closed or open loop in which it takes the advantage of the prediction and pattern recognition of the AI algorithms.

The AI algorithms have the ability to predict and recognize patterns precisely. That can improve the most challenging issue in most of the SGSE studies which is the network configuration since it is greatly changeable in the SG. The ANN algorithm has been proposed in different approaches and shows a promising results and field of interest for the future works. Moreover, the ANN is a powerful optimization technique which can present an effective alternative method to overcome the SGSE since the SE problem is formulated as an optimization problem. Another field of interest is the pseudo measurement which can fulfill the lack of measurement in SG by generating high quality pseudo measurement. The learning based methods could improve the pseudo measurement greatly as shown in sect. IV in which it uses the load profiles to learn from and then present a more realistic data to the SGSE. It also uses the learning method to improve itself frequently to produce the best measurements to the system.

Eventually, the recent improvement in employing the measurement devices in the SG will have a huge impact on the SGSE. Taking the advantage of having the real time measurements from these devices will be a study of interest since many companies have applied them widely. The ANN seems to be preferable since it learns from less number of samples which will be provided from the measurement devices and then it can learn the topological changes to provide the best state estimator as well as it could reduce the computational cost.

References

1. J. Sakhnini, H. Karimipour, A. Dehghantanha, R. M. Parizi, and G. Srivastava, "Security aspects of internet of things aided smart Grids: a bibliometric survey," *arXiv*. 2020.
2. F. Ahmad, A. Rasool, E. Ozsoy, R. Sekar, A. Sabanovic, and M. Elitaş, "Distribution system state estimation-A step towards smart grid," *Renewable and Sustainable Energy Reviews*. 2018.
3. H. Karimipour and H. Leung, "Relaxation-based anomaly detection in cyber-physical systems using ensemble kalman filter," *IET Cyber-Physical Syst. Theory Appl.*, 2020.
4. M. Majdoub, A. Belfqih, J. Boukherouaa, O. Sabri, B. Cheddadi, and T. Haidi, "A review on distribution system state estimation techniques," *Proc. 2018 6th Int. Renew. Sustain. Energy Conf. IRSEC 2018*, pp. 1–6, 2019.

5. K. Dehghanpour, Z. Wang, J. Wang, Y. Yuan, and F. Bu, "A survey on state estimation techniques and challenges in smart distribution systems," *IEEE Trans. Smart Grid*, vol. 10, no. 2, pp. 2312–2322, 2019.
6. B. Hayes and M. Prodanović, "State estimation techniques for electric power distribution systems," *Proc.—UKSim-AMSS 8th Eur. Model. Symp. Comput. Model. Simulation, EMS 2014*, pp. 303–308, 2014.
7. J. Zhao *et al.*, "Power System Dynamic State Estimation: Motivations, Definitions, Methodologies, and Future Work," *IEEE Trans. Power Syst.*, vol. 34, no. 4, pp. 3188–3198, 2019.
8. R. E. Larson, W. F. Tinney, and J. Peschon, "State Estimation in Power Systems .1. Theory and Feasibility," *IEEE Trans. Power Appar. Syst.*, vol. Pa89, no. 3, pp. 345, 1970.
9. S. Bhela, V. Kekatos, and S. Veeramachaneni, "Enhancing observability in distribution grids using smart meter data," *IEEE Trans. Smart Grid*, vol. 9, no. 6, pp. 5953–5961, 2018.
10. G. R. Gray, J. Simmins, G. Rajappan, G. Ravikumar, and S. A. Khaparde, "Making Distribution Automation Work: Smart Data Is Imperative for Growth," *IEEE Power Energy Mag.*, vol. 14, no. 1, pp. 58–67, 2016.
11. H. Karimipour and V. Dinavahi, "Extended Kalman Filter-Based Parallel Dynamic State Estimation," *IEEE Trans. Smart Grid*, vol. 6, no. 3, pp. 1539–1549, 2015.
12. A. Abur and A. G. Expósito, *Power System State Estimation: Theory and Implementation*. 2004.
13. F. C. Schweppe, "Power System Static-State," no. 1, pp. 120–125, 1970.
14. A. Jain and N. R. Shivakumar, "Power system tracking and dynamic state estimation," *2009 IEEE/PES Power Syst. Conf. Expo. PSCE* 2009, pp. 1–8, 2009.
15. M. B. Do Coutto Filho and J. C. Stacchini de Souza, "Forecasting-aided state estimation—Part I: Panorama," *IEEE Trans. Power Syst.*, vol. 24, no. 4, pp. 1667–1677, 2009.
16. F. Shabaninia, M. Vaziri, M. Amini, M. Zarghami, and S. Vadhava, "Kalman-filter algorithm and PMUs for state estimation of distribution networks," *Proc. 2014 IEEE 15th Int. Conf. Inf. Reuse Integr. IEEE IRI 2014*, pp. 868–873, 2014.
17. U. Kuhar, M. Pantos, G. Kosec, and A. Svigelj, "The Impact of Model and Measurement Uncertainties on a State Estimation in Three-Phase Distribution Networks," *IEEE Trans. Smart Grid*, 2019.
18. J. S. Thorp, K. J. Karimi, and A. G. Phadke, "Real Time Voltage-Phasor Measurements for Static State Estimation," *IEEE Power Eng. Rev.*, 1985.
19. A. S. Zamzam and N. D. Sidiropoulos, "Physics-Aware Neural Networks for Distribution System State Estimation," *IEEE Trans. Power Syst.*, pp. 1–1, 2020.
20. H. Karimipour and V. Dinavahi, "Parallel domain decomposition based distributed state estimation for large-scale power systems," *IEEE Trans. Ind. Appl.*, vol. 2015, no. July, 2015.
21. L. Mili, M. Cheniae, and P. Rousseeuw, "Robust state estimation of electric power systems," *Circuits Syst. I Fundam. Theory Appl. IEEE Trans.*, vol. 41, no. 5, pp. 349–358, 2002.
22. M. Göl and A. Abur, "LAV based robust state estimation for systems measured by PMUs," *IEEE Trans. Smart Grid*, 2014.
23. J. Zhao, G. Zhang, M. La Scala, and Z. Wang, "Enhanced robustness of state estimator to bad data processing through multi-innovation analysis," *IEEE Trans. Ind. Informatics*, 2017.
24. R. E. Kalman and R.Buey, "A new approach to linear filtering and prediction theory.," *Trans. ASME, J. Basic Eng.*, vol. 83, pp. 95–108, 1961.
25. H. Tebianian and B. Jeyasurya, "Dynamic state estimation in power systems using Kalman filters," *2013 IEEE Electr. Power Energy Conf. EPEC 2013*, pp. 1–5, 2013.
26. E. Ghahremani and I. Kamwa, "PMU analytics for decentralized dynamic state estimation of power systems using the Extended Kalman Filter with Unknown Inputs," *IEEE Power Energy Soc. Gen. Meet.*, vol. 2015-Septe, pp. 1–5, 2015.
27. E. Ghahremani and I. Kamwa, "Dynamic state estimation in power system by applying the extended kalman filter with unknown inputs to phasor measurements," *IEEE Trans. Power Syst.*, vol. 26, no. 4, pp. 2556–2566, 2011.

28. F. Shabani, M. Seyedyazdi, M. Vaziri, M. Zarghami, and S. Vadhva, "State Estimation of a Distribution System Using WLS and EKF Techniques," *Proc.—2015 IEEE 16th Int. Conf. Inf. Reuse Integr. IRI 2015*, pp. 609–613, 2015.
29. H. Karimipour and V. Dinavahi, "On false data injection attack against dynamic state estimation on smart power grids," in *2017 5th IEEE International Conference on Smart Energy Grid Engineering, SEGE 2017*, 2017.
30. H. Karimipour and V. Dinavahi, "Robust Massively Parallel Dynamic State Estimation of Power Systems Against Cyber-Attack," *IEEE Access*, 2017.
31. A. Brüggemann, K. Görner, and C. Rehtanz, "Evaluation of extended Kalman filter and particle filter approaches for quasi-dynamic distribution system state estimation," *CIRED—Open Access Proc. J.*, vol. 2017, no. 1, pp. 1755–1758, 2017.
32. A. Yazdinejad, H. HaddadPajouh, A. Dehghantanha, R. M. Parizi, G. Srivastava, and M.-Y. Chen, "Cryptocurrency malware hunting: A deep recurrent neural network approach," Applied Soft Computing, vol. 96, p. 106630, 2020.
33. E. Nowroozi, A. Dehghantanha, R. M. Parizi, and K.-K. R. Choo, "A survey of machine learning techniques in adversarial image forensics," Computers & Security, vol. 100, p. 102092, 2021.
34. A. Yazdinejad, R. M. Parizi, A. Dehghantanha, G. Srivastava, S. Mohan, and A. M. Rababah, "Cost optimization of secure routing with untrusted devices in software defined networking," Journal of Parallel and Distributed Computing, vol. 143, pp. 36-46, 2020
35. A. N. Jahromi, S. Hashemi, A. Dehghantanha, R. M. Parizi and K. -K. R. Choo, "An Enhanced Stacked LSTM Method With No Random Initialization for Malware Threat Hunting in Safety and Time-Critical Systems," in IEEE Transactions on Emerging Topics in Computational Intelligence, vol. 4, no. 5, pp. 630–640, Oct. 2020, doi: https://doi.org/10.1109/TETCI.2019.2910243
36. H. S. Hippert, C. E. Pedreira, and R. C. Souza, "Neural networks for short-term load forecasting: a review and evaluation Full Text as PDF Full Text in HTML," *IEEE Trans. Power Syst.*, vol. 16, no. 1, p. 4333, 2014.
37. D. W. Patterson, *Artificial Neural Networks: Theory and Applications*, 1st ed. NJ, USA, 1998.
38. V. D. A. Sánchez, "Advanced support vector machines and kernel methods," *Neurocomputing*, vol. 55, no. 1–2, pp. 5–20, 2003.
39. N. Cristianini and J. Shawe-Taylor, *An Introduction to Support Vector Machines and Other Kernel-based Learning Methods*, 1st ed. Cambridge, 2000.
40. Y. C. Li and K. Gao, "A KPCA and SVR based dynamic state estimation method for power system," *ICCMS 2010—2010 Int. Conf. Comput. Model. Simul.*, vol. 2, pp. 493–497, 2010.
41. S. Xu and C. Yuan, "A Probabilistic Fuzzy Principle Component Analysis Model for Scene Denoising," *Proc.—9th Int. Conf. Intell. Human-Machine Syst. Cybern. IHMSC 2017*, vol. 1, pp. 276–279, 2017.
42. A. A. Abod, A. H. Abdullah, and M. K. Abd, "Support vector machine based approach for state estimation of Iraqi super grid network," *Proc.—2008 Work. Power Electron. Intell. Transp. Syst. PEITS 2008*, pp. 252–256, 2008.
43. S. Geris and H. Karimipour, "Joint State Estimation and Cyber-Attack Detection Based on Feature Grouping," in *Proceedings of 2019 the 7th International Conference on Smart Energy Grid Engineering, SEGE 2019*, 2019.
44. D. E. Holmes and L. C. Jain, *A Tutorial on Learning with Bayesian Networks*. Springer, Berlin, Heidelberg, 2008.
45. S. Francisco *et al.*, "Graphical models: Foundations of neural computation," vol. 4, no. 510, pp. 2012–2015, 2017.
46. T. Data, "Journal of Statistical Software_R," *Network*, vol. 59, no. 10, pp. 1–5, 2014.
47. R. Singh, E. Manitsas, B. C. Pal, and G. Strbac, "A recursive bayesian approach for identification of network configuration changes in distribution system state estimation," *IEEE Trans. Power Syst.*, 2010.

48. A. K. Ghosh, D. L. Lubkeman, M. J. Downey, and R. H. Jones, "Distribution circuit state estimation using a probabilistic approach," *IEEE Power Eng. Rev.*, vol. 17, no. 2, pp. 46–47, 1997.
49. N. Petra, C. G. Petra, Z. Zhang, E. M. Constantinescu, and M. Anitescu, "A Bayesian Approach for Parameter Estimation with Uncertainty for Dynamic Power Systems," *IEEE Trans. Power Syst.*, vol. 32, no. 4, pp. 2735–2743, 2017.
50. Y. Yuan, W. Zhou, H. T. Zhang, Z. Ping, and O. Ardakanian, "Sparse Bayesian Harmonic State Estimation," *2018 IEEE Int. Conf. Commun. Control. Comput. Technol. Smart Grids, SmartGridComm 2018*, pp. 1–6, 2018.
51. P. A. Pegoraro *et al.*, "Bayesian Approach for Distribution System State Estimation with Non-Gaussian Uncertainty Models," *IEEE Trans. Instrum. Meas.*, vol. 66, no. 11, pp. 2957–2966, 2017.
52. K. R. Mestav, J. Luengo-Rozas, and L. Tong, "Bayesian State Estimation for Unobservable Distribution Systems via Deep Learning," *IEEE Trans. Power Syst.*, pp. 1–1, 2019.
53. Y. Hu, A. Kavcic, and A. Kuh, "A Belief Propagation Based Power Distribution System State Estimator," *IEEE Comput. Intell. Mag.*, vol. 112, no. 483, pp. 36–46, 2011.
54. R. B. Darlington and A. F. Hayes, *Regression Analysis and Linear Models: Concepts, Applications, and Implementation*. 2016.
55. J. Nazarko and W. Zalewski, "Application of the Fuzzy Regression Analysis," pp. 2–5, 1996.
56. J. Nazarko and W. Zalewski, "The fuzzy regression approach to peak load estimation in power distribution systems," *IEEE Trans. Power Syst.*, vol. 14, no. 3, pp. 809–814, 1999.
57. M. Hassanzadeh and C. Y. Evrenosoglu, "A regression analysis based state transition model for power system dynamic state estimation," *NAPS 2011—43rd North Am. Power Symp.*, 2011.
58. M. Hassanzadeh and C. Y. Evrenosoglu, "Power system state forecasting using regression analysis," *IEEE Power Energy Soc. Gen. Meet.*, pp. 12–17, 2012.
59. M. Hassanzadeh and C. Y. Evrenosoglu, "Use of PMUs in regression-based power system dynamic state estimation," *2012 North Am. Power Symp. NAPS 2012*, pp. 1–5, 2012.
60. M. Haque and A. Kashtiban, "Application of neural networks in power systems; a review," *Power*, vol. 2005, 2000.
61. B. P. Hayes, J. K. Gruber, and M. Prodanovic, "A Closed-Loop State Estimation Tool for MV Network Monitoring and Operation," *IEEE Trans. Smart Grid*, 2015.
62. R. W. K. Chan, J. K. K. Yuen, E. W. M. Lee, and M. Arashpour, "Application of Nonlinear-Autoregressive-Exogenous model to predict the hysteretic behaviour of passive control systems," *Eng. Struct.*, vol. 85, pp. 1–10, 2015.
63. C. Workshop, "Paper 0098 SHORT-TERM OPERATIONAL PLANNING AND STATE ESTIMATION IN POWER DISTRIBUTION NETWORKS Overview of Methodology CIRED Workshop—Rome, 11–12 June 2014 Paper 0098," no. 0098, pp. 11–12, 2014.
64. M. Abdel-Nasser, K. Mahmoud, and H. Kashef, "A Novel Smart Grid State Estimation Method Based on Neural Networks," *Int. J. Interact. Multimed. Artif. Intell.*, vol. inPress, no. inPress, p. 1, 2018.
65. J. Wu, Y. He, and N. Jenkins, "A robust state estimator for medium voltage distribution networks," *IEEE Trans. Power Syst.*, vol. 28, no. 2, pp. 1008–1016, 2013.
66. E. Manitsas, R. Singh, B. C. Pal, and G. Strbac, "Distribution system state estimation using an artificial neural network approach for pseudo measurement modeling," *IEEE Trans. Power Syst.*, vol. 27, no. 4, pp. 1888–1896, 2012.
67. A. Angioni, T. Schlösser, F. Ponci, and A. Monti, "Impact of pseudo-measurements from new power profiles on state estimation in low-voltage grids," *IEEE Trans. Instrum. Meas.*, 2016.
68. D. F. U. P., I. Suryawati, O. Penangsang, A. Suprijanto, and M. Syai'in, "Online State Estimator for Three Phase Active Distribution Networks Displayed on Geographic Information System," *J. Clean Energy Technol.*, vol. 2, no. 4, pp. 357–362, 2014.
69. A. Onwuachumba, Y. Wu, and M. Musavi, "Reduced model for power system state estimation using artificial neural networks," *IEEE Green Technol. Conf.*, no. 1, pp. 407–413, 2013.

70. A. Onwuachumba and M. Musavi, "New Reduced Model approach for Power System State Estimation Using Artificial Neural Networks and Principal Component Analysis," *Proc.—2014 Electr. Power Energy Conf. EPEC 2014*, pp. 15–20, 2014.
71. F. Shabaninia, H. Sadeghi, M. Vaziri, and S. Vadhva, "PMU-based recursive state estimation and its performance with neural network," *IEEE Power Energy Soc. Gen. Meet.*, pp. 1–5, 2012.
72. J. Watitwa and K. Awodele, "A Review on Active Distribution System State Estimation," *Proc.—2019 South. African Univ. Power Eng. Conf. Mechatronics/Pattern Recognit. Assoc. South Africa, SAUPEC/RobMech/PRASA 2019*, pp. 726–731, 2019.
73. O. Ivanov and M. Garvrilas, "State estimation for power systems with multilayer perceptron neural networks," *11th Symp. Neural Netw. Appl. Electr. Eng. 2012—Proc.*, pp. 243–246, 2012.
74. J. Sakhnini, H. Karimipour, and A. Dehghantanha, "Smart Grid Cyber Attacks Detection Using Supervised Learning and Heuristic Feature Selection," in *Proceedings of 2019 the 7th International Conference on Smart Energy Grid Engineering, SEGE 2019*, 2019.
75. A. Al-Abassi, H. Karimipour, A. Dehghantanha, and R. M. Parizi, "An ensemble deep learning-based cyber-attack detection in industrial control system," *IEEE Access*, 2020.
76. A. N. Jahromi, H. Karimpour, J. Sakhnini, and A. Dehghantanha, "A deep unsupervised representation learning approach for effective cyber-physical attack detection and identification on highly imbalanced data," in *CASCON 2019 Proceedings—Conference of the Centre for Advanced Studies on Collaborative Research—Proceedings of the 29th Annual International Conference on Computer Science and Software Engineering*, 2020.
77. H. Karimipour, A. Dehghantanha, R. M. Parizi, K. K. R. Choo, and H. Leung, "A Deep and Scalable Unsupervised Machine Learning System for Cyber-Attack Detection in Large-Scale Smart Grids," *IEEE Access*, 2019.
78. M. Huang, Z. Wei, G. Sun, and H. Zang, "Hybrid State Estimation for Distribution Systems With AMI and SCADA Measurements," *IEEE Access*, vol. 7, pp. 120350–120359, 2019.
79. "Pecan Street Inc. Dataport." [Online]. Available: https://dataport.pecanstreet.org/.
80. M. Farajollahi, A. Shahsavari, and H. Mohsenian-Rad, "Topology Identification in Distribution Systems Using Line Current Sensors: An MILP Approach," *IEEE Trans. Smart Grid*, 2020.
81. R. D. Zimmerman, C. E. Murillo-Sánchez, and R. J. Thomas, "MATPOWER: Steady-state operations, planning, and analysis tools for power systems research and education," *IEEE Trans. Power Syst.*, 2011.
82. "Data." [Online]. Available: https://www.kaggle.com/c/global-energy-forecasting-competition-2012-load-forecasting/data.
83. L. Zhang, G. Wang, and G. B. Giannakis, "Real-Time Power System State Estimation and Forecasting via Deep Unrolled Neural Networks," *IEEE Trans. Signal Process.*, vol. 67, no. 15, pp. 4069–4077, 2019.
84. H. K. Shahrzad Hadayeghparast, Amir Namavar Jahromi, "A Hybrid Deep Learning-Based Power System State Forecasting," *IEEE Int. Conf. Syst. MAN, Cybern.*, 2020.
85. I. Goodfellow, Y. Bengio, and A. Courville, "Deep Learning— An MIT Press book," *MIT Press*, 2016.
86. A. Yazdinejad, R. M. Parizi, A. Bohlooli, A. Dehghantanha, and K.-K. R. Choo, "A high-performance framework for a network programmable packet processor using P4 and FPGA," Journal of Network and Computer Applications, vol. 156, p. 102564, 2020.

Cyber Security of Smart Manufacturing Execution Systems: A Bibliometric Analysis

Amir Hossein Bahrami and Hossein Mohammadi Rouzbahani

1 Introduction

A manufacturing execution system is a computerized structure that tracks and monitors all the elements involved in the process of production while documents the crucial information [1]. Figure 1 shows a general functionality of smart MES.

MES is an efficient way of gathering and documenting data from the production process. It is undeniable that by the rapid improvements in technology, specifically in the industrial sector, organizations by any size are obligated to move towards computerized tools for managing their business [2, 3]. The industrial internet of things (IIOT) made this change possible by using MES as an essential tool that helps us decrease mistakes of human intervention and make better decisions for future steps.

While IIoT makes the automation process more comfortable, there are many challenges as well as security concerns [4]. As machines operate, they generate data, which is valuable for the SMES and attractive for attackers [5, 6].

A practical MES is a system that takes control over "data acquisition," "process modelling," "documentation," and "decision making" by itself, not by using various licensed software among them. Robust and accurate decision-making ability is the key that separates an ordinary MES from a forceful one. This is the part where Machine Learning (ML) plays a critical role.

Machine learning (ML) is an important method that can be used for data analysis to make sense of the data generated from the production by building analytical models and cyber-attack detection. ML is already known to serve the purpose of

A. H. Bahrami
Peter the Great St. Petersburg Polytechnic University, St. Petersburg, Russia

H. M. Rouzbahani (✉)
University of Guelph, Guelph, ON, Canada
e-mail: hmoham15@uoguelph.ca

H. Karimipour, F. Derakhshan (eds.), *AI-Enabled Threat Detection and Security Analysis for Industrial IoT*, https://doi.org/10.1007/978-3-030-76613-9_6

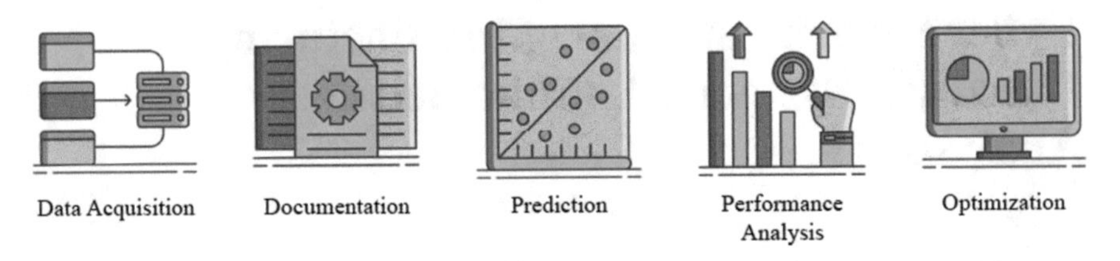

Fig. 1 General functionality of a smart MES with predictive analysis

knowledge synthesis in engineering automation (Lu 1990*). MES can perform learning to apply on a wide range of production data, including optimization of individual behavior, optimization across one or more production lines [5]. ML helps MES to export useful information from the production modules analyzing with a modern computing capability using learning algorithms. This will consequently result in exploring new opportunities, business models and solving challenges that were not possible before [7]. With the aid of similar systems, not only will the enterprise obtains the benefit of Manufacturing Execution Systems (MES), but also the advantages of utilizing Artificial Neural Networks (ANN) [8, 9].

ML can also be used in order to detect abnormality and defects in production [10]. There are several use cases of ML in smart manufacturing.

ML can be used to obtain better business plans or optimized scheduling and quality improvement. The best results will be obtained when a comprehensive MES will be created that first conducts the optimization, and second, will decrease the amount of paid and licensed software in the process. This will focus on what matters in solving the problems. Furthermore, Enterprises will have more control in personalizing the MES for their own purpose.

Cyber-attacks have various derivations, such as Denial of Service (DoS), scan, etc. ML can perform a significant role in addressing cyber-attacks, as well as their classification. Mohammadi et al. [11] conducted research in Intrusion Detection Systems (IDS) based on feature selection and clustering algorithms and presented high accuracy of 95.03% and a detection rate of 95.23% with their proposed model. Rouzbahani et al. [12] presented research on using ML algorithms for the classification of False Data Injection (FDI) attacks in CPS. Karimipour et al. [13] proposed a scalable anomaly detection engine for Cyber-Attack Detection in Large-Scale Smart Grids. The result of their deep unsupervised machine learning system demonstrated high accuracy of 99% in addressing cyber-attacks. Sakhnini et al. [14] studied cyber-attacks detection of smart grids by proposing three supervised learning techniques combined with three feature selection techniques. They concluded that supervised Learning combined with heuristic feature selection methods performs better in detecting False Data Injection (FDI) attacks. Rouzbahani et al. [15] conducted research and performed cyber-attack detection in smart cyber-physical grids by using different ML algorithms combined with Random forest and K-Nearest Neighbor (KNN). Jahromi et al. [16] proposed a deep unsupervised learning approach for cyber-physical attack detection and identification on highly imbalanced data. Their supervised stacked autoencoder converts the raw features to a low-dimensional new feature. They used this feature learning technique in different methods, and the resulting accuracy, and F1-score was promising.

Rouzbahani et al. [17] proposed an Ensemble Deep Convolutional Neural Network (EDCNN) model for electricity theft detection in smart grids. They used an unbalanced dataset, compared the results with other models, and concluded that EDCNN could detect electricity theft in smart grids with an accuracy of 0.981, indicating that the model is precise. Rouzbahani et al. proposed an Incentive-based Demand Response Optimization (IDRO) model in order to efficiently schedule household appliances for minimum usage during peak hours [18, 19], which demonstrates noticeable improvements in power factor and cost-saving during peak hours for individual households.

These examples and many others demonstrate that the research investigations conducted in this field are noteworthy. However, no bibliometric analysis has been done to report the impacts and effects of such investigations and research.

Bibliometric is a statistical analysis that helps researchers analyze parameters such as citations, publications, and results. It allows researchers to understand the characteristics, structure, and patterns of research activities. The research activities are also combined into a realistic trend of a research domain by this statistical analysis. This involves literature studies of scientific activities in different contexts such as publications, authors, institutions, citations, and countries. Moreover, this method reports on the comprehensive evaluation of the expansion of research fields [20–23].

This study aims to investigate the achievements in Cyber Security of SMES. It is necessary to know the trending topics and organizations in the field of study before conducting scientific research work, as well as well-known researchers. This paper intends to find the strengths and weaknesses, and trends in this area of science by conducting a bibliometric analysis on relevant papers published in Web of Science between 2010 and 2020 and help researchers in their path of performing productive research.

In this paper, the methodology is described in Sect. 2. Section 3 presents findings and information about the smart manufacturing execution system. Section 4 is devoted to the conclusion of this paper.

2 Methodology

There are two main types of data extraction for bibliometric analysis, Content or citation analysis [24]. These methods have benefited dramatically from computerized data treatment, and in recent years there has been an immense increase in the number of publications within the field. This is partly because of computerized methods, but also a bibliometric method has to include a certain volume of data in order to be statistically reliable [25].

Citation analysis is probably the most traditional method applied in bibliometrics as an approximate measure of scientific quality, particularly in the case of individual researchers, rankings of universities and institutions ([26, 27], as cited in Ellegaard

[14]). Content analysis can also provide quantitative measures through harvesting of keywords [28], like Song and Zhao's research in forest ecology [29].

In order to conduct a bibliometric analysis, researchers often use three main tools WoS, Scopus and Google Scholar. As it has been shown in the work of A. M. Martín [30], none of these tools are perfect, but the citation tool in WoS is considered to be more accurate because it is an older tool and covers older citations as opposed to google scholar and Scopus which were created in 2004. This bibliometric tool has over 12,000 titles of journals since 1900–present covers 45 Languages and provides citation analysis by author, country, document type, institution, language, publication year, source title, subject area and funding information. WoS contains citation maps that assist with visualizing the result of the citing references. The cited reference search in WoS is a unique feature that cannot be found in any other database [31]. Besides, 94% of Scopus's highest impact factor journals were indexed in WoS [28].

WoS has been used for this bibliometric analysis; some keywords need to be selected to start searching for documents related to the topic. There may be other keywords related to the articles on this topic, but the main ones have been considered in this research to expand this search in all types of documents in any category. The query of search in this paper is as follows: TOPIC: (((Security) AND (IIOT OR Industry 4.0) AND (Manufacturing Execution System OR Smart Manufacturing Execution System)))), Timespan: 2010–2020. Databases: WOS, KJD, MEDLINE, RSCI, SCIELO.

As it can be observed from the query, the period was selected from 2010 to 2020. The databases Web of Science Core Collection (WOS), KCI-Korean Journal Database (KJD), MEDLINE, Russian Science Citation Index (RSCI), SciELO Citation Index (SCIELO) were selected to cover as many documents as possible. As the result, 149 records have been achieved. Figure 2 shows the diagram of the data collection process . The criteria of this bibliometric analysis are: (a) productivity, (b) research areas, (c) institutions, (d) authors, (e) Impact Publishers, (f) highly cited articles and (g) keyword frequency. It should be noted, and no results have been found before 2014 for this topic. So, in the rest of this research, presented results will be limited to 2014–2019.

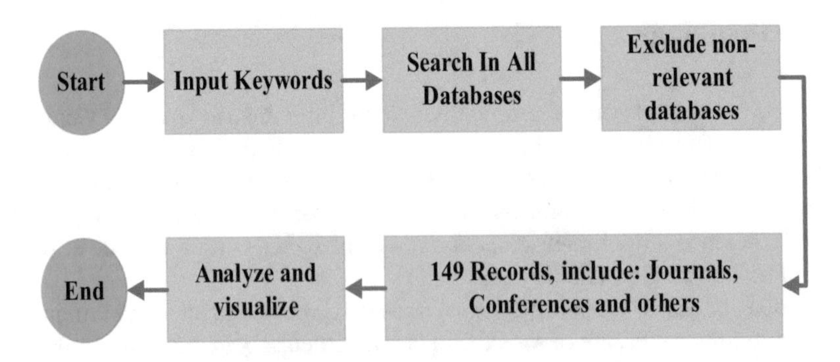

Fig. 2 The schematic of the data collection process

3 Findings

In this section, the bibliometric analysis results for Smart Manufacturing Execution System will be demonstrated and analyzed. Because the criteria for inclusion in Web of Science are based on Scholarly and quality criteria determined by the literature review committee, the record must also fit the subject matter. The results will be high-quality materials to assist researchers in their research.

In the following subsections, the results will be presented, which are productivity, research areas, institutions, impact journals, authors, highly cited articles and keyword frequency. Figure 3 illustrates the number of publications between 2014 and 2019.

In Fig. 3, categories consist of meetings, articles, and others (early access, reviews, and editorials). The meeting category has the highest number of total publications by 49.12%. In 2016 the number of published articles was higher than the meetings category. After 2017 the records of meeting documents have been dropped significantly. This change can be observed after 2016 in the article category.

Citation analysis is an approximate measure of scientific quality for both researchers and journals. In this study's findings, the result of citation analysis in the time period of 2014 to 2020 is presented in Fig. 4. In 2018 and 2019, the number of citations has increased dramatically, and by the time of the conduct of this study 15th of November 2020, the numbers are less than in 2019, which is not predicted to become more than 2019 by the end of the year. Citation is a method of providing evidence in investigations, and it can demonstrate the number of activities in a specific topic of research in publications. The older publication has a higher chance of being cited in recent publications, and one factor that affects such probability can be earlier studies are more general and can be cited in various category of researches.

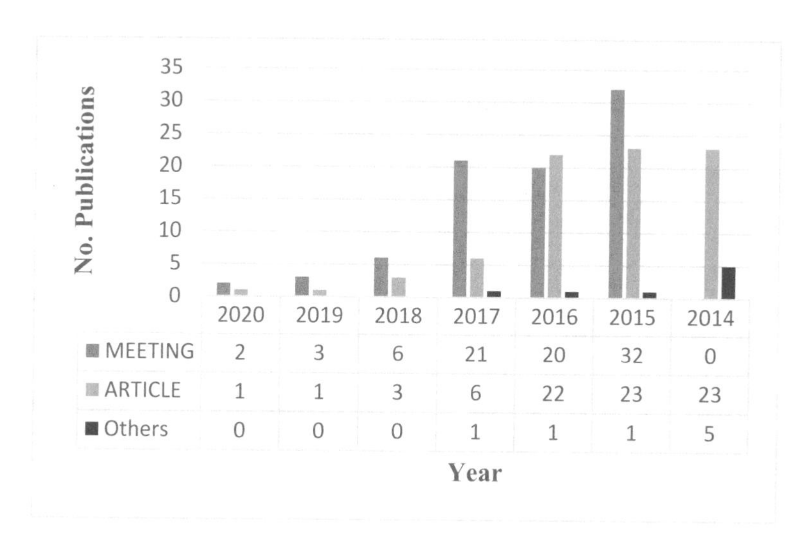

	2020	2019	2018	2017	2016	2015	2014
MEETING	2	3	6	21	20	32	0
ARTICLE	1	1	3	6	22	23	23
Others	0	0	0	1	1	1	5

Fig. 3 The number of publications

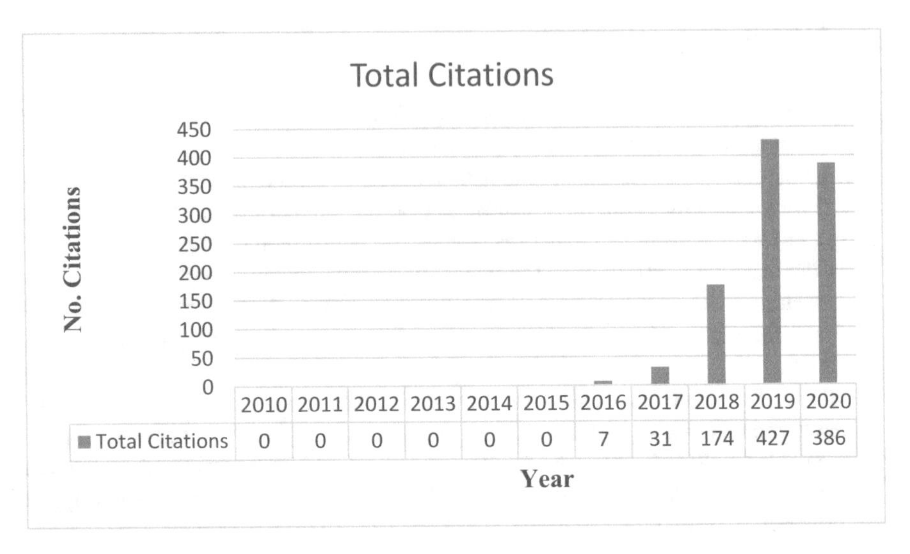

Fig. 4 The number of citations

3.1 Productivity

In this section, countries' productivity in conducting studies and the number of their publications will be discussed. It will show the strength of each country's publications and research in smart manufacturing execution systems and how these countries are moving forward in this area. Investment and development of these countries in the field are higher, and the best organizations and institutions for research and funding can be found there.

Figure 5 shows that Germany, China, and Italy are the lead countries in the field of SMES, and they have most of the publications in this area combined. Other countries can be observed in Table 1. This observation can also be discussed in much more detail; for instance, the countries concentrating more on different aspects and areas of industry 4.0 are developing their manufacturing industry quicker than other countries. Therefore, they will have a bigger share of the world's market in various areas in the nearest future.

3.2 Research Areas

The application of a particular topic will not be bound to one area. For instance, one topic can emerge from the engineering area, but scientists and researchers will discover other applications of the topic in biology or the food industry over time. Hence, the necessity of having an analysis that will illustrate the application of a research topic in all categories of science is vital for the researchers. Research area analysis will demonstrate a logical way of understanding how each field can

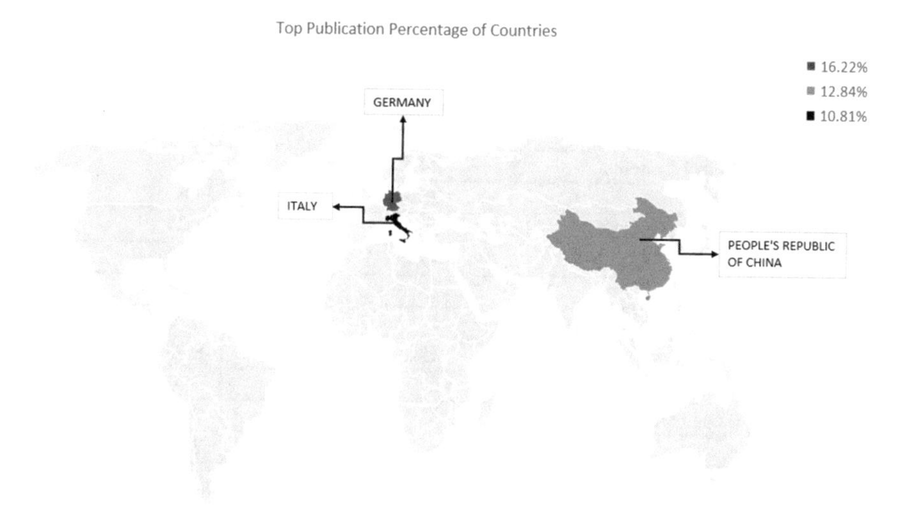

Fig. 5 The most productive countries

Table 1 Productivity

Countries	Publication No.	Publication %
Germany	24	16.22
People's R China	19	12.84
Italy	16	10.81
China	10	6.76
England	8	5.41
France	8	5.41
Spain	8	5.41
Portugal	7	4.73
Brazil	6	4.05
Greece	6	4.05
Taiwan	6	4.05
USA	6	4.05
Austria	5	3.38
INDIA	5	3.38
New Zealand	5	3.38
UK	5	3.38

connect and affect others as well as the industries. Table 2 shows more details about the research areas.

It can be observed in Table 2, that the majority of the publications fall under the category of engineering and computer science. Accordingly, computer science and engineering are the two main research areas for SMES.

Table 2 Research areas

Research areas	Publication No.	Publication %
Engineering	128	86.49
Computer Science	104	70.27
Automation Control Systems	39	26.35
Business Economics	29	19.60
Operations Research Management Science	17	11.49
Robotics	17	11.49
Telecommunications	15	10.14
Instruments Instrumentation	13	8.78
Mathematics	13	8.78
Communication	6	4.05
Science Technology Other Topics	5	3.38
Chemistry	4	2.70
Education Educational Research	4	2.70
Energy Fuels	4	2.70
Environmental Sciences Ecology	4	2.70
Materials Science	4	2.70
Metallurgy Metallurgical Engineering	4	2.70
Geography	3	2.03
Transportation	3	2.03
Physics	2	1.35
Construction Building Technology	1	0.68

3.3 Institutions

In this section, leading institutions in the field will be demonstrated, based on their efforts in conducting studies and publications in the field. As well as their number of publications and their countries, institutions can be observed from Table 3. Based on Table 3, Fraunhofer-Gesellschaft and the University of Auckland have the highest number of publications. Noted that this is the individual work of the institutions, in fact, we learned from Table 1 that Germany and China are the leading countries by the total number of publications. Nevertheless, from this table, we can recognize the leading institutions from each country in the area. The rest of the leading institutions can be observed in Table 3.

3.4 Authors

This section is devoted to the recognition of the most productive authors in the field. Results are shown in Table 4. As it can be implied from the number of publications, SMES is a relatively new topic, and researchers can still conduct revolutionary studies in the field.

Table 3 Institution

Institutions	Publication (No)	Publication (%)	Country
Fraunhofer Gesellschaft	5	3.38	Germany
University of Auckland	5	3.38	New Zealand
Polytechnic University of Turin	4	2.70	Italy
Royal Institute of Technology	4	2.70	Sweden
Siemens AG	4	2.70	Germany
Aalborg University	3	2.03	Denmark
Chinese Academy of Sciences	3	2.03	China
National Central University	3	2.03	Taiwan
Natl Cent Univ	3	2.03	Taiwan
South China University of Technology	3	2.03	China
Technische Universitat Wien	3	2.03	Austria
Universidade Federal De Santa Catarina UFSC	3	2.03	Brazil
University of Basque Country	3	2.03	Spain
University of Hong Kong	3	2.03	Hong Kong
University of Split	3	2.03	Croatia
University of Stuttgart	3	2.03	Germany
CEA	2	1.35	–
Centre National De La Recherche Scientifique CNRS	2	1.35	France

3.5 Publishers

It is significantly important to recognize the lead journals in any category of science to publish studies. In this way, the quality of the researcher's work will be guaranteed, but the chance of research citation will be higher. This section analyzed the publishers who published the most research in this field. Results of such analysis can be found in Table 5, where there are two journals listed. IEEE Access journal has roughly all proportion of the publications, ACM journals with only one publication were added in the table. This result shows that the IEEE journal is the leading publisher in the field.

Other results, such as total cites, cited documents, and citations per document, can be observed in this table. By comparing these results, we can imply that researchers in the area of cybersecurity in SMES tend to publish their work in journals, specifically in IEEE.

Table 4 List of authors

Authors	Publication (No)	Publication (%)
Zhong RY	5	3.38
Xu X	4	2.70
Zhong RY	4	2.70
Di LI	3	2.03
Frazzon EM	3	2.03
Gjeldum N	3	2.03
Huang GQ	3	2.03
Li D	3	2.03
Mantravadi S	3	2.03
Moller C	3	2.03
Tsai WH	3	2.03
Wang LH	3	2.03
Xun XU	3	2.03
Aljinovic A	2	1.35
Bettayeb B	2	1.35
Bratukhin A	2	1.35
Bruno G	2	1.35
Cadavid J	2	1.35
Chen X	2	1.35
Chengliang LIU	2	1.35
Crnjac M	2	1.35
Faccio M	2	1.35
Fumagalli L	2	1.35
Furstner I	2	1.35
Gamberi M	2	1.35

Table 5 List of publishers

Title	Type	P	TC	CD	CPD
IEEE	Journal	31	76	72	2.4
ACM	Journal	1	3	3	3

P publication no, *TC* total cites, *CD* cited documents, *CPD* citations per document (2010–2020)

3.6 Highly Cited Articles

In this section analysis of the top 15 articles and their received number of citations was conducted. As we understood earlier in this paper, citation demonstrates the quality and influence of an article, and the higher number of citations means better quality of the article and more significant influence of it on the area of study.

The top article in this category is "Intelligent Manufacturing in the Context of Industry 4.0: A Review," which discusses new concepts and ideas in the fourth industrial revolution. The date of publishing of this paper is 2017, and it shows the proof of the hypothesis that was mentioned before, the chance of a paper being cited is higher if the paper is older and is in the database for much longer than others. In such a way, it has a higher chance to be shown in other research. The results are listed in Table 6.

3.7 Keywords Frequency

This section analyzed keyword frequency in all the articles which have been indexed in WOS. This analysis will show the most frequent keyword related to articles whose contexts are related to the cyber security of SMES. In 149 articles, 61 unique keywords have been detected, which have been used in these articles 2055 times in total. The results can be observed in Table 7, which implies that the most frequent title and the keyword is automation and security. From this table, it can be noted that keywords and titles such as "Control," "CPS," "IoT," "IIoT," "MES," "Attack," and "digital twin" have been frequently used as well.

Figure 6 provides a desirable observation. This figure is a world map created from Table 7. It will demonstrate how titles and keywords are connected to each other. In this figure, keywords are divided into two clusters based on their relevance

Table 6 Top 15 highly cited publications

Title	Times cited
Intelligent Manufacturing in the Context of Industry 4.0: A Review	395
A dynamic model and an algorithm for short-term supply chain scheduling in the smart factory industry 4.0	176
Smart manufacturing systems for Industry 4.0: Conceptual framework, scenarios, and future perspectives	123
An event-driven manufacturing information system architecture for Industry 4.0	101
A critical investigation of Industry 4.0 in manufacturing: theoretical operationalization framework	74
Part data integration in the Shop Floor Digital Twin: Mobile and cloud technologies to enable a manufacturing execution system	37
Defining and assessing industry 4.0 maturity levels - case of the defense sector	32
Context-Aware Cloud Robotics for Material Handling in Cognitive Industrial Internet of Things	31
IoT-enabled Smart Factory Visibility and Traceability using Laser-scanners	30
A Fog Computing Based Cyber-Physical System for the Automation of Pipe-Related Tasks in the Industry 4.0 Shipyard	27
RFID-based Production Data Analysis in an IoT-enabled Smart Job-shop	21
Review of digital twin applications in manufacturing	16
A Mobile Cloud-Based Scheduling Strategy for Industrial Internet of Things	15
Towards Industry 4.0: Gap Analysis between Current Automotive MES and Industry Standards using Model-Based Requirement Engineering	14
Green Production Planning and Control for the Textile Industry by Using Mathematical Programming and Industry 4.0 Techniques	13

Table 7 Frequency of keywords in titles and abstracts

Titles	Frequency	Keywords	Frequency
Automation	151	Security	57
CPS	98	Control	47
IOT	98	Attack	39
Industrial internet	96	Manufacturing execution system	37
Cloud	95	Factory	34
Execution system	93	Level	33
IIOT	88	Smart factory	30
Smart manufacturing	84	Network	24
Application	65	Algorithm	22
Task	58	Algorithm	22
Environment	50	Cyber Security	22
Manufacturing	50	RFID	21
Thing	48	Sensor	20
Device	47	Production planning	19
Analysis	47	Control system	17
Manufacturing system	35	MES	16
Problem	35	Shop floor	15
Service	35	Communication	14
Internet	33	Enterprise	14
Cyber physical system	31	Manufacturing process	14
Resource	22	Vision	14
Scheduling	22	Enterprise resource planning	13
Machine	18	ERP	13
Robot	18	Production process	13
Real time	17	CPPS	12
Simulation	15	Cyber physical production system	11
Digital twin	14	ISA	11
Interaction	13	Radio frequency identification	9

and frequency. The size of their related circles demonstrates the higher frequency of the keywords.

4 Conclusions

In this bibliometric research, WoS was used as the search engine and discovered all publications in the cybersecurity area of SMES between the years 2010 and 2020, from all databases. It was found that the publications started to appear from 2014, and before that, there was no publication, but the rate of growth of the publications from 2014 until 2020 was immense. All publications were analyzed based on several criteria such as "productivity," "research areas," "institutions," "authors,"

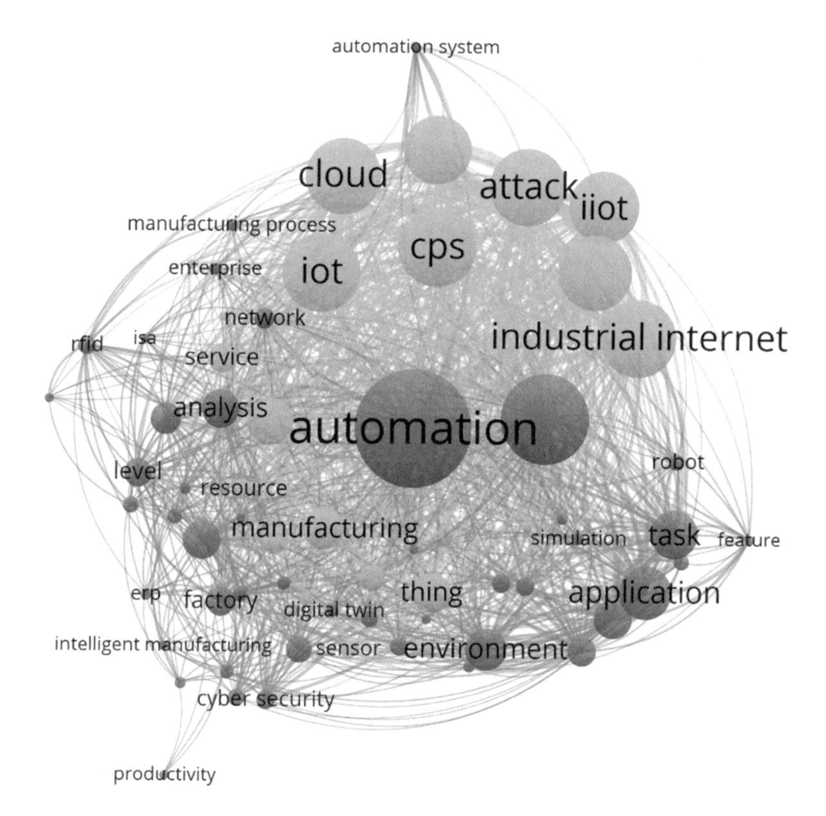

Fig. 6 Keywords map

"impact publishers," "highly cited articles," and "keyword frequency." From this analysis, trends, best publishers and institutions, and leading countries in the field were discovered.

Germany, China and Italy were the leading countries in the field, but the University of Auckland in New Zealand and the Royal Institute of Technology from Sweden were productive in this research field. Among publishers, IEEE Journals had the highest number of publications in this area.

Lastly, top authors in this field and trends and keywords for better direction in the future studies in the field of cybersecurity of SMES were recognized.

References

1. V. V Potekhin, A. H. Bahrami, and B. Katalinič, "Developing manufacturing execution system with predictive analysis," *IOP Conf. Ser. Mater. Sci. Eng.*, vol. 966, p. 12117, 2020, doi: https://doi.org/10.1088/1757-899x/966/1/012117.

2. S. Mantravadi and C. Møller, "An Overview of Next-generation Manufacturing Execution Systems: How important is MES for Industry 4.0?," *Procedia Manuf.*, vol. 30, pp. 588–595, 2019, doi: https://doi.org/10.1016/j.promfg.2019.02.083.
3. A. Yazdinejad, G. Srivastava, R. M. Parizi, A. Dehghantanha, H. Karimipour, and S. R. Karizno, "SLPoW: Secure and Low Latency Proof of Work Protocol for Blockchain in Green IoT Networks," in 2020 IEEE 91st Vehicular Technology Conference (VTC2020-Spring), 2020, pp. 1–5: IEEE.
4. M. F. Ayub, S. Shamshad, K. Mahmood, S. H. Islam, R. M. Parizi and K. -K. R. Choo, "A Provably Secure Two-Factor Authentication Scheme for USB Storage Devices," in *IEEE Transactions on Consumer Electronics*, vol. 66, no. 4, pp. 396–405, Nov. 2020, doi: https://doi.org/10.1109/TCE.2020.3035566
5. S. M. Tahsien, H. Karimipour, P. Spachos, "Machine Learning Based Solutions for Security of Internet of Things (IoT): A Survey", Journal of Network and Computer Applications, vol. 161, pp. 1–18, April. 2020. https://doi.org/10.1016/j.jnca.2020.102630
6. A. Alabasi, H. Karimipour, A. Dehghantanha, R. M. Parizi "An Ensemble Deep Learning-based Cyber-Attack Detection in Industrial Control System", IEEE Access, vol. 8, pp. 83965-83973, April. 2020. doi: https://doi.org/10.1109/ACCESS.2020.2992249
7. S. Mantravadi., C. Li., and C. Møller., "Multi-agent Manufacturing Execution System (MES): Concept, Architecture & ML Algorithm for a Smart Factory Case," in *Proceedings of the 21st International Conference on Enterprise Information Systems—Volume 1: ICEIS,* 2019, pp. 477–482, doi: https://doi.org/10.5220/0007768904770482.
8. H. H. Pajouh, A. Dehghantanha, R. Parizi, H. Karimipour, "A Survey on Internet of Things Security: Requirements, Challenges, and Solutions", Internet of Things Journal, pp. 1–16, Oct. 2019. https://doi.org/10.1016/j.iot.2019.100129.
9. R. Y. Zhong, X. Xu, E. Klotz, and S. T. Newman, "Intelligent Manufacturing in the Context of Industry 4.0: A Review," *Engineering*, vol. 3, no. 5, pp. 616–630, 2017, doi: https://doi.org/10.1016/J.ENG.2017.05.015.
10. K. Taehoon, L. J. Hyuk, C. Hyunchang, C. Sungzoon, L. Wounjoo, and L. Miji, "Machine learning-based anomaly detection via integration of manufacturing, inspection and after-sales service data," *Ind. Manag. Data Syst.*, vol. 117, no. 5, pp. 927–945, Jan. 2017, doi: https://doi.org/10.1108/IMDS-06-2016-0195.
11. S. Mohammadi, H. Mirvaziri, M. Ghazizadeh-Ahsaee, and H. Karimipour, "Cyber intrusion detection by combined feature selection algorithm," *J. Inf. Secur. Appl.*, vol. 44, pp. 80–88, 2019, doi: https://doi.org/10.1016/j.jisa.2018.11.007.
12. H. Mohammadi Rouzbahani, H. Karimipour, A. Rahimnejad, A. Dehghantanha, and G. Srivastava, "Anomaly Detection in Cyber-Physical Systems Using Machine Learning BT—Handbook of Big Data Privacy," K.-K. R. Choo and A. Dehghantanha, Eds. Cham: Springer International Publishing, 2020, pp. 219–235.
13. H. Karimipour, A. Dehghantanha, R. M. Parizi, K. R. Choo, and H. Leung, "A Deep and Scalable Unsupervised Machine Learning System for Cyber-Attack Detection in Large-Scale Smart Grids," *IEEE Access*, vol. 7, pp. 80778–80788, 2019, doi: https://doi.org/10.1109/ACCESS.2019.2920326.
14. J. Sakhnini, H. Karimipour, and A. Dehghantanha, "Smart Grid Cyber Attacks Detection Using Supervised Learning and Heuristic Feature Selection," in *2019 IEEE 7th International Conference on Smart Energy Grid Engineering (SEGE)*, 2019, pp. 108–112, doi: https://doi.org/10.1109/SEGE.2019.8859946.
15. H. M. Rouzbahani, Z. Faraji, M. Amiri-Zarandi, and H. Karimipour, "AI-Enabled Security Monitoring in Smart Cyber Physical Grids BT—Security of Cyber-Physical Systems: Vulnerability and Impact," Springer International Publishing, 2020, pp. 145–167.
16. A. N. Jahromi, J. Sakhnini, H. Karimpour, and A. Dehghantanha, "A Deep Unsupervised Representation Learning Approach for Effective Cyber-Physical Attack Detection and Identification on Highly Imbalanced Data," in *Proceedings of the 29th Annual International Conference on Computer Science and Software Engineering*, 2019, pp. 14–23.

17. H. M. Rouzbahani, H. Karimipour, and L. Lei, "An Ensemble Deep Convolutional Neural Network Model for Electricity Theft Detection in Smart Grids," in *2020 IEEE International Conference on Systems, Man, and Cybernetics (SMC)*, 2020, pp. 3637–3642, https://doi.org/10.1109/SMC42975.2020.9282837.
18. H. M. Ruzbahani and H. Karimipour, "Optimal incentive-based demand response management of smart households," in *2018 IEEE/IAS 54th Industrial and Commercial Power Systems Technical Conference (I&CPS)*, 2018, pp. 1–7, https://doi.org/10.1109/ICPS.2018.8369971.
19. H. M. Ruzbahani, A. Rahimnejad, and H. Karimipour, "Smart Households Demand Response Management with Micro Grid," in *2019 IEEE Power & Energy Society Innovative Smart Grid Technologies Conference (ISGT)*, 2019, pp. 1–5, https://doi.org/10.1109/ISGT.2019.8791595.
20. M. F. A. Razak, N. B. Anuar, R. Salleh, and A. Firdaus, "The rise of 'malware': Bibliometric analysis of malware study," *J. Netw. Comput. Appl.*, vol. 75, pp. 58–76, 2016, doi: https://doi.org/10.1016/j.jnca.2016.08.022.
21. A. Yazdinejadna, R. M. Parizi, A. Dehghantanha, and M. S. Khan, "A kangaroo-based intrusion detection system on software-defined networks," *Computer Networks,* vol. 184, p. 107688, 2021.
22. R. M. Parizi, S. Homayoun, A. Yazdinejad, A. Dehghantanha, and K.-K. R. Choo, "Integrating privacy enhancing techniques into blockchains using sidechains," in *2019 IEEE Canadian Conference of Electrical and Computer Engineering (CCECE)*, 2019, pp. 1–4: IEEE.
23. H. Haddadpajouh, A. Azmoodeh, A. Dehghantanha and R. M. Parizi, "MVFCC: A Multi-View Fuzzy Consensus Clustering Model for Malware Threat Attribution," *IEEE Access*, vol. 8, pp. 139188-139198, 2020, doi: https://doi.org/10.1109/ACCESS.2020.30129
24. J. A. Wallin, "Bibliometric Methods: Pitfalls and Possibilities," *Basic Clin. Pharmacol. Toxicol.*, vol. 97, no. 5, pp. 261–275, Nov. 2005, doi: https://doi.org/10.1111/j.1742-7843.2005.pto_139.x.
25. O. Ellegaard and J. A. Wallin, "The bibliometric analysis of scholarly production: How great is the impact?," *Scientometrics*, vol. 105, no. 3, pp. 1809–1831, 2015, doi: https://doi.org/10.1007/s11192-015-1645-z.
26. L. Waltman *et al.*, "The Leiden ranking 2011/2012: Data collection, indicators, and interpretation," *J. Am. Soc. Inf. Sci. Technol.*, vol. 63, no. 12, pp. 2419–2432, Dec. 2012, doi: https://doi.org/10.1002/asi.22708.
27. P. Weingart, "Impact of bibliometrics upon the science system: Inadvertent consequences?," *Scientometrics*, vol. 62, no. 1, pp. 117–131, 2005, doi: https://doi.org/10.1007/s11192-005-0007-7.
28. C. López-Illescas, F. de Moya-Anegón, and H. F. Moed, "Coverage and citation impact of oncological journals in the Web of Science and Scopus," *J. Informetr.*, vol. 2, no. 4, pp. 304–316, 2008, doi: https://doi.org/10.1016/j.joi.2008.08.001.
29. Y. Song and T. Zhao, "A bibliometric analysis of global forest ecology research during 2002–2011," *Springerplus*, vol. 2, no. 1, p. 204, 2013, doi: https://doi.org/10.1186/21931801-2-204.
30. A. Martín-Martín, E. Orduna-Malea, M. Thelwall, and E. Delgado López-Cózar, "Google Scholar, Web of Science, and Scopus: A systematic comparison of citations in 252 subject categories," *J. Informetr.*, vol. 12, no. 4, pp. 1160–1177, 2018, doi: https://doi.org/10.1016/j.joi.2018.09.002.
31. J. Li, J. F. Burnham, T. Lemley, and R. M. Britton, "Citation Analysis: Comparison of Web of Science®, Scopus™, SciFinder®, and Google Scholar," *J. Electron. Resour. Med. Libr.*, vol. 7, no. 3, pp. 196–217, Aug. 2010, doi: https://doi.org/10.1080/15424065.2010.505518.

The Role of Machine Learning in IIoT Through FPGAs

Behzad Joudat and Mina Zolfy Lighvan

1 Introduction

The most significant advances in industrial manufacturing have been accomplished during the last 300 years [1]. As the focus of mankind was about the use of steam and water to come up with mechanical innovations in the first industrial revolution, in the second one the focus was about advanced machine tools and electrifying them and as it continued, the focus changed on the scope of improving the output production. The third industrial revolution starting from the 1950s, adopted the use of semiconductors and communication networks. Artificial intelligence (AI) and machine learning (ML) for enabling the efficiency in processes reducing waste and material consumption, better quality and productivity, and safer environments have been introduced in the last decade [2–7].

The ongoing and fourth industrial revolution (industry 4.0) is being implemented by global industries. It is a combination of the internet of things model and service-oriented concepts of industrial manufacturing, leading to complete integrated systems [8].

In our daily life, technology plays an important role which is the same for every other business, agency, and industry. Technologies combined with the Industrial Internet of Things can help industries in many ways like object identification, monitoring, automation, etc. [9]. In general, they help people to have access to the services around the clock and make services available even during natural disasters. IIoT, also known as industry 4.0 has been in the background of industrial mutation for the best. There has been a dramatic increase in intensity and extensive applications of information-based technologies, and supply chains and businesses [10].

B. Joudat · M. Z. Lighvan (✉)
Faculty of Electrical and Computer Engineering, University of Tabriz, Tabriz, Iran
e-mail: Mzolfy@tabrizu.ac.ir

H. Karimipour, F. Derakhshan (eds.), *AI-Enabled Threat Detection and Security Analysis for Industrial IoT*, https://doi.org/10.1007/978-3-030-76613-9_7

With the industrial internet of things managers would have comprehensive control at hand over their industries. Through IIoT many concepts like machine learning, artificial intelligence, machine-to-machine communications, distributed computing, cloud computing, edge computing, and data analytics are getting implanted in factory machines, materials, and methods [11]. Therefore, IIoT is the combination of the aforementioned technologies [12].

Looking at this background, following the challenges faced, machine learning solutions help us in achieving automated processes.

1.1 Industrial Internet of Things (IIoT)

With recent advances in domains such as industrial wireless networks (IWNs), big data, and Cloud/Fog computing, etc., greater opportunities for contributing to industrial improvements arose, granting the definition of the fourth industrial revolution referred to as Industry 4.0 known as Industrial internet of things (IIoT). Industry 4.0 goal is to provide industries with intelligent, interoperable, real-time autonomous manufacturing environments [13, 14].

To achieve these goals, industry 4.0 is based on the Internet of Things(IoT) and cloud computing [15], therefore an IIoT system contains all subset devices and definitions related to IoT and cloud computing such as sensors and actuators, intelligent system applications, microcontrollers, etc., [6, 16, 17]. Thus attracting research interest of academy and industry in a plethora applications, Industry 4.0 is perceived to provide competency, flexibility, self-optimization, automation, and complete physical and digital complex tasks in the scope of quality requirements [18]. It is concluded that almost 81% of global industries consider the critical aspect for the success of industries in the future is adopting the industrial internet of things [19].

Looking at the above information, industry 4.0 will be the definition that combines the internet with the ability to control physical systems directly, in the subset of the industry. IIoT implementation in the real world will change present industrial organizations in a way every person related to IIoT will benefit from it, either costumers or organizations. Key elements of the Industrial Internet of Things are shown in Fig. 1.

1.2 Challenges of IIot

As technology takes over our world and transforms it into a digitally connected one, devices and end-users will be in continuous communication with each other. Nowadays most organizations in the industry are eager to implement IIoT in their businesses, As IIoT in definition makes industries more efficient, improves performance and productivity [20]. However, this introduces new challenges for scientists

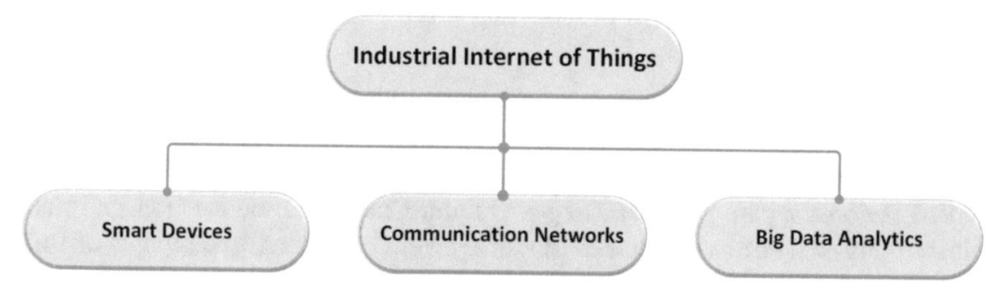

Fig. 1 Key Elements of Industrial Internet of Things

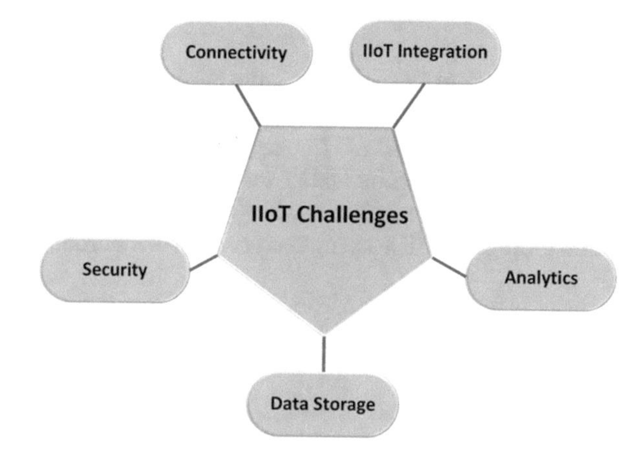

Fig. 2 Challenges of IIoT

and business owners. In [2] there are some challenges referred to which are shown in Fig. 2 and discussed as follows:

1.2.1 Security

Security challenges in IIoT arise from two parts of its architecture as it is connecting physical devices and industrial systems to the internet and challenges about security in IoT in a fact are a challenge in industrial IoT [21]. In [22] they categorized IoT security challenges into three layers of Perception, Transportation, and Application. The complete discussion and analysis of every security issue in each layer have been provided by [23]. The biggest challenge that IIoT faces is security since the smallest warning to components of its architecture could rupture the complete business [24]. Due to increased cost and maintenance requirements of solving issues related to security in IIoT [25], until a rigid security scheme is implemented in IIoT businesses won't risk it putting IIoT in practice in their firms.

1.2.2 Connectivity

In IIoT it is crucial to make sure all industry's machines and devices are working and are connected to each other and surveillance machines to increase the manufacturing outcome, therefore with unacceptable or poor connectivity critical challenges in IIoT become apparent. In cases of power failure, internet disconnection, and technical or physical errors, difficulties of management between different units of IIoT machines may arise [26].

1.2.3 IIoT Integration

One of the other challenges IIoT encounters is the integration of information technology. For providing integration between IIoT and information technology connectivity and synchronization is critical. To execute processes in IIoT systems, the data gathered by machines in industries and IT sectors are needed to coordinate with each other, and inner organizations [27, 28]. In [29] IoT, IIoT, and cyber-physical systems integration through SEPT learning factory has been introduced.

1.2.4 Data Storage

Another important challenge in IIoT is data storage for companies and/or businesses. Secure data storage is a fundamental part to adopt IIoT in industries. In [30] authors have talked about the challenges of storing data and how to secure produced data as to use them later, concluded frameworks for searching and storing data in IIoT, and designed secure data storage and retrieval systems based on fog and cloud computing. Investigation about challenges of data security in edge-based IIoT, such as fog in which nodes are not trusted has been investigated [31], and a basic cloud-fog-device data storage framework for solving corresponding challenges has been proposed.

1.2.5 Analytics Challenges

As to execute processes in IoT architecture it is crucial for data analytics to include data processing, cleansing, and representation at the same time [32, 33]. Authors in [34] investigated research studies on IIoT related to big data analytics and presented frameworks, covered several challenges and opportunities of big data analytics, and also case studies of big data analytics used in IIoT systems.

In [35] the use of edge computing architecture in IIoT is discussed and therefore the challenges of data analytics in edge computing have been provided, moreover, some machine learning regression algorithms for data analytics in state-of-the-art research have been discussed. Data analytics plans and the platform's goals are to eliminate related challenges by transferring part of its processes from the cloud to the edge, as to obtain good real-time benefits of data analysis.

2 Machine Learning

Looking at what has been said before, such as the amount of massive data produced in IIoT and the challenges it brings, we will discuss machine learning techniques, some most used algorithms, and their effects on IIoT in this chapter.

In [36] the importance of machine learning and deep learning for increasing the potentiality of big data analytics and IoT platforms have been mentioned, giving worth to each of the sections. Furthermore, three types of data that are dealt with in industrial IoT are shown as Raw data (untouched and unstructured), Metadata (info about data), and Transformed data (valued-added data). Machine learning algorithms are used for identifying, categorizing, and decision-making purposes. Using machine learning in industries will provide solutions for challenges IIoT faces and will bring forward improvements in operations, production, and services. Leading the businesses to better adaptation with technology [37, 38].

By using machine learning models and algorithms in many businesses the cost of procedures and manufacturing has been reduced. In industries for production, maintenance, software testing, and pattern imaging analytics, machine learning and deep learning are being used [39].

Machine learning as a subset of artificial intelligence is divided into sorts such as supervised learning, unsupervised learning, reinforcement learning, and deep learning [40, 41]. Which are discussed as following:

- Supervised learning: is mainly used for classification and regression which in this practice known results are put into an algorithm for particular inputs for training purposes, in which the most common algorithms are artificial neural networks (AANs) and support vector machines (SVMs) [40, 42–44]. Therefore, in schemes with tagged data, usually supervised learning is employed.
- Unsupervised learning: is where the ML algorithm finds patterns in unknown data sets, thus unlabeled data sets are used for this type of ML [40, 42, 44, 45]. One of the most common algorithms in unsupervised learning is principal component analysis (PCA) which is used for surveillance objectives [44].
- Reinforcement learning: is for specific performance metrics by examining unsupervised learning operations in which actions result rewards [40]. In RL actions are sequential and from their results, the best possible fit for the problem at hand is selected.
- Deep learning: is used to make intelligent decisions by building an ANN through making use of multiple layers, thus without any human intervention, large amounts of data could be handled [40, 46, 47]. Convolutional networks (CNNs), restricted Boltzmann machine (RBM), and autoencoders (AE) are some deep learning algorithms [47, 48].

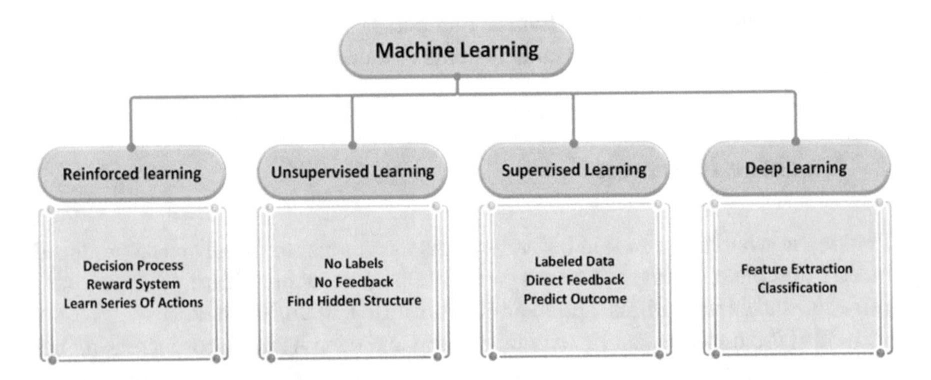

Fig. 3 Machine Learning Classification

Various classifications of ML with their crucial features are summarized in Fig. 3.

To improve reliability, efficiency, and productivity within industries machine learning technologies are used in forms of mobile intelligence and automated guided vehicles [49, 50]. Also, machine learning algorithms are used in IIoT for feature selection [51], on the data gathered and stored in the cloud from sensors through manufacturing processes and products for evaluation and prediction [52].

For smart decision, datasets created by IIoT are trained and tested by ML algorithms and models [53]. To have structured and potent businesses now machine learning has become a crucial part of IIoT.

Machine learning has been successful in implementations such as image or video recognition, natural language process, etc. [54]. It has the ability to handle the challenges of IIoT systems in security, accuracy, real-time response, and big data.

To make use of machine learning models in IIoT, it is necessary to select machine learning frameworks with open source ml frameworks and their implementation processes in industrial IoT are discussed in [55]. Also, it is necessary to have access to all data collected by sensors and other equipment from manufacturing, control, and monitoring systems. Availability of data increases by coming forth of IIoT, and cloud computing getting used more extensively. Leading to the need for immense computational need for processing large amounts of data, like multicore central processing unit (CPU) architectures, graphic processing units (GPUs), and DL libraries [1].

Thus, bearing in mind limitations of power consumption, flexibility in processing different types of data (binary, ternary &, etc.), better performance for DNNs, and being more potent than GPUs Field-programmable gate arrays (FPGAs) are much more desirable [56].

To use machine learning algorithms and its frameworks there is a need for a good processing unit. Therefore, using FPGAs will have a positive effect on the efficiency of ML algorithms implemented on embedded systems, in which the processes should take place and if that place would be near to edge or in middle frames of IIoT. Thus, it would be much more efficient and would help have better results, reduce the amount of data sent to other levels of IIoT, and solve some problems related to IIoT by doing so. In the next chapter, we will discuss more about the FPGAs and then present a dynamic reconfigurable architecture research and the use FPGAs near the edge of IIoT and fog so to overcome these challenges.

3 FPGAs

Field Programmable Gate Arrays (FPGAs) are general-purpose multi-level programmable logic devices customized by end-users [57]. Enabling users to change its logic anytime for different processing purposes.

To provide distributed intelligence, towards overcoming challenges like; acting in time, working offline, serving many, decreasing the size of transferred data to Cloud and improving flexibility, new definitions and methods have been defined, like sensor function virtualization (SFV) [58–61].

For any required processes or algorithms written in VHDL/Verilog programming languages, FPGAs provide us with the ability to dynamically reconfigure its fabric to the selected processes on run time. So, in different hierarchies of IIoT looking at the need for any special process over data obtained, these previously written codes could be utilized to configure the FPGA. Moreover, it is possible to change the configuration of FPGA during run-time [62–66].

By using FPGAs in different places of IIoT architecture like in monitoring systems, next to sensors on the production line, gateways, access points, etc., ML algorithms could be implemented and processed independent of the cloud and its processing resources. Thus, raw data on edge devices could be preprocessed and the extracted knowledge sent to other levels. Therefore, in higher levels of IIoT infrastructure other processes make use of the knowledge gained instead of raw data. This way, we could tackle some difficulties in IIoT.

It is possible to use FPGAs in monitoring systems of IIoT with ML or DL algorithms for the live feed of cameras, etc. As in [67], a way to implement a partial FPGA configuration for video applications has been provided. In this research, using the partial FPGA configuration feature, a video input has been processed by two image processing modules, and a partial reconfiguration was performed on the board. The increase in flexibility and speed of the implementation for varied processes of video streams was the main purpose of this work.

The approach for improving managements of algorithms used in various configurations and a library to use and store them and an approach to automate the necessary automation of operations have been discussed in [63]. consequently, the essential activities performed manually with human interaction would be prevented and partial configuration processes are executed faster. Giving us the idea of trying to automate decision making in IIoT systems using FPGAs.

In [68] different ways to achieve partial dynamic reconfiguration on the FPGA at runtime and its methods are presented, resulting in speeding up the implementation of the FPGA's partial and dynamic configuration with the help of an embedded software processor.

Looking at the internet of things architecture, the required services, and processes, the ability of communication between all connected devices, low processing power on edge devices, and the challenges faced by IoT (e.g. the need for immediate response to services, the offline use of devices, …), in [69] they have tried to implement distributed intelligence in IoT to overcome some existing challenges. However, due to the limitation of the device's communication and processing capabilities in IoT, the distribution of necessary processes on all IoT devices was not attainable. Leading them to suggest distribution of knowledge and intelligence to routers, gateways, adjacent devices with high-connectivity and processing capabilities, and then to Cloud. Approaches to implementing distributed intelligence in the Internet of things, applications required, and locations of the appropriate distribution are demonstrated in this document.

In [70] a use case of FPGAs in IIoT for baseband signal processing as a solution for wireless communication challenges of industrial IoT is presented. In which solutions for multi-user support, high data throughput with low latency, etc. are covered.

To the ends of processing power in edge nodes of the industrial internet of things in [71] the idea of a smart sensor with characteristics like self-configuration, self-optimization, and self-protection is implemented by using FPGAs.

For difficulties faced in IIoT in the reliable data collection process, integration, and analysis of data, FPGA-based hardware and software have been introduced, resulting in a significant increase in performance and resource utilization decrease up to 55% in [72].

In this chapter, with the information said before. It is proposed to use reconfigurable hardware within various hierarchies of the industrial Internet of things, gateways, monitoring systems, edge nodes, and access networks to provide a good processing unit and also a flexible solution for different processes. To this end, by utilizing FPGAs, it is feasible to distribute required algorithms for particular implementations into modules, and allocating them to crucial sections. This way, obtained data in the first layers of architecture would be processed and the results of processes as knowledge gained would be sent to the next layer. Naturally causing a decrease in sent data size rather than the raw data, decreasing the amount of traffic on the communication network. As shown in Fig. 4.

In the logical part of FPGA, multiple reconfigurable partitions could be implemented. Each partition for different modules, ML, DL, or any other needed processes. For the FPGA, the data could be sent through wireless or connected networks. After allocating devices, with the use of Microblaze processor implemented in Xilinx FPGAs, considering the data received from inputs, chooses which modules of algorithms are chosen to be configured. Following, the processor acquires generated modules of algorithms for the partial configuration of the FPGA, then on the selected segment of FPGA fabric configures it for data processing, Fig. 5.

In this work, it is tried to propose the use of FPGAs dynamic partial reconfiguration to overcome problems such as instantaneous response to services, respond to several simultaneous needs, acting in multi-user and multiprocessing situations, and a way to automate decision making by implementing ML in FPGAs. Other

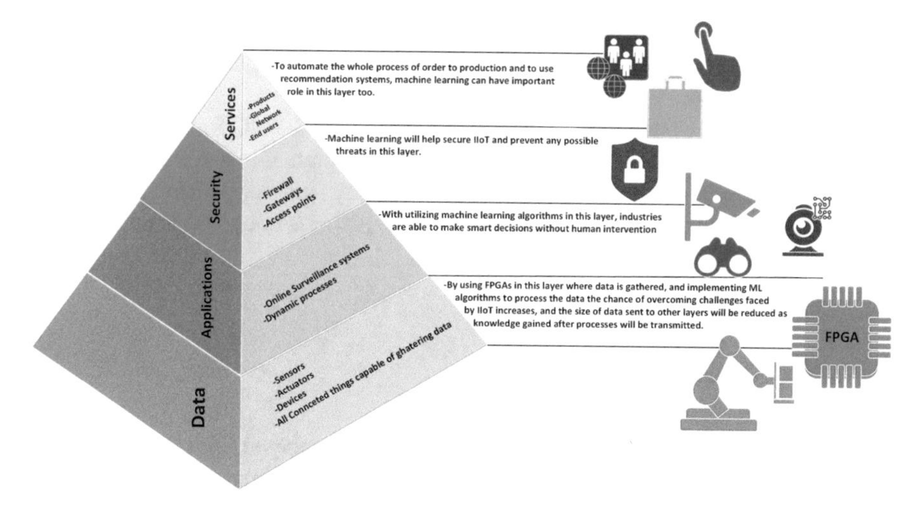

Fig. 4 Traffic on the communication network

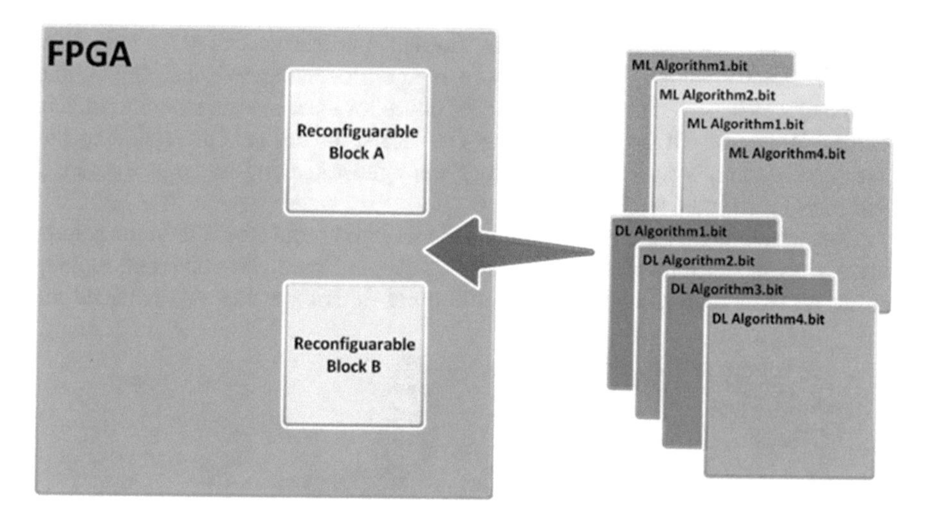

Fig. 5 Different modules to implement in reconfigurable partitions of FPGA

researches like the concept of SFV and distributed intelligence trying to distribute the processes to various parts of the Internet of things to resolve these problems and eliminate the dependency of the processes to the Cloud for IoT made lead us to the use of FPGAs and distribution of processes in IIoT. That's why in this work, by presenting the use of available features of reconfigurable hardware (FPGAs), a new architecture based on Use of FPGAs in the industrial Internet of Things in order to overcome obstacles of previous works is proposed. In this architecture, multiple components of the reconfigurable partition could be used to implement various modules, such as image processing, ML, DL, and computational modules. Moreover, during the execution of various algorithms on need-based biases, the FPGA fabric could be reconfigured and replaced by required modules instead of the old ones that are not used anymore.

In papers [73, 74] dynamic partial reconfiguration of PYNQ for different applications has been covered. As PYNQ is a field-programmable gate array with the ability to work with the python programming language, it's easy to use implementation for data scientists, and the availability of ML libraries in python makes it interesting to use in IIoT.

4 Case Study

A case study for the use of FPGAs reconfigurability merged with the internet of things and image processing was conducted and modules of image processing filters were implemented in FPGA. Two reconfigurable partitions were created on FPGA logic so to implement image processing filters and numerical computations the "filter" and "normal". Sobel and Gaussian image processing filters were to be implemented in the "filter" partition. And the "normal" partition was used for numerical computational modules, Fig. 6. The two image processing modules, Sobel, and Gaussian were selected regarding [75]. The codes for these modules were written in Verilog language and then with the help of ISE software, the corresponding bitstreams for configuring FPGAs Partitions were generated and stored in a compact flash drive connected to FPGA.

Data was fed into the FPGA board via the inquired serial input. Then the Microblaze processor (Embedded on FPGA Development Board), At this point, looking at the received data associating the serial input, looking at the data received the

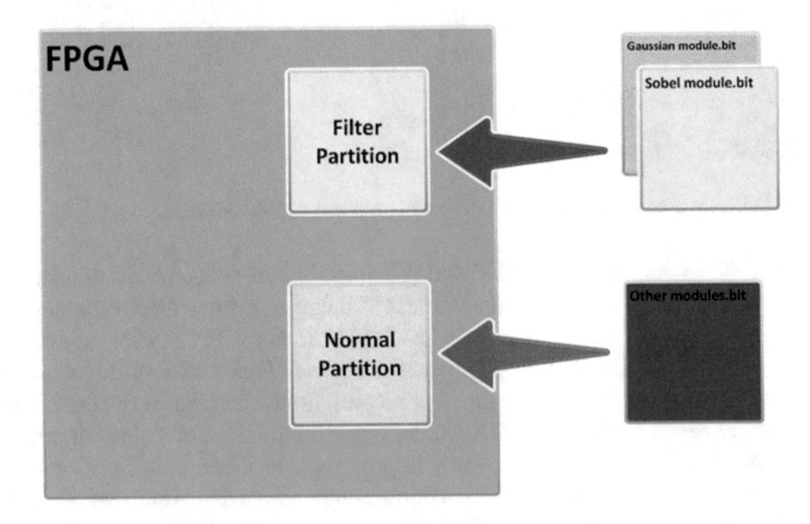

Fig. 6 Reconfigurable partitions: filter & normal

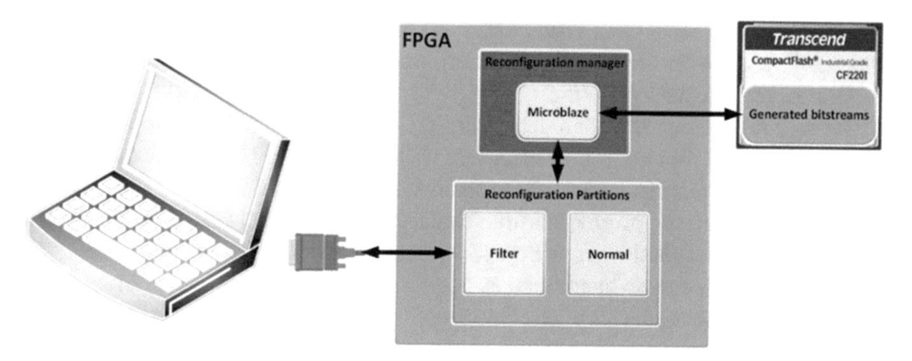

Fig. 7 The Case Study

selects the processing module which must be configured (Done manually here by giving commands to micro-blaze through a terminal connected to pc). Afterward, the generated Bitstream information for the partial configuration of the module is acquired by the processor and then configured on the specified partition. After performing processes, results are sent to the serial output, Fig. 7.

In this approach combining the previous architectures, by use of Microblaze processor FPGA configuration is executed automatically. Time-consuming manual operations necessary to load bitstreams of various modules to related partitions are terminated from the configuration steps, thus speeding up the configuration process. Moreover, there is a decrease in power consumption, according to the shorter time it takes for configuring FPGA comparing to manual configuration processes. For evaluating the proposed approach, it was obvious that the need to process images in MATLAB (Cloud) by the use of FPGA will be eliminated, resulting in computability near the edge devices for better efficiency. After execution of processes over data the knowledge afterward will be sent to the cloud if needed, resulting in data size reduction. Also, the processes done at the hardware level are executed much faster than what occurs with a software like MATLAB.

Resulting in a better understanding of the concept of distributed intelligence, and overcoming some obstacles faced by IoT at a small and experimental scale. In referred work, we were able to respond to different requests at the same time by using FPGAs reconfigurability characteristic, eliminate the dependency on other software like MATLAB for processing images, and reduce the size of the data sent after processing level to other layers.

5 Challenges and Open Issues

In this section, we will talk about challenges and open issues existing in IIoT that could be tackled by researchers. One of the first challenges faced in implementing industrial IoT assets is the high investment cost for businesses which rises from applying IIoT architecture over the industry and all connected devices to it, which has maintenance costs from hiring staff and the surveillance of all operations. So, there is a chance to create an automated architecture using ML algorithms and FPGAs to reduce the costs [76].

Using Machine learning and deep learning algorithms and FPGAs to implement them in IIoT will transform industries and will lead them to possible outcomes like self-sufficient industries, reducing the faults made by human intervention, securing the connected devices of any possible threat in dynamic and real-time situations. But there still exist challenges of how to implement and use all these coherently and practically, which can be mentioned as open issues to be tackled in the future.

As to implement Machine learning algorithms we need to have access to data collected by sensors and other devices in the IIoT network and this data needs massive computational processing power to be processed. Therefore, the need for hardware with massive computational ability is realized. The influential multi-core central processing units (CPUs), graphical processing units (GPUs) are indeed good answers giving us parallel and fast computing capability [77]. However, FPGAs have shown better processing competencies over GPUs and CPUs in real-time processing scenarios with restrictions faced like the amount of power consumption, flexibility to process different types of data, and good performance achieved. Making use of FPGAs in embedded systems and implement ML and DL algorithms on them is also an open issue to tackle.

To make FPGAs work with ml algorithms in a fast and efficient manner it is necessary to make libraries of Machine Learning algorithms in hardware level language so to be implemented on FPGAs and used with hardware accelerators. Moreover, designs of the necessary peripherals in FPGAs architecture are needed to be taken care of. To implement ml and FPGAs near edge devices namely fog computing idea there are significant limitations as for limited existing libraries and platforms. And as fog nodes are considered with moderate computational processing powers, it is necessary to come up with ideas of using FPGA-based devices near edge nodes looking at the fact they consume less power to process data comparing to other processing units. To have the ability to use ML near edge devices.

Also, for future work, the big data and data analysis can be a good research scope for the use of Machine Learning algorithms. Machine Learning algorithms are going to play an important role in the automation of decision making in industry reducing the mistakes made by human intervention in decision making. As technology advances the threats to security advance too with it. So, regarding all knowledge gained about security using ml models, it is necessary to have advanced ml algorithms to keep up with the threats and provide industry safety.

We can summarize open issues as following but not limited to:

- Big data analytics
- Providing effective training data
- Enhanced ML algorithms altered to hardware level languages
- Capable fog devices for processing near edge node
- Designing, implementing and embedding FPGAs in IIoT infrastructure
- Providing Security for software and hardware connected to IIoT

6 Conclusion

ML plays a key role in IIoT, enabling new possibilities to improve the quality of businesses and therefore our lives, by optimizing processes, reducing transferred data size, smart decision making, and quality of production. Therefore, in this chapter challenges of IIoT, ML solutions, and at the end the use of FPGAs for implementing machine learning algorithms in IIoT presented. The complete implementation of FPGAs in IIoT as a potential solution for some challenges has a lot to be done, so it is yet an open issue that can be addressed alongside the challenges and open issues mentioned in Sect. 5 of this chapter.

References

1. Angelopoulos A, Michailidis E, Nomikos N, Panagiotis T, Hatziefremidis A, Voliotis S, Zahariadis T (2019) Tackling Faults in the Industry 4.0 Era—A Survey of Machine-Learning Solutions and Key Aspects. Sensors 20:109. doi:https://doi.org/10.3390/s20010109
2. Chen B, Wan J, Shu L, Li P, Mukherjee M, Yin B (2018) Smart Factory of Industry 4.0: Key Technologies, Application Case, and Challenges. IEEE Access 6:6505–6519. doi:https://doi.org/10.1109/ACCESS.2017.2783682
3. Li B-h, Hou B-c, Yu W-t, Lu X-b, Yang C-w (2017) Applications of artificial intelligence in intelligent manufacturing: a review. Frontiers of Information Technology & Electronic Engineering 18 (1):86–96. doi:https://doi.org/10.1631/FITEE.1601885
4. Lu Y (2017) Industry 4.0: A survey on technologies, applications and open research issues. Journal of Industrial Information Integration 6:1–10. doi: https://doi.org/10.1016/j.jii.2017.04.005
5. Muhuri PK, Shukla AK, Abraham A (2019) Industry 4.0: A bibliometric analysis and detailed overview. Engineering Applications of Artificial Intelligence 78:218–235. doi: https://doi.org/10.1016/j.engappai.2018.11.007
6. Xu LD, Xu EL, Li L (2018) Industry 4.0: state of the art and future trends. International Journal of Production Research 56 (8):2941–2962. doi:https://doi.org/10.1080/00207543.2018.1444806
7. Tahsien SM, Karimipour H, Spachos P (2020) Machine learning based solutions for security of Internet of Things (IoT): A survey. Journal of Network and Computer Applications 161:102630. doi: https://doi.org/10.1016/j.jnca.2020.102630
8. Thoben K, Wiesner S, Wuest T (2017) "Industrie 4.0" and Smart Manufacturing—A Review of Research Issues and Application Examples. Int J Autom Technol 11:4–16
9. Sakhnini J, Karimipour H, Dehghantanha A, Parizi RM, Srivastava G (2019) Security aspects of Internet of Things aided smart grids: A bibliometric survey. Internet of Things:100111. doi: https://doi.org/10.1016/j.iot.2019.100111
10. Davis J, Edgar T, Porter J, Bernaden J, Sarli M (2012) Smart manufacturing, manufacturing intelligence and demand-dynamic performance. Comput Chem Eng 47:145–156
11. Karimipour H, Dehghantanha A, Parizi RM, Choo KR, Leung H (2019) A Deep and Scalable Unsupervised Machine Learning System for Cyber-Attack Detection in Large-Scale Smart Grids. IEEE Access 7:80778–80788. doi:https://doi.org/10.1109/ACCESS.2019.2920326
12. Madakam S, Uchiya T Industrial Internet of Things (IIoT): Principles, Processes and Protocols. In, 2019.
13. Qin J, Liu Y, Grosvenor R (2016) A Categorical Framework of Manufacturing for Industry 4.0 and Beyond. Procedia CIRP 52:173–178. doi: https://doi.org/10.1016/j.procir.2016.08.005

14. Wan J, Tang S, Shu Z, Li D, Wang S, Imran M, Vasilakos A (2016) Software-Defined Industrial Internet of Things in the Context of Industry 4.0. IEEE Sensors Journal 16:7373–7380
15. Zhong RY, Xu X, Klotz E, Newman ST (2017) Intelligent Manufacturing in the Context of Industry 4.0: A Review. Engineering 3 (5):616–630. doi: https://doi.org/10.1016/J.ENG.2017.05.015
16. Monostori L, Kádár B, Bauernhansl T, Kondoh S, Kumara S, Reinhart G, Sauer O, Schuh G, Sihn W, Ueda K (2016) Cyber-physical systems in manufacturing. CIRP Annals 65 (2):621–641. doi: https://doi.org/10.1016/j.cirp.2016.06.005
17. Al-Abassi A, Karimipour H, HaddadPajouh H, Dehghantanha A, Parizi RM (2020) Industrial Big Data Analytics: Challenges and Opportunities. In: Choo K-KR, Dehghantanha A (eds) Handbook of Big Data Privacy. Springer International Publishing, Cham, pp. 37–61. doi:https://doi.org/10.1007/978-3-030-38557-6_3
18. Oztemel E, Gursev S (2020) Literature review of Industry 4.0 and related technologies. Journal of Intelligent Manufacturing 31 (1):127–182. doi:https://doi.org/10.1007/s10845-018-1433-8
19. Jeschke S, Brecher C, Song H, Rawat DB (2017) Industrial Internet of Things - Cybermanufacturing Systems. Springer Series in Wireless Technology, vol RWTH-2016-08282. Springer International Publishing, Cham. https://doi.org/10.1007/978-3-319-42559-7
20. HaddadPajouh H, Dehghantanha A, M. Parizi R, Aledhari M, Karimipour H (2019) A survey on internet of things security: Requirements, challenges, and solutions. Internet of Things:100129. doi: https://doi.org/10.1016/j.iot.2019.100129
21. Al-Abassi A, Karimipour H, Dehghantanha A, Parizi RM (2020) An Ensemble Deep Learning-Based Cyber-Attack Detection in Industrial Control System. IEEE Access 8:83965–83973. doi:https://doi.org/10.1109/ACCESS.2020.2992249
22. Frustaci M, Pace P, Aloi G, Fortino G (2018) Evaluating Critical Security Issues of the IoT World: Present and Future Challenges. IEEE Internet of Things Journal 5 (4):2483–2495. doi:https://doi.org/10.1109/JIOT.2017.2767291
23. Jing Q, Vasilakos A, Wan J, Lu J, Qiu D (2014) Security of the Internet of Things: Perspectives and challenges. Wireless Networks 20:2481–2501. doi:https://doi.org/10.1007/s11276-014-0761-7
24. Himanshu J, Nikhil S, Rajinder S (2020) Evolution of IoT to IIoT: Applications & Challenges. Proceedings of the International Conference on Innovative Computing & Communications (ICICC). doi: https://doi.org/10.2139/ssrn.3603739
25. Park H, Kim H, Joo H, Song J (2016) Recent advancements in the Internet-of-Things related standards: A oneM2M perspective. ICT Express 2 (3):126–129. doi: https://doi.org/10.1016/j.icte.2016.08.009
26. Sharma A, Singh A, Sharma N, Kaushik I, Bhushan B Security Countermeasures in Web Based Application. In: 2019 2nd International Conference on Intelligent Computing, Instrumentation and Control Technologies (ICICICT), 5–6 July 2019 2019. pp 1236–1241. doi:https://doi.org/10.1109/ICICICT46008.2019.8993141
27. Sun Y, Zhang L, Feng G, Yang B, Cao B, Imran MA (2019) Blockchain-Enabled Wireless Internet of Things: Performance Analysis and Optimal Communication Node Deployment. IEEE Internet of Things Journal 6 (3):5791–5802. doi:https://doi.org/10.1109/JIOT.2019.2905743
28. Sakhnini J, Karimipour H (2020) AI and Security of Cyber Physical Systems: Opportunities and Challenges. In: Karimipour H, Srikantha P, Farag H, Wei-Kocsis J (eds) Security of Cyber-Physical Systems: Vulnerability and Impact. Springer International Publishing, Cham, pp. 1–4. doi:https://doi.org/10.1007/978-3-030-45541-5_1
29. Singh I, Centea D, Elbestawi M (2019) IoT, IIoT and Cyber-Physical Systems Integration in the SEPT Learning Factory. Procedia Manufacturing 31:116–122. https://doi.org/10.1016/j.promfg.2019.03.019
30. Fu J, Liu Y, Chao H, Bhargava BK, Zhang Z (2018) Secure Data Storage and Searching for Industrial IoT by Integrating Fog Computing and Cloud Computing. IEEE Transactions on Industrial Informatics 14 (10):4519–4528. doi:https://doi.org/10.1109/TII.2018.2793350

31. Yu Y, Chen R, Li H, Li Y, Tian A (2019) Toward Data Security in Edge Intelligent IIoT. IEEE Network 33 (5):20–26. doi:https://doi.org/10.1109/MNET.001.1800507
32. Astarloa A, Bidarte U, Jimenez J, Zuloaga A, Lázaro J (2016) Intelligent gateway for Industry 4.0-compliant production. doi:https://doi.org/10.1109/IECON.2016.7793890
33. Yousefi S, Derakhshan F, Karimipour H (2020) Applications of Big Data Analytics and Machine Learning in the Internet of Things. In: Choo K-KR, Dehghantanha A (eds) Handbook of Big Data Privacy. Springer International Publishing, Cham, pp. 77–108. doi:https://doi.org/10.1007/978-3-030-38557-6_5
34. ur Rehman MH, Yaqoob I, Salah K, Imran M, Jayaraman PP, Perera C (2019) The role of big data analytics in industrial Internet of Things. Future Generation Computer Systems 99:247–259. doi: https://doi.org/10.1016/j.future.2019.04.020
35. Qiu T, Chi J, Zhou X, Ning Z, Atiquzzaman M, Wu DO (2020) Edge Computing in Industrial Internet of Things: Architecture, Advances and Challenges. IEEE Communications Surveys & Tutorials 22 (4):2462–2488. doi:https://doi.org/10.1109/COMST.2020.3009103
36. Ambika P (2020) Chapter Thirteen—Machine learning and deep learning algorithms on the Industrial Internet of Things (IIoT). In: Raj P, Evangeline P (eds) Advances in Computers, vol 117. Elsevier, pp 321–338. doi: https://doi.org/10.1016/bs.adcom.2019.10.007
37. Namavar Jahromi A, Hashemi S, Dehghantanha A, Choo K-KR, Karimipour H, Newton DE, Parizi RM (2020) An improved two-hidden-layer extreme learning machine for malware hunting. Computers & Security 89:101655. doi: https://doi.org/10.1016/j.cose.2019.101655
38. Mohammadi S, Mirvaziri H, Ghazizadeh-Ahsaee M, Karimipour H (2019) Cyber intrusion detection by combined feature selection algorithm. Journal of Information Security and Applications 44:80–88. doi: https://doi.org/10.1016/j.jisa.2018.11.007
39. Khan A, Al-Mulla Y (2019) Unmanned Aerial Vehicle in the Machine Learning Environment. Procedia Computer Science 160:46–53. doi:https://doi.org/10.1016/j.procs.2019.09.442
40. Wuest T, Weimer D, Irgens C, Thoben K-D (2016) Machine learning in manufacturing: advantages, challenges, and applications. Production & Manufacturing Research 4 (1):23–45. doi:https://doi.org/10.1080/21693277.2016.1192517
41. Jahromi AN, Sakhnini J, Karimpour H, Dehghantanha A (2019) A deep unsupervised representation learning approach for effective cyber-physical attack detection and identification on highly imbalanced data. Paper presented at the Proceedings of the 29th Annual International Conference on Computer Science and Software Engineering, Toronto, Ontario, Canada
42. Xu LD, Duan L (2019) Big data for cyber physical systems in industry 4.0: a survey. Enterprise Information Systems 13 (2):148–169. doi:https://doi.org/10.1080/17517575.2018.1442934
43. Kim D-H, Kim TJY, Wang X, Kim M, Quan Y-J, Oh JW, Min S-H, Kim H, Bhandari B, Yang I, Ahn S-H (2018) Smart Machining Process Using Machine Learning: A Review and Perspective on Machining Industry. International Journal of Precision Engineering and Manufacturing-Green Technology 5 (4):555–568. doi:https://doi.org/10.1007/s40684-018-0057-y
44. Ge Z, Song Z, Ding SX, Huang B (2017) Data Mining and Analytics in the Process Industry: The Role of Machine Learning. IEEE Access 5:20590–20616. doi:https://doi.org/10.1109/ACCESS.2017.2756872
45. Al-Abassi A, Sakhnini J, Karimipour H Unsupervised Stacked Autoencoders for Anomaly Detection on Smart Cyber-physical Grids. In: 2020 IEEE International Conference on Systems, Man, and Cybernetics (SMC), 11–14 Oct. 2020 2020. pp 3123–3129. doi:https://doi.org/10.1109/SMC42975.2020.9283064
46. Sonntag D, Zillner S, van der Smagt P, Lörincz A (2017) Overview of the CPS for Smart Factories Project: Deep Learning, Knowledge Acquisition, Anomaly Detection and Intelligent User Interfaces. In: Jeschke S, Brecher C, Song H, Rawat DB (eds) Industrial Internet of Things: Cybermanufacturing Systems. Springer International Publishing, Cham, pp. 487–504. doi:https://doi.org/10.1007/978-3-319-42559-7_19
47. Wang J, Ma Y, Zhang L, Gao RX, Wu D (2018) Deep learning for smart manufacturing: Methods and applications. Journal of Manufacturing Systems 48:144–156. doi: https://doi.org/10.1016/j.jmsy.2018.01.003

48. Darbandi F, Jafari A, Karimipour H, Dehghantanha A, Derakhshan F, Raymond Choo K-K (2020) Real-time stability assessment in smart cyber-physical grids: a deep learning approach. IET Smart Grid 3 (4):454–461. doi: https://doi.org/10.1049/iet-stg.2019.0191
49. Hassanzadeh A, Modi S, Mulchandani S Towards effective security control assignment in the Industrial Internet of Things. In: 2015 IEEE 2nd World Forum on Internet of Things (WF-IoT), 14–16 Dec. 2015 2015. pp 795–800. doi:https://doi.org/10.1109/WF-IoT.2015.7389155
50. Duan Y, Luo Y, Li W, Pace P, Aloi G, Fortino G (2018) A collaborative task-oriented scheduling driven routing approach for industrial IoT based on mobile devices. Ad Hoc Networks 81:86–99. doi: https://doi.org/10.1016/j.adhoc.2018.07.022
51. Zhao L, Dong X (2018) An Industrial Internet of Things Feature Selection Method Based on Potential Entropy Evaluation Criteria. IEEE Access 6:4608–4617. doi:https://doi.org/10.1109/ACCESS.2018.2800287
52. Ma M, He D, Kumar N, Choo KR, Chen J (2018) Certificateless Searchable Public Key Encryption Scheme for Industrial Internet of Things. IEEE Transactions on Industrial Informatics 14 (2):759–767. doi:https://doi.org/10.1109/TII.2017.2703922
53. Li Y (2018) An Integrated Platform for the Internet of Things Based on an Open Source Ecosystem. Future Internet 10:105
54. Liang F, Yu W, Liu X, Griffith D, Golmie N (2020) Toward Edge-Based Deep Learning in Industrial Internet of Things. IEEE Internet of Things Journal 7 (5):4329–4341. doi:https://doi.org/10.1109/JIOT.2019.2963635
55. Khan AI, Al-Badi A (2020) Open Source Machine Learning Frameworks for Industrial Internet of Things. Procedia Computer Science 170:571–577. doi: https://doi.org/10.1016/j.procs.2020.03.127
56. Nurvitadhi E, Venkatesh G, Sim J, Marr D, Huang R, Hock JOG, Liew YT, Srivatsan K, Moss D, Subhaschandra S, Boudoukh G (2017) Can FPGAs Beat GPUs in Accelerating Next-Generation Deep Neural Networks? Paper presented at the Proceedings of the 2017 ACM/SIGDA International Symposium on Field-Programmable Gate Arrays, Monterey, California, USA,
57. Trimberger SM (2012) Field-Programmable Gate Array Technology. Springer US,
58. Mohammed FH, Esmail2 DR (2013) Survey on IoT Services: Classifications and Applications. International Journal of Science and Research (IJSR)
59. Azzarà A, Alessandrelli D, Bocchino S, Petracca M, Pagano P (2014) PyoT, a macroprogramming framework for the Internet of Things. doi:https://doi.org/10.1109/SIES.2014.6871193
60. Atzori L, Iera A, Morabito G (2010) The Internet of Things: A Survey. Computer Networks:2787-2805. doi:https://doi.org/10.1016/j.comnet.2010.05.010
61. Alessandrelli D, Petracca M, Pagano P (2013) T-Res: Enabling Reconfigurable In-network Processing in IoT-based WSNs. doi:https://doi.org/10.1109/DCOSS.2013.75
62. Najmabadi SM, Wang Z, Baroud Y, Simon S A self-adaptive dynamic partial reconfigurable architecture for online data stream compression. In: 2016 International Conference on FPGA Reconfiguration for General-Purpose Computing (FPGA4GPC), 9–10 May 2016 2016. pp 19–24
63. Mao F, Zhang W, He B Towards automatic partial reconfiguration in FPGAs. In: 2014 International Conference on Field-Programmable Technology (FPT), 10–12 Dec. 2014 2014. pp 286–287
64. Hauck S, DeHon A (2010) Reconfigurable computing: the theory and practice of FPGA-based computation. Elsevier,
65. Griese B, Vonnahme E, Porrmann M, Rückert U Hardware Support for Dynamic Reconfiguration in Reconfigurable SoC Architectures. In, Berlin, Heidelberg, 2004. Field Programmable Logic and Application. Springer Berlin Heidelberg, pp. 842–846
66. Choi C-S, Lee H (2007) A Self-Reconfigurable Adaptive FIR Filter System on Partial Reconfiguration Platform. IEICE Transactions 90–D:1932–1938. doi:https://doi.org/10.1093/ietisy/e90-d.12.1932

67. Leray P, Nafkha A, Moy C (2011) Implementation Scenario for Teaching Partial Reconfiguration of FPGA.
68. Borkute CV, Deshmukh AY (2013) RUN TIME DYNAMIC PARTIAL RECONFIGURATION USING MICROBLAZE SOFT CORE PROCESSOR FOR DSP APPLICATIONS.
69. Van den Abeele F, Hoebeke J, Teklemariam GK, Moerman I, Demeester P (2015) Sensor Function Virtualization to Support Distributed Intelligence in the Internet of Things. Wireless Personal Communications 81 (4):1415–1436. doi:https://doi.org/10.1007/s11277-015-2481-4
70. Hao M, Karsthof L, Rust J, Demel J, Bockelmann C, Dekorsy A, Houry AA, Mackenthun F, Paul S FPGA-based Baseband Solution for High Performance Industrial Wireless Communication. In: 2018 IEEE 23rd International Conference on Digital Signal Processing (DSP), 19–21 Nov. 2018 2018. pp 1–5. doi:https://doi.org/10.1109/ICDSP.2018.8631662
71. Abbas SSA, Priya KL Self Configurations, Optimization and Protection Scenarios with wireless sensor networks in IIoT. In: 2019 International Conference on Communication and Signal Processing (ICCSP), 4–6 April 2019 2019. pp 0679–0684. doi:https://doi.org/10.1109/ICCSP.2019.8697973
72. Al Azzawi AKY, Ercan T (2019) Design of an FPGA-based Intelligent Gateway for Industrial IoT.
73. Janßen B, Zimprich P, Hübner M A dynamic partial reconfigurable overlay concept for PYNQ. In: 2017 27th International Conference on Field Programmable Logic and Applications (FPL), 4–8 Sept. 2017 2017. pp 1–4. doi:https://doi.org/10.23919/FPL.2017.8056786
74. Kästner F, Janßen B, Kautz F, Hübner M, Corradi G Hardware/Software Codesign for Convolutional Neural Networks Exploiting Dynamic Partial Reconfiguration on PYNQ. In: 2018 IEEE International Parallel and Distributed Processing Symposium Workshops (IPDPSW), 21–25 May 2018 2018. pp 154–161. doi:https://doi.org/10.1109/IPDPSW.2018.00031
75. K. Lakshmi SG (2017) Implementation of IoT with Image processing in plant growth monitoring system. Journal of Scientific and Innovative Research; 6(2): 80–83
76. Munirathinam S (2020) Chapter Six—Industry 4.0: Industrial Internet of Things (IIOT). In: Raj P, Evangeline P (eds) Advances in Computers, vol 117. Elsevier, pp 129–164. doi: https://doi.org/10.1016/bs.adcom.2019.10.010
77. Kan C, Yang H, Kumara S (2018) Parallel computing and network analytics for fast Industrial Internet-of-Things (IIoT) machine information processing and condition monitoring. Journal of Manufacturing Systems 46:282–293. doi: https://doi.org/10.1016/j.jmsy.2018.01.010

Deep Representation Learning for Cyber-Attack Detection in Industrial IoT

Amir Namavar Jahromi, Hadis Karimipour, Ali Dehghantanha, and Reza M. Parizi

1 Introduction

Cyber-attack is an attempt to expose, alter, destroy, steal, or gain unauthorized access to or make unauthorized use of an asset that targets computer information systems, infrastructures, computer networks, or personal computers. Connecting the Industrial Control System (ICS) networks to the internet in Industry 4.0 defined the Industrial Internet of Things (IIoT), made these systems accessible on the internet, and made them a new target for attackers [1, 2]. Based on the SonicWall security report, 34.3 million IoT malware attacks were detected in 2019, which shows a 4.8% increase compared to 2018 [3].

ICS was initially designed to increase performance, reliability, and safety by reducing manual monitoring, controlling, and management. Traditionally, the ICS security was provided by physical obscurity, or a so-called air gap [4]. By introducing Industry 4.0, ICS information is routed to sophisticated applications across enterprises through the local area network and the internet; and this is where security by obscurity is no longer a valid security solution to protect the system.

A. N. Jahromi (✉) · H. Karimipour
Department of Electrical and Software Engineering, University of Calgary, Calgary, AB, Canada
e-mail: anamavar@uoguelph.ca; hadis.karimipour@ucalgary.ca

A. Dehghantanha
School of Computer Sciecne, University of Guelph, Guelph, ON, Canada
e-mail: adehghan@uoguelph.ca

R. M. Parizi
Department of Software Engineering and Game Development, Kennesaw State University, Kennesaw, GA, USA
e-mail: rparizi1@kennesaw.edu

H. Karimipour, F. Derakhshan (eds.), *AI-Enabled Threat Detection and Security Analysis for Industrial IoT*, https://doi.org/10.1007/978-3-030-76613-9_8

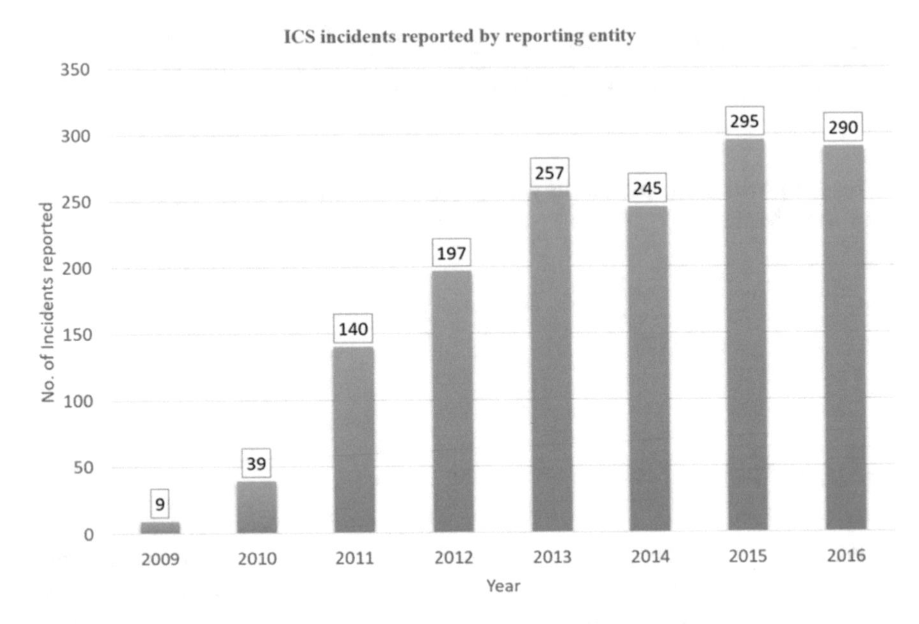

Fig. 1 The number of reported ICS incidents per year from 2009 to 2016 [4]

Although the integration of the IoT and communication network increases the efficiency and agility of ICS, it significantly increases the attack surface and the possibility of cyber-attacks [5–7]. Based on the ICS Computer Emergency Readiness Team (ICS-CERT) reports, the number of reported ICS incidents increased significantly from 2009 to 2016. Figure 1 shows the number of reported ICS attacks in each year from 2009 to 2016 [4].

For example, Stuxnet attacked Iranian centrifuges for nuclear enrichment in 2010, causing severe damage to equipment [8, 9]. Zero-day exploits were mounted on a USB drive and injected malicious code into Siemens Programmable Logic Controller (PLC) to spin centrifuges at their natural frequencies, causing their wear rates to be much higher than expected. A BlackEnergy malware was used in 2015 to target the power grids in Ukraine, causing an industrial power outage, which affected about 230 thousand people [10, 11]. Another major cyber-attack was reported in April 2018 by three US gas pipeline firms claiming a shutdown of electronic customer communication systems for several days [12].

Aside from the cyber layer, the physical part of the IIoT network also could be a target of a cyber-attack. Operational Technology (OT) devices in IIoT have no built-in security, as they were designed to be installed on a network that conveys any threat [13]. However, with the convergence of Information Technology (IT) and OT networks, devices are now exposed to many types of threats. For example, the attacker can manipulate the physical layer device such as a sensor or an actuator to unstable the system. Therefore, system-level security methods are required to analyze the physical behavior and maintain the system's reliable operation [8, 14]. While IT security solutions are mature, they cannot be directly implemented in the OT environment due to different reasons. The OT security goals are prioritized in the order of availability, integrity, and confidentiality, while IT systems are prioritized in the order of confidentiality, integrity, and availability [15, 16].

Moreover, due to close coupling between variables of the feedback control loop and physical processes, IIoT cyber-attacks have potentially fatal and environmentally damaging effects. In contrast, IT cyber-attacks may cause business continuity. Consequently, IIoT and ICS require extremely robust safety and security measurements to detect and prevent intrusions [8]. So, security solutions are required to secure connected devices spread across both OT and IT environments.

While security solutions developed for IT are mature, they are insufficient due to differences between ICS/IIoT and IT [6, 15]. Most of the available IT cybersecurity solutions (e.g., antivirus and firewalls) are oriented to monitor and protect the cyber part of the ICS, such as network and device layers. However, IIoT networks are different from ordinary IT networks [17]. Despite the IT networks mainly focusing on managing the high throughput of the network, IIoT networks have to carry out tasks reliable and punctual. Moreover, the IIoT network often has redundant assets to recover the system if it fails and continuously carries out the processes. When failure occurs, the IIoT network cannot merely be rebooted, like a typical computer network [4].

On the other side, IIoT introduces more security vulnerability due to the tight integration between the controlled physical environment and the cyber system [14]. Most of the industrial control protocols were not designed with inherent stiff security requirements. These protocols were implemented under the assumption that they will be used in a secure network. So, their operational environments were not designed with secured access control. Besides, old variations of industrial network protocols such as Supervisory Control And Data Acquisition (SCADA) suffer from common security issues such as the absence of authentication, lack of protection or security measures for data traversing over the link, and insufficient control measures to avoid default broadcast approaches [4]. These protocols were designed to perform safely in isolated, self-contained systems or zones. However, they require special considerations while using in a public network such as a Wide Area Network (WAN). Private firms may not originate from such industrial protocols, and their operational methods and structure are readily available in public. Thus, malicious actors may use these protocols to hack the control systems for either reconnaissance or attack, leading a normal functioning systems operation to an undesirable state. Another difference between IT and ICS/IIoT network is that incidents may cause severe impacts like an explosion, blackout, and leaking hazardous materials that threaten human life and/or the environment.

To mitigate the above challenges, the Canadian government, like the others, considers the ICS/IIoT cyber-security as the first goal of the *National Cyber Security Action Plan (2019–2024)* [18], which shows the importance of this subject in this era.

Considering the vital role of ICS/IIoT in critical infrastructural operations, there is a significant need for sophisticated security solutions specifically designed for IIoT to address the challenges mentioned above.

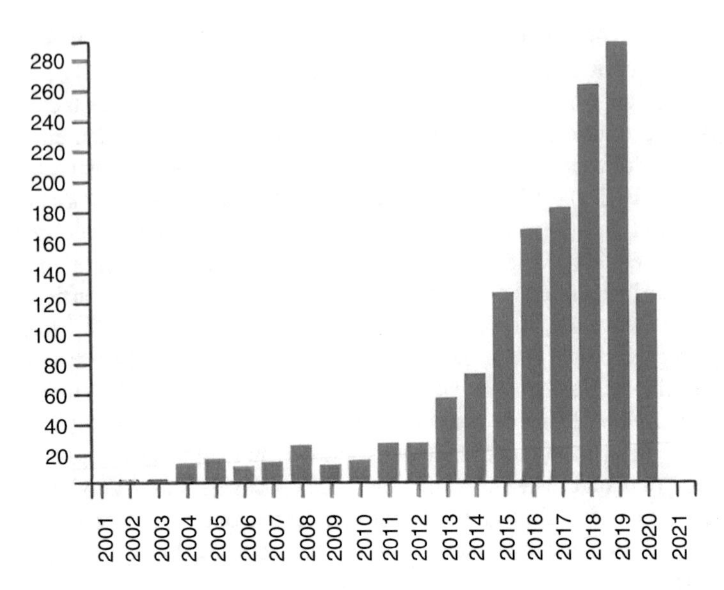

Fig. 2 The number of papers with ICS/IIoT cyber-security title in each year (Source: Web of Science)

Due to the number of assets in IIoT networks, the number of features is significantly high. A large number of features makes the manual and traditional statistical analysis of the IIoT data time-consuming and even impossible for the large real-world IIoT networks. On the other hand, ML-based techniques can analyze this high dimensional data autonomously by extracting abstract patterns from the data.

To show the importance of the subject, the Web of Science was used to analyze the literature review, the cyber-security in ICS/IIoT. As illustrated in Fig. 2, before introducing Industry 4.0, the number of researches on ICS and IIoT security is a few and changed linearly. However, after introducing Industry 4.0, this number increased significantly from about 30 papers in 2011 to about 300 papers in 2019. This trend shows that by integrating the internet with the ICS network, the importance of security in this domain has significantly increased.

From 2000, 9165 pieces of research with 76,467 citations were found by the Web of Science tool on the ICS and IIoT cyber-attack detection. Except for the years between 2010 and 2014, the research trend focused on the attack and intrusion detection increased (see Fig. 3a), which shows the importance of this subject in cyber-security research. Attack detection in ICS/IIoT research was started in 2014, and 57 ICS/IIoT attack detection researchers were found using Web of Science. Comparing the trend of ICS/IIoT attack detection and attack detection charts show that working on the ICS/IIoT attack detection methods is introduced recently. It is in an era that the challenges are almost identified (like the 2010–2014 era of attack detection). The number of researches will be increased by resolving these challenges (see Fig. 3b. Based on Fig. 3 and the Canadian and other governments

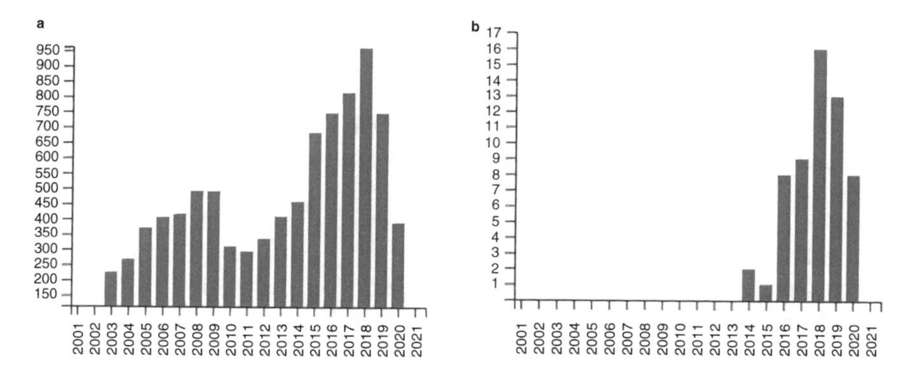

Fig. 3 The number of papers with the title of (**a**) attack detection, threat detection, or intrusion detection, (**b**) IIoT and ICS attack detection, threat detection, or intrusion detection in each year (Source: Web of Science)

considering the ICS/IIoT security as a priority in their roadmap, this subject will be a field of interest in the following years.

Considering the vital role of ICS/IIoT in critical infrastructural operations, there is a significant need for sophisticated security solutions specifically designed for ICS to address the challenges mentioned above.

The rest of this chapter is organized as follows. Section 2 defines the cyber-attack detection in IIoT and review some recent literature. Section 3 presents machine learning (ML) and explains some conventional ML techniques and challenges. In Sect. 4, an ML-based attack detection technique is proposed. Section 5 presents the experimental setup and evaluation results. This chapter will end with the conclusion in Sect. 6.

2 Cyber-Attack Detection

Cyber-threat detection answers the question of "is the incoming packet or flow a known (previously seen) cyber-attack?" To mitigate the effect of cyber-attack on ICS, attack detection techniques, including signature-based and anomaly-based detection systems, are proposed in the literature [19–23]. Signature-based methods use databases, and fixed signatures, making them unreliable in detecting unknown or new attacks [24–26]. Alternatively, anomaly-based approaches aim to identify process patterns or habits that improve the ability to deal with any new or unexpected intrusions [27]. Traditionally, attack detection has been widely used as a useful security tool to prevent cyber-attacks in CPS. The hybrid attack detection combines signature-based and anomaly-based detection, which combines the accuracy of signature-based approaches for known attacks with anomaly-based systems' generalizability [23]. While these approaches effectively detect unusual activates, they are not reliable due to frequent upgrades in the network, resulting in different

Intrusion Detection System (IDS) typologies [28]. Besides that, traditional attack detection techniques mainly rely on network metadata analysis, including IP addresses, transmission ports, traffic duration, and packet intervals. To overcome these issues, ML-based attack detection is proposed.

Moreover, attack detection can also be categorized according to network-based or host-based approaches [29]. Supervised clustering [26], single-class or multi-class Support Vector Machine (SVM) [30, 31], fuzzy logic [28, 32, 33], Deep Neural Networks (DNNs) [29, 32, 34] and deep learning [35] are among commonly used techniques for attack detection on network traffic. These techniques analyze real-time traffic data to detect malicious attacks promptly. However, attack detection that considers the only network and host data may fail to detect sophisticated attacks and insider exploits. Unsupervised models that incorporate process/physical data must monitor the system complementary without relying on detailed knowledge of the exploit. In general, a sophisticated attacker with sufficient knowledge and time can easily bypass even a robust attack detection [15].

ML-based attack detection techniques work based on a moving target to continually evolve and learn new vulnerabilities rather than identify the attack signatures or the network's normal pattern [23, 36]. There are different ML algorithms available in the literature to detect cyber-attacks compromising data integrity [32, 37, 38], availability [39], and confidentiality [33].

In 2016, Shang et al. [36] proposed a one-class SVM (OCSVM) based model to detect attacks from a sequence of Modbus function codes. One year later, Ashfaq et al. [40] proposed a fuzzy neural network method for attack detection. They did not try their method on the ICS data, but their proposed method worked in a semi-supervised environment. The research proposed by [41] used the DNN algorithm to detect false data injection attacks in power systems as a type of ICS. They have tested their proposed method on two datasets and reported 91.80% accuracy for their method. Wang et al. [42] proposed a deep learning state estimation method for attack detection. In this method, stacked autoencoders were used for state estimation, and an attack was detected by comparing the predicted value with the actual one. In 2019, [43] compared several ML algorithms, such as K-Nearest Neighbor (KNN), Random Forest (RF), Decision Tree (DT), Logistic Regression (LR), DNN, Naïve Bayes (NB), and SVM, to detect backdoor, command, and SQL injection attacks in the water storage system. This paper's results showed that the RF algorithm is the best attack detection model with the recall of 0.9744, while the DNN is the fifth algorithm with the recall of 0.8718, and the LR is the worst algorithm with the recall of 0.4744. They also reported that the DNN could not detect 12.82% of the attacks but consider 0.03% of the normal samples as an attack. Based on the results, LR, SVM, and KNN consider many attack samples normal without labeling so many normal samples as an attack. It shows that these ML algorithms are sensitive to imbalanced data and are not suitable for ICS attack detection. Moreover, Dovom et al. [44] proposed a fuzzy-based technique to detect IoT malware samples using their OpCodes. They compared several fuzzy-based techniques in detection time, accuracy, and f-measure [45]. suggested a KNN algorithm to detect cyber-attacks in the gas pipeline dataset. To decrease the effect of using an imbalanced dataset on the algorithm, they oversampled the data and balanced it. Using the KNN on the balanced dataset, they reported an accuracy of 97%, the precision of 0.98, recall of 0.92, and the f-measure of 0.95 for the effective eavesdropping attacks detection. Sakhnini et al. [46] compared SVM, KNN, and DNN techniques for IIoT false injection attack detection in another research. Moreover, they used three feature selection techniques and reported that the KNN could better detect attack samples even with less selected features. However, SVM had the best accuracy for all the test cases. Wang et al. [47] proposed a state forecasting-based attack detection method. They modeled the ICS and compared the model's states with the normal situation to detect an attack. In the same year, Li et al. [48] proposed a cost-sensitive online learning approach for attack detection. They compared their proposed method with their online learning algorithms to show its advantages. Also, Abokifa et al. [49] proposed a prediction-based method for attack detection. In this work, a predictive neural network was trained over the normal data. Then, the system's states were predicted using the trained model, and the attack was detected by comparing the predicted values and the real ones. Haddadpajouh et al. [50] proposed a multi-kernel SVM for IoT malware hunting. They used this technique to hunt malware samples using OpCode and ByteCode of IoT samples and reported an accuracy of 99.63% and an f-measure of 0.996. Fard et al. [51] proposed a multi-view sparse representation classifier method to hunt IoT malware samples and reported an accuracy of 99.2%. Yang et al. [52] proposed a combination of a long short-term memory (LSTM) and a convolutional neural network (CNN) to detect an attack based on the network flow fingerprint in ICS.

Most of the existing work on the attack detection in ICSs addresses the challenges raised by the ICS's datasets' imbalanced nature by ignoring the minority class or balancing the dataset that both have significant drawbacks. We will discuss this challenge in Sect. 3.

3 Machine Learning (ML)

In this section, DNN and autoencoders, as an unsupervised DNN, will be presents. Moreover, classification methods that were used in the experiments will be introduced.

3.1 Deep Neural Network (DNN)

Until recently, most machine learning and signal processing techniques had exploited shallow-structured architecture [53]. These architectures typically contain at most one or two layers of nonlinear feature transformation. Examples of the shallow architectures are Gaussian mixture models (GMMs), linear or nonlinear dynamic systems, conditional random fields (CRFs), maximum entropy (MaxEnt) models, support vector machines, logistic regression, kernel regression, multi-layer perceptron (MLP) with a single hidden layer including extreme learning machine (ELM) [54, 55].

Deep neural network methods try to learn features hierarchically, in the way that features of higher levels are formed by the composition of lower-level features [56, 57]. Automatic learning of features at multiple levels allows the model to learn more complex systems directly from data, without depending on human-crafted features [57, 58].

The performance of machine learning methods depends heavily on the choice of data representation (or features). Therefore, much of the actual effort to deploy machine learning algorithms go into designing preprocessing pipelines and data transformations representing the data supporting effective machine learning. Such feature engineering is essential but labor-intensive. It highlights the inability of current learning algorithms to extract and organize the discriminative information from the data. Feature engineering is a way to take advantage of human ingenuity and prior knowledge to compensate for that weakness. To expand the scope and ease of applicability of machine learning, it would be highly desirable to make learning algorithms less dependent on feature engineering so that novel applications could be constructed faster, and more importantly, to make progress towards Artificial Intelligence (AI). These algorithms are called representation learning (also called deep learning or feature learning) [59].

Deep learning solves the central problem in representation learning by introducing representations expressed in terms of other more straightforward representations. Deep learning enables the computer to build complex concepts out of more straightforward concepts [59].

3.1.1 Autoencoder

An autoencoder is a three-layer neural network trained to attempt to copy its input to its output. Internally, it has a hidden layer h that describes a code used to represent the input. The network may be viewed as consisting of two parts: an encoder function h (see Eq. (1)) and a decoder that produces a reconstruction $\hat{x}$. This architecture is presented in Fig. 4. If an autoencoder succeeds in merely learning to set

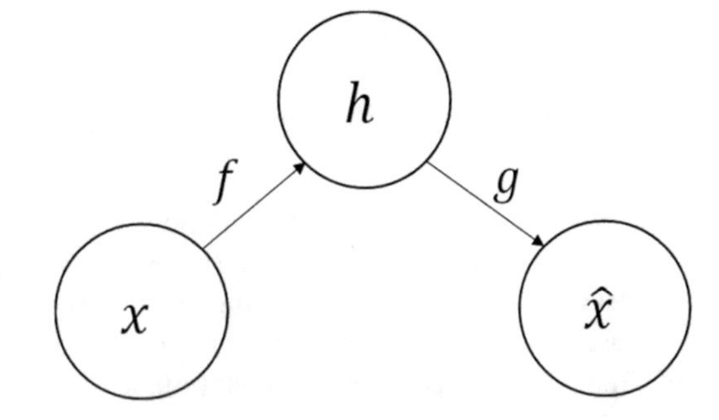

Fig. 4 The general structure of an autoencoder, mapping an input x to an output (called reconstruction) $\hat{x}$ through an internal representation or code h [59]

$g(f(x)) = x$ everywhere, then it is not especially useful. Instead, autoencoders are designed to be unable to learn to copy correctly. Usually, they are restricted in ways that allow them to copy only approximately and to copy only input that resembles the training data. Because the model is forced to prioritize which aspects of the input should be copied, it often learns useful properties of the data [60].

As mentioned before, each autoencoder consists of two parts, the encoder and the decoder. Equation (1) represents the encoder function of an autoencoder.

$$h = f(x) = \sigma(W_1 x + b_1) \tag{1}$$

where σ is the sigmoid function ($\sigma(x) = \frac{1}{1+e^{-x}}$) that is used to make the function nonlinear.

Also, Eq. (2) shows the decoder function.

$$\hat{x} = g(h) = \sigma(W_2 h + b_2) \tag{2}$$

Traditionally, autoencoders were used for dimensionality reduction or feature learning. Recently, theoretical connections between autoencoders and latent variable models have brought autoencoders to the forefront of generative modeling and representation learning [57, 58]. An autoencoder can capture the most salient features of the training data. The learning process is described as minimizing a loss function as Eq. (3) [61]. Figure 5 shows an autoencoder model with an encoder and decoder layer that encodes the data to the new representation and reconstructs it from the new representation.

$$L(x, g(f(x))) = L(x, \hat{x}) \tag{3}$$

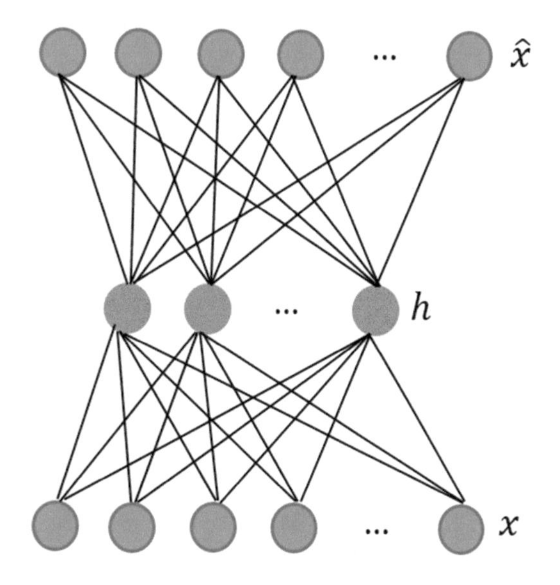

Fig. 5 Architecture of an autoencoder

Where L is a loss function penalizing $g(f(x))$ for being dissimilar from x, such as the mean squared error.

When the decoder is linear, and L is the mean squared error, an undercomplete autoencoder learns to span the same subspace as *Principal Component Analysis (PCA)*. In this case, an autoencoder trained to perform the copying task has learned the training's principal substance as a side effect [60]. However, autoencoders with nonlinear encoder function f, and nonlinear decoder functions g can thus learn a more powerful nonlinear generalization of PCA [60].

A stacked autoencoder comprises multiple autoencoders that make the architecture deep to find highly nonlinear and intricate data patterns [62]. Figure 6 illustrates the stacked autoencoder model in which multiple encoders and decoders exist. In this figure, data encodes from the input to the first latent space, h_1, and then it encodes to the more abstract representation, h_2. Then, it passes through the decoder layer to reconstruct the later layer $\hat{h}_1$ and it passes another decoder to reconstruct the input $\hat{x}$.

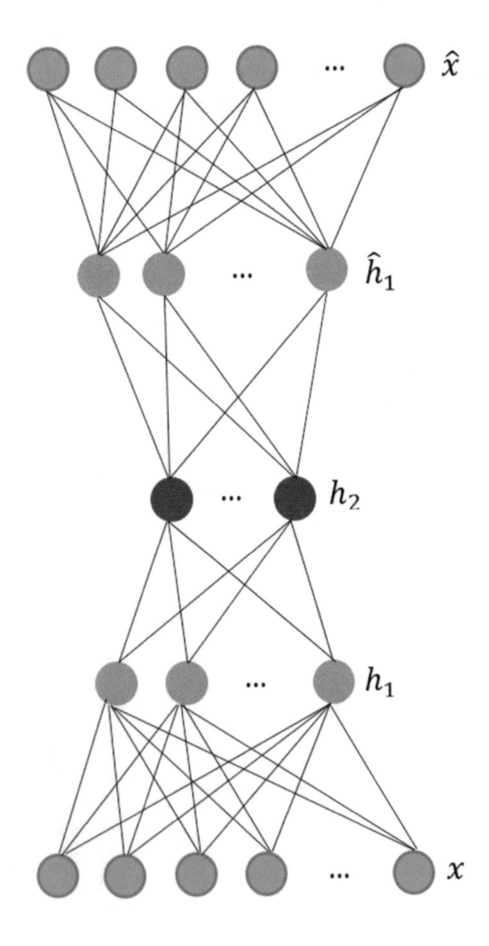

Fig. 6 Architecture of stacked autoencoder with two encoders and decoders

3.1.2 Long Short-Term Memory (LSTM)

Long-short-term memory (LSTM) networks are a special kind of *Recurrent Neural Network (RNN)* and can learn long-term dependencies. Hochreiter and Schmidhuber introduced LSTM to avoid the vanishing gradient problem of previous RNNs [63]. LSTM was refined and popularized by many people.

3.2 Decision Tree (DT)

A decision tree algorithm begins with a set of cases or examples and creates a tree data structure that can be used to classic new cases. Each case is described by a set of values or symbolic values [64].

3.3 K-Nearest Neighbors (KNN)

K-nearest neighbor algorithm is classified according to the distance between feature values. The formula for calculating distance mainly includes the Euclidean distance or Manhattan distance formula [65].

3.4 Random Forest (RF)

Random forest, developed by Breiman in 2001, is a tree-based ensemble classifier [66]. Random forest is widely used and exhibits exceptionally high performance [67].

3.5 Support Vector Machine (SVM)

An SVM is an ML technique for separating data in the feature space using an optimal hyperplane. SVM finds the optimal hyperplane using *support vectors,* the data points near that hyperplane [68].

3.6 Naïve Bayes (NB)

Naïve Bayes is a probabilistic model that assumes all features are mutually independent, given a particular variable called the label, and tries to solve the joint distribution problem [69].

3.7 Challenges of Applying ML on IIoT Data

Despite promising results, most existing ML algorithms suffer from the curse of dimensionality due to the large data volume generated in real-world ICS. Therefore, feature engineering must reduce the number of features or generate a new representation of the features to reduce computational overhead. Moreover, an imbalanced dataset of the ICS is another challenge that should be considered. Researchers have attempted to resolve this issue using oversampling/undersampling, as well as ignoring attack samples and building algorithms using normal samples.

Furthermore, most of the existing approaches ignore ICS data's imbalanced property by modeling only a system's normal behavior and reporting deviations from normal behavior as anomalies. This is, perhaps, due to limited attack samples in existing datasets and real-world scenarios. Although using majority class samples is a reasonable solution to avoid issues due to imbalanced datasets, the trained model will not view the attack samples' patterns. In other words, such an approach fails to detect unseen attacks and suffers from a high false-positive rate [70]. Thus, there have been attempts to utilize DNN approaches, for example, to facilitate automated feature (representation) learning to model complex concepts from simpler ones [60] without depending on human-crafted features [59].

Using a conventional unsupervised DNN on an imbalanced dataset yielded a DNN model that mainly learned majority class patterns and missed minority class characteristics. Most researchers have tried to address this challenge by generating new samples or removing specific samples to make the dataset balanced and then passing the data to a DNN. However, in ICS/IIoT security applications, generating or removing samples are not reasonable solutions. Due to the ICS/IIoT systems' sensitivity, generated samples should be validated in a real network, which is impossible since the generated attack samples may be harmful to the network and cause severe impacts on the environment or human life. Besides, validation of the generated samples is time-consuming. Moreover, removing the normal data from a dataset is not the right solution since the number of attack samples in ICS/IIoT datasets is usually less than 10% of the dataset, and most of the dataset knowledge is discarded by removing 80% of the dataset.

To avoid the above mentioned problems in handling imbalanced datasets, this study proposed an unsupervised deep representation learning method to make the ML techniques able to handle imbalanced datasets without changing, generating, or removing samples.

4 The Proposed ML-Based Detection Method

To analyze the effect of an unsupervised deep representation learning on the imbalanced IIoT and ICS data and extend our previous experiment using deep unsupervised representation learning in [64], a more complex deep representation learning consists of two stacked autoencoders will be proposed in this section. This method consists of four main steps, including (1) data engineering, (2) data splitting, (3) training the model, and (4) evaluating it using several machine learning classification algorithms. The first three are covered in this section, but the last one will be described in the next section.

4.1 Data Engineering

In the first step of the implementation process, data were analyzed, and useless features were deleted. These features are ones with constant values for all samples or having many missing values. Moreover, data should be normalized or standardized to avoid biasing ML techniques on features. We used normalization, which means scaling features values between 0 and 1 based on each features' maximum and minimum. Equation (4) shows the transformation function.

$$X_{scale} = \frac{x - \min(x)}{\max(x) - \min(x)} \tag{4}$$

4.2 Data Splitting

To have enough data for the training phase and have unseen data for the test phase, datasets were split into two categories, 90% of data was used for the training, and the remaining 10% was kept unseen for the test phase. Since our datasets are highly imbalanced, we split the dataset so that data distribution is the same in both sets. This process was done ten times to shape the ten-fold cross-validation.

4.3 Training the Proposed Method

In this proposed method, to handle the imbalanced IIoT data without ignoring the minority data or balancing the dataset, attack and normal samples were separated. Each passed through a stacked autoencoder to build its unique representation. These representations then fused and formed a super-vector. The resulting super-vector passed a PCA for feature extraction and dimensionality reduction. A DT was placed

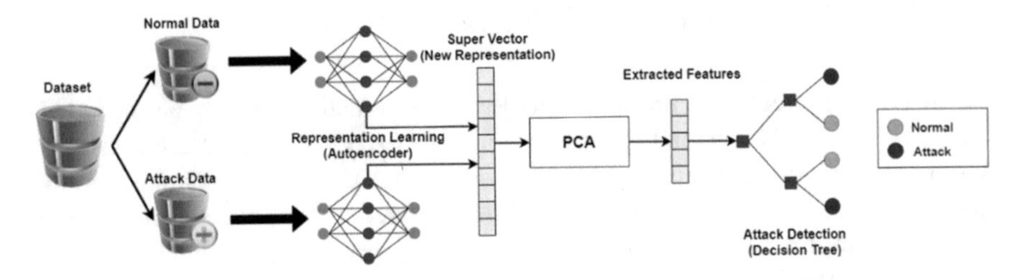

Fig. 7 Structure of the proposed IIoT attack detection method

at the end and made the final decision. Figure 7 shows the structure of the proposed method.

The proposed attack detection consists of two parts, the representation learning, and the detection phase. DNN by itself cannot perform well on imbalanced data and learn the patterns of the majority class and detects most of the minority class samples as the majority ones, which leads to a low detection rate (recall). A method consists of two autoencoder-based representation learning is proposed in this section to handle the mentioned challenge. As mentioned before, each autoencoder is responsible for learning the representation learning of one class without considering the samples of other classes, so each autoencoder's output represents its input class well. The used autoencoder had three encoders and three decoders. After training the autoencoders, all normal and attack samples passed through both autoencoders and fused into a super-vector to make the new representation of all samples.

In the second phase, the super-vector passed the PCA for feature extraction and dimensionality reduction. The extracted features are then passed to a DT classifier to make the final decision. DT has been chosen as the final classifier since it is fast and robust to the imbalanced data and worked well on IIoT data, based on our previous experiments [71]. Algorithm 1 shows the procedure of the proposed detection method.

Algorithm 1: The proposed two-phase ensemble AD method

Data: Dataset including *Normal* and *Attack* samples (X) and their labels($y = \{0, 1\}$)

Training Phase:

$X = z(X)$: $z = \dfrac{x - min(x)}{max(x) - min(x)}$;

$X_{attack} = X[y == 1]$;

$X_{normal} = X[y == 0]$;

♯ Training Representation Learning Models:

for *number of epochs* **do**

  for *number of batches in Normal set* **do**

    Train the Normal autoencoder (AE_{normal}):

     $\min \mathcal{L}(X_{normal}, \hat{X}_{normal})$;

  end

  for *number of batches in Attack set* **do**

    Train the Attack autoencoder (AE_{attack}):

     $\min \mathcal{L}(X_{attack}, \hat{X}_{attack})$;

  end

end

♯ Fusion Layer:

$newRep_{normal} = AE_{normal}.predict(X)$;

$newRep_{attak} = AE_{attack}.predict(X)$;

$X_{superVector} =$
 $concat(newRep_{normal}, newRep_{attack})$;

♯ Detection Model:

Feature selection using PCA:

$Selected\_Features(X_{superVector}) =$
 $PCA(X_{superVector})$;

Train a DT using the new features:

$DT = Train_DT(Selected\_Features)$

Testing Phase:

$x_{test} = z(x_{text})$;

$newRep_{normal} = AE_{normal}(x_{test})$;

$newRep_{attack} = AE_{attack}(x_{test})$;

$superVector =$
 $concat(newRep_{normal}, newRep_{attack})$;

$\hat{x}_{test} = Selected\_Features(superVector)$;

$\hat{y} = DT\hat{x}_{test}()$;

Output: Normal/Attack Label ($\hat{y}$)

5 Experimental Setup and Evaluation Results

5.1 Dataset

We used Ton_IoT datasets [72], collected at the University of New South Wales, to evaluate the proposed IIoT attack detection method. These datasets contain *telemetry data of IoT/IIoT services*, *operating system logs*, and *network traffics*. As illustrated in Fig. 8, the testbed used for collecting these datasets was designed based on the IIoT network elements' interaction with the three layers of the *Edge*, *Fog*, and *Cloud* to simulate a real-world IIoT network [72].

TON_IoT datasets contain normal data and nine types of attack, including scanning, Denial of Service (DoS), Distributed Denial of Service (DDoS), ransomware, backdoor, data injection, Cross-site Scripting (XSS), password cracking, and Man-in-The-Middle (MITM) were launched against various IoT and IIoT networks. Moreover, TON_IoT contains seven datasets: *fridge*, *garage door*, *GPS tracker*, *Modbus*, *motion light*, *thermostat*, and *weather*.

5.2 Attack Scenario

As mentioned before, TON_IoT contains nine types of attack [72].

- *Scanning*—is the first step of the attack. In this type of attack, the attacker gathers information about the target before launching the actual attack. The useful information for the attacker are opening ports and available services.
- *Denial of Service (DoS)*—is a flooding attack in which an attacker launches several malicious attempts to disrupt access to services.
- *Distributed Denial of Services (DDoS)*—is a type of DoS launched by a large number of compromised devices known as bots.
- *Ransomware*—is a complex type of malware that encrypts user files and limits the user access to a system or a service by encrypting them and sell the decryption key. IIoT devices and applications are victims of IoT ransomware since they often carry out critical tasks.
- *Backdoors*—is an attack that allows the attacker to gain unauthorized remote access to an IIoT device.
- *Injection attack*—tries to execute malicious codes or inject malicious data into IIoT servers and applications.
- *Cross-site scripting (XSS)*—tries to run malicious commands on a Web server in the IIoT applications by HTML or JavaScript codes.
- *Password cracker*—tries to guess IIoT devices and servers using various brute force or dictionary attacks.
- *Man-in-the-middle (MITM)*—is a network attack that can intercept the communication channel between tao devices and manipulate the data.

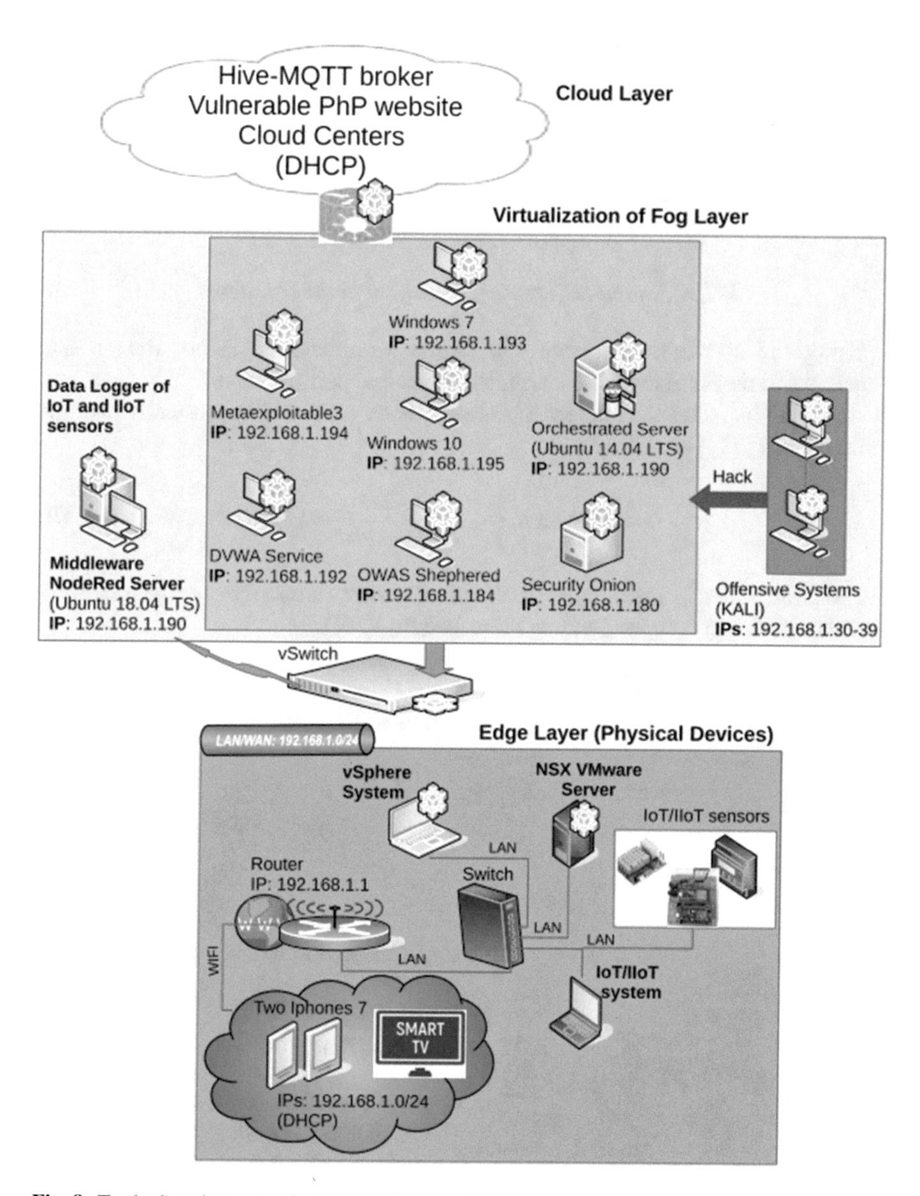

Fig. 8 Testbed environment for TON_IOT dataset [72]

5.3 Evaluation Metrics

In IIoT attack detection, the primary concerns are the high detection rate and the ability to avoid false alarms. Therefore, the performance of the proposed method is analyzed based on the True Positives (TP), the True Negatives (TN), the False

Positives (FP), and the False Negatives (FN), which are defined in Eqs. (5)–(8), respectively.

$$TP_i = \sum samples\ correctly\ classified\ as\ calss\ i \tag{5}$$

$$TN_i = \sum samples\ correctly\ classified\ as\ other\ classes \tag{6}$$

$$FP_i = \sum samples\ wrongly\ classified\ as\ class\ i \tag{7}$$

$$FN_i = \sum samples\ wrongly\ classified\ as\ other\ calsses \tag{8}$$

Using the above metrics, we can define Accuracy, Precision, Recall, and F-measure to measure machine learning algorithms' performance.

Accuracy indicates the number of samples that were classified correctly over the entire dataset [73] (see Eq. (9)).

$$Accuracy = \frac{TP + TN}{TP + TN + FP + FN} \tag{9}$$

Precision indicates the number of classified samples correctly over the total samples classified as the corresponding class (see Eq. (10)).

$$Precision = \frac{TP}{TP + FP} \tag{10}$$

Recall indicates the number of classified samples correctly over the total samples of the corresponding class (see Eq. (11)).

$$Recall = \frac{TP}{TP + FN} \tag{11}$$

Also, f-measure is the harmonic value of precision and recall (see Eq. (12)).

$$f - measure = \frac{2 \times Precision \times Recall}{Precision + Recall} \tag{12}$$

As mentioned before, our task is a three-class classification. Our data is hugely imbalanced, so we used the f-measure metric suitable for multiclass classification and the classification of imbalanced data by considering TP, FP, and FN of each class.

5.4 Evaluation Results

To evaluate the proposed method and determine the effect of the representation learning on imbalanced IIoT datasets, the proposed method was compared with the same algorithm without using the representation learning (DT row of Table 1) and the evaluation results reported in [72]. Table 1 illustrates the evaluation results.

Table 1 Evaluation results of the proposed method, DT, and reported results in [57]

Dataset	Method	Accuracy	Precision	Recall	f-measure
Fridge sensor	Proposed method	100	1.0	1.0	1.0
	DT	85	0.73	0.85	0.78
	KNN [57]	99	0.99	0.99	0.99
	RF [57]	97	0.97	0.97	0.97
	NB [57]	50	0.53	0.51	0.51
	SVM [57]	81	0.86	0.82	0.80
	LSTM [57]	100	1.0	1.0	1.0
Modbus	Proposed method	99	0.99	0.99	0.99
	DT	95	0.95	0.95	0.95
	KNN [57]	77	0.77	0.78	0.77
	RF [57]	97	0.98	0.98	0.98
	NB [57]	67	0.46	0.68	0.55
	SVM [57]	67	0.46	0.68	0.55
	LSTM [57]	68	0.47	0.68	0.55
Light motion	Proposed method	95	0.95	0.95	0.95
	DT	86	0.74	0.86	0.80
	KNN [57]	54	0.34	0.59	0.43
	RF [57]	58	0.34	0.59	0.43
	NB [57]	58	0.34	0.59	0.43
	SVM [57]	58	0.34	0.59	0.43
	LSTM [57]	59	0.35	0.59	0.44
Garage door	Proposed method	100	1.0	1.0	1.0
	DT	100	1.0	1.0	1.0
	KNN [57]	100	1.0	1.0	1.0
	RF [57]	100	1.0	1.0	1.0
	NB [57]	100	1.0	1.0	1.0
	SVM [57]	100	1.0	1.0	1.0
	LSTM [57]	100	1.0	1.0	1.0
GPS sensor	Proposed method	100	1.0	1.0	1.0
	DT	92	0.92	0.92	0.92
	KNN [57]	88	0.89	0.88	0.88
	RF [57]	85	0.85	0.85	0.85
	NB [57]	84	0.86	0.85	0.86
	SVM [57]	86	0.88	0.87	0.87
	LSTM [57]	87	0.89	0.88	0.88
Thermostat	Proposed method	92	0.92	0.92	0.92
	DT	81	0.78	0.81	0.79
	KNN [57]	60	0.56	0.61	0.57
	RF [57]	66	0.59	0.66	0.53
	NB [57]	66	0.44	0.66	0.53
	SVM [57]	66	0.44	0.66	0.53
	LSTM [57]	66	0.45	0.67	0.54

(continued)

Table 1 (continued)

Dataset	Method	Accuracy	Precision	Recall	f-measure
Weather	Proposed method	93	0.93	0.93	0.93
	DT	81	0.81	0.81	0.81
	KNN [57]	81	0.81	0.81	0.81
	RF [57]	84	0.84	0.84	0.84
	NB [57]	69	0.72	0.69	0.67
	SVM [57]	63	0.68	0.63	0.55
	LSTM [57]	82	0.82	0.81	0.80

As shown in Table 1, the proposed method outperformed all other methods, including the DT, which has a similar procedure except the representation learning, which shows the power of the proposed representation learning on IIoT imbalanced datasets. The proposed method has the best recall metric compared to the other competitors, which means a higher detection rate. Moreover, it has the best precision metric among the compared techniques, which shows fewer false alarms.

6 Conclusion

This chapter focused on cyber-attack detection as a part of IIoT cybersecurity. In this chapter, an unsupervised deep representation learning approach was proposed to handle the IIoT imbalanced data without manipulating it. The proposed approach consists of two phases, representation learning, and detection. In the representation learning phase, two stacked autoencoders were used to learn independent representations from normal and attack samples and made the data ready for the detection phase. In the detection phase, a decision tree was used to separate normal and attack data based on the learned representation. Compared to six conventional ML techniques, the proposed model shows its superior performance in accuracy, precision, recall, and f-measure. The higher recall of the proposed method shows its higher detection rate, while the higher precision shows the fewer false alarms of the proposed method over the other techniques.

The proposed cyber-attack detection approach will be completed by a cyber-attack hunting component that makes a two-stage detection and hunting framework to detect the known attack in the first step and hunt the previously unseen attacks in the hunting stage.

References

1. HaddadPajouh H, Dehghantanha A, M. Parizi R, et al (2019) A survey on internet of things security: Requirements, challenges, and solutions. Internet of Things 100129. https://doi.org/10.1016/j.iot.2019.100129

2. Sakhnini J, Karimipour H, Dehghantanha A, et al (2019) Security aspects of Internet of Things aided smart grids: A bibliometric survey. Internet of Things 100111. https://doi.org/10.1016/j.iot.2019.100111
3. SonicWall (2020) 2020 SonicWall cyber threat report
4. Singh S, Karimipour H, HaddadPajouh H, Dehghantanha A (2020) Artificial Intelligence and Security of Industrial Control Systems. In: Choo K-KR, Dehghantanha A (eds) Handbook of Big Data Privacy. Springer International Publishing, Cham, pp 121–164
5. Karimipour H, Dehghantanha A, Parizi RM, et al (2019) A Deep and Scalable Unsupervised Machine Learning System for Cyber-Attack Detection in Large-Scale Smart Grids. IEEE Access 7:80778–80788. https://doi.org/10.1109/ACCESS.2019.2920326
6. Yan W, Mestha LK, Abbaszadeh M (2019) Attack Detection for Securing Cyber Physical Systems. IEEE Internet Things J 6:8471–8481. https://doi.org/10.1109/JIOT.2019.2919635
7. Cui Z, Xue F, Cai X, et al (2018) Detection of Malicious Code Variants Based on Deep Learning. IEEE Trans Ind Informatics 1–1. https://doi.org/10.1109/TII.2018.2822680
8. Zhang F, Kodituwakku HADE, Hines JW, Coble J (2019) Multilayer Data-Driven Cyber-Attack Detection System for Industrial Control Systems Based on Network, System, and Process Data. IEEE Trans Ind Informatics 15:4362–4369. https://doi.org/10.1109/TII.2019.2891261
9. Ma R, Cheng P, Zhang Z, et al (2019) Stealthy Attack Against Redundant Controller Architecture of Industrial Cyber-Physical System. IEEE Internet Things J 6:9783–9793. https://doi.org/10.1109/JIOT.2019.2931349
10. CISA (2016) Cyber-Attack Against Ukrainian Critical Infrastructure. https://www.us-cert.gov/ics/alerts/IR-ALERT-H-16-056-01
11. Falco G, Caldera C, Shrobe H (2018) IIoT Cybersecurity Risk Modeling for SCADA Systems. IEEE Internet Things J 5:4486–4495. https://doi.org/10.1109/JIOT.2018.2822842
12. Higgins KJ (2010) Security Incidents Rise In Industrial Control Systems. https://www.darkreading.com/attacks-breaches/security-incidents-rise-in-industrial-control-systems-/d/d-id/1133388
13. Karimipour H, Srikantha P, Farag H, Wei-Kocsis J (2020) Security of Cyber-Physical Systems. Springer International Publishing, Cham
14. Al-Abassi A, Karimipour H, HaddadPajouh H, et al (2020) Industrial Big Data Analytics: Challenges and Opportunities. In: Choo K-KR, Dehghantanha A (eds) Handbook of Big Data Privacy. Springer International Publishing, Cham, pp 37–61
15. Yang J, Zhou C, Yang S, et al (2018) Anomaly Detection Based on Zone Partition for Security Protection of Industrial Cyber-Physical Systems. IEEE Trans Ind Electron 65:4257–4267. https://doi.org/10.1109/TIE.2017.2772190
16. Singh S, Karimipour H, Pajooh H, Dehghantanha A (2019) Artificial Intelligence and Security of Industrial Control Systems. In: Handbook of Big Data and Privacy. pp 1–32
17. Tahsien SM, Karimipour H, Spachos P (2020) Machine learning based solutions for security of Internet of Things (IoT): A survey. J Netw Comput Appl 161:102630. https://doi.org/10.1016/j.jnca.2020.102630
18. Public Safety Canada (2018) National Cyber Security Action Plan (2019-2024). https://www.publicsafety.gc.ca/cnt/rsrcs/pblctns/ntnl-cbr-scrt-strtg-2019/index-en.aspx
19. Ntalampiras S (2015) Detection of integrity attacks in cyber-physical critical infrastructures using ensemble modeling. IEEE Trans Ind Informatics 11:104–111. https://doi.org/10.1109/TII.2014.2367322
20. Mohammadi S, Mirvaziri H, Ghazizadeh-Ahsaee M, Karimipour H (2019) Cyber intrusion detection by combined feature selection algorithm. J Inf Secur Appl 44:80–88. https://doi.org/10.1016/j.jisa.2018.11.007
21. E. Nowroozi, A. Dehghantanha, R. M. Parizi, and K.-K. R. Choo, "A survey of machine learning techniques in adversarial image forensics," Computers & Security, vol. 100, p. 102092, 2021.
22. A. Yazdinejad, G. Srivastava, R. M. Parizi, A. Dehghantanha, H. Karimipour, and S. R. Karizno, "SLPoW: Secure and Low Latency Proof of Work Protocol for Blockchain in Green IoT

Networks," in 2020 IEEE 91st Vehicular Technology Conference (VTC2020-Spring), 2020, pp. 1–5: IEEE.
23. Ponomarev S, Atkison T (2016) Industrial Control System Network Intrusion Detection by Telemetry Analysis. IEEE Trans Dependable Secur Comput 13:252–260. https://doi.org/10.1109/TDSC.2015.2443793
24. A. Yazdinejadna, R. M. Parizi, A. Dehghantanha, and M. S. Khan, "A kangaroo-based intrusion detection system on software-defined networks," *Computer Networks,* vol. 184, p. 107688, 2021.
25. A. N. Jahromi, S. Hashemi, A. Dehghantanha, R. M. Parizi and K. -K. R. Choo, "An Enhanced Stacked LSTM Method With No Random Initialization for Malware Threat Hunting in Safety and Time-Critical Systems," in *IEEE Transactions on Emerging Topics in Computational Intelligence*, vol. 4, no. 5, pp. 630–640, Oct. 2020, https://doi.org/10.1109/TETCI.2019.2910243.
26. Chi-Ho Tsang, Kwong S (2005) Multi-agent intrusion detection system in industrial network using ant colony clustering approach and unsupervised feature extraction. In: 2005 IEEE International Conference on Industrial Technology. pp 51–56
27. Pang Z, Liu G, Zhou D, et al (2016) Two-Channel False Data Injection Attacks Against Output Tracking Control of Networked Systems. IEEE Trans Ind Electron 63:3242–3251. https://doi.org/10.1109/TIE.2016.2535119
28. Clemente JF (2018) No CYBER SECURITY FOR CRITICAL ENERGY INFRASTRUCTURE. Naval Postgraduate School
29. Gao W, Morris T (2014) On Cyber Attacks and Signature Based Intrusion Detection for Modbus Based Industrial Control Systems. J Digit Forensics, Secur Law. https://doi.org/10.15394/jdfsl.2014.1162
30. Maglaras LA, Jiang J (2014) Intrusion detection in SCADA systems using machine learning techniques. In: 2014 Science and Information Conference. pp 626–631
31. Luo Y (2013) Research and design on intrusion detection methods for industrial control system. Zhejiang University
32. He Y, Mendis GJ, Wei J (2017) Real-Time Detection of False Data Injection Attacks in Smart Grid: A Deep Learning-Based Intelligent Mechanism. IEEE Trans Smart Grid 8:2505–2516. https://doi.org/10.1109/TSG.2017.2703842
33. Krawczyk B (2016) Learning from imbalanced data: open challenges and future directions. Prog Artif Intell 5:221–232. https://doi.org/10.1007/s13748-016-0094-0
34. Linda O, Manic M, Vollmer T, Wright J (2011) Fuzzy logic based anomaly detection for embedded network security cyber sensor. In: 2011 IEEE Symposium on Computational Intelligence in Cyber Security (CICS). pp 202–209
35. Vollmer T, Manic M (2009) Computationally efficient Neural Network Intrusion Security Awareness. In: 2009 2nd International Symposium on Resilient Control Systems. pp 25–30
36. Javaid A, Niyaz Q, Sun W, Alam M (2016) A Deep Learning Approach for Network Intrusion Detection System. In: Proceedings of the 9th EAI International Conference on Bio-Inspired Information and Communications Technologies (Formerly BIONETICS). ICST (Institute for Computer Sciences, Social-Informatics and Telecommunications Engineering), Brussels, BEL, pp 21–26
37. Jahromi AN, Hashemi S, Dehghantanha A, et al (2020) An Improved Two-Hidden-Layer Extreme Learning Machine for Malware Hunting. Comput Secur 89:101655. https://doi.org/10.1016/j.cose.2019.101655
38. Karimipour H, Leung H (2020) Relaxation-based anomaly detection in cyber-physical systems using ensemble kalman filter. IET Cyber-Physical Syst Theory Appl 5:49–58
39. Karimipour H, Dinavahi V (2017) On false data injection attack against dynamic state estimation on smart power grids. In: 2017 IEEE International Conference on Smart Energy Grid Engineering (SEGE). pp 388–393

40. Zolanvari M, Teixeira MA, Gupta L, et al (2019) Machine Learning-Based Network Vulnerability Analysis of Industrial Internet of Things. IEEE Internet Things J 6:6822–6834. https://doi.org/10.1109/JIOT.2019.2912022
41. Shang W, Zeng P, Wan M, et al (2016) Intrusion detection algorithm based on OCSVM in industrial control system. Secur Commun Networks 9:1040–1049. https://doi.org/10.1002/sec.1398
42. (2017) Fuzziness based semi-supervised learning approach for intrusion detection system. Inf Sci (Ny) 378:484–497. https://doi.org/10.1016/j.ins.2016.04.019
43. Yu JJQ, Hou Y, Li VOK (2018) Online False Data Injection Attack Detection With Wavelet Transform and Deep Neural Networks. IEEE Trans Ind Informatics 14:3271–3280. https://doi.org/10.1109/TII.2018.2825243
44. Wang H, Ruan J, Wang G, et al (2018) Deep Learning-Based Interval State Estimation of AC Smart Grids Against Sparse Cyber Attacks. IEEE Trans Ind Informatics 14:4766–4778. https://doi.org/10.1109/TII.2018.2804669
45. Dovom EM, Azmoodeh A, Dehghantanha A, et al (2019) Fuzzy pattern tree for edge malware detection and categorization in IoT. J Syst Archit 97:1–7. https://doi.org/10.1016/j.sysarc.2019.01.017
46. Khan IA, Pi D, Khan ZU, et al (2019) HML-IDS: A Hybrid-Multilevel Anomaly Prediction Approach for Intrusion Detection in SCADA Systems. IEEE Access 7:89507–89521. https://doi.org/10.1109/ACCESS.2019.2925838
47. Sakhnini J, Karimipour H, Dehghantanha A (2019) Smart Grid Cyber Attacks Detection Using Supervised Learning and Heuristic Feature Selection. In: 2019 IEEE 7th International Conference on Smart Energy Grid Engineering (SEGE). IEEE, pp 108–112
48. Wang H, Ruan J, Zhou B, et al (2019) Dynamic Data Injection Attack Detection of Cyber Physical Power Systems with Uncertainties. IEEE Trans Ind Informatics 15:5505–5518. https://doi.org/10.1109/TII.2019.2902163
49. Li D, Chen D, Jin B, et al (2019) MAD-GAN: Multivariate Anomaly Detection for Time Series Data with Generative Adversarial Networks. Lect Notes Comput Sci (including Subser Lect Notes Artif Intell Lect Notes Bioinformatics) 11730 LNCS:703–716. https://doi.org/10.1007/978-3-030-30490-4_56
50. Abokifa AA, Haddad K, Lo C, Biswas P (2019) Real-Time Identification of Cyber-Physical Attacks on Water Distribution Systems via Machine Learning-Based Anomaly Detection Techniques. J Water Resour Plan Manag 145:4018089. https://doi.org/10.1061/(ASCE)WR.1943-5452.0001023
51. Haddadpajouh H, Mohtadi A, Dehghantanaha A, et al (2020) A Multi-Kernel and Metaheuristic Feature Selection Approach for IoT Malware Threat Hunting in the Edge Layer. IEEE Internet Things J 1–1. https://doi.org/10.1109/JIOT.2020.3026660
52. Fard SMH, Karimipour H, Dehghantanha A, et al (2020) Ensemble sparse representation-based cyber threat hunting for security of smart cities. Comput Electr Eng 88:106825. https://doi.org/10.1016/j.compeleceng.2020.106825
53. Yang K, Li Q, Lin X, et al (2020) iFinger: Intrusion Detection in Industrial Control Systems via Register-Based Fingerprinting. IEEE J Sel Areas Commun 38:955–967
54. Bengio Y (2009) Learning Deep Architectures for AI. Found Trends® Mach Learn 2:1–127. https://doi.org/10.1561/2200000006
55. Huang G Bin, Zhu QY, Siew CK (2004) Extreme learning machine: A new learning scheme of feedforward neural networks. In: IEEE International Conference on Neural Networks—Conference Proceedings. IEEE, pp 985–990
56. Huang G-B, Zhu Q-Y, Siew C-K (2006) Extreme learning machine: Theory and applications. Neurocomputing 70:489–501. https://doi.org/10.1016/j.neucom.2005.12.126
57. Bourlard H, Kamp Y (1988) Auto-association by multilayer perceptrons and singular value decomposition. Biol Cybern 59:291–294. https://doi.org/10.1007/BF00332918

58. Hinton GE, Zemel RS (1994) Autoencoders, Minimum description length and helmholtz free energy. In: Cowan JD, Tesauro G, Alspector J (eds) Advances in Neural Information Processing Systems 6. Morgan-Kaufmann, pp 3–10
59. Bengio Y, Courville A, Vincent P (2013) Representation Learning: A Review and New Perspectives. IEEE Trans Pattern Anal Mach Intell 35:1798–1828
60. Goodfellow I, Bengio Y, Courville A (2016) Deep learning. MIT Press
61. Ng A, Ngiam J, Foo CY, et al (2013) Unsuoervised feature and deep learning (UFLDL). In: Stanford Univ.
62. Karnouskos S (2011) Stuxnet worm impact on industrial cyber-physical system security. In: IECON 2011—37th Annual Conference of the IEEE Industrial Electronics Society. pp 4490–4494
63. Hochreiter S, Schmidhuber J (1997) Long Short-Term Memory. Neural Comput 9:1735–1780. https://doi.org/10.1162/neco.1997.9.8.1735
64. Fayyad UM, Irani KB (1992) On the handling of continuous-valued attributes in decision tree generation. Mach Learn 8:87–102. https://doi.org/10.1007/BF00994007
65. Wang D, Wang X, Zhang Y, Jin L (2019) Detection of power grid disturbances and cyber-attacks based on machine learning. J Inf Secur Appl 46:42–52. https://doi.org/10.1016/j.jisa.2019.02.008
66. Breiman L (2001) Random Forests. Mach Learn 45:5–32. https://doi.org/10.1023/A:1010933404324
67. Genuer R, Poggi J-M, Tuleau-Malot C, Villa-Vialaneix N (2017) Random Forests for Big Data. Big Data Res 9:28–46. https://doi.org/10.1016/j.bdr.2017.07.003
68. Sebald DJ, Bucklew JA (2000) Support Vector Machine Techniques for Nonlinear Equalization. IEEE Trans Signal Process 48:3217–3226. https://doi.org/10.1109/78.875477
69. Lowd D, Domingos P (2005) Naive Bayes Models for Probability Estimation. In: Proceedings of the 22nd international conference on Machine learning—ICML '05. ACM Press, New York, New York, USA, pp 529–536
70. Bellinger C, Sharma S, Japkowicz N (2012) One-Class versus Binary Classification: Which and When? In: 2012 11th International Conference on Machine Learning and Applications. pp 102–106
71. Namavar Jahromi A, Sakhnini J, Karimpour H, Dehghantanha A (2019) A Deep Unsupervised Representation Learning Approach for Effective Cyber-Physical Attack Detection and Identification on Highly Imbalanced Data. In: Proceedings of the 29th Annual International Conference on Computer Science and Software Engineering. IBM Corp., pp 14–23
72. Alsaedi A, Moustafa N, Tari Z, et al (2020) TON_IoT Telemetry Dataset: A New Generation Dataset of IoT and IIoT for Data-Driven Intrusion Detection Systems. IEEE Access 8:165130–165150. https://doi.org/10.1109/ACCESS.2020.3022862
73. Ramirez AG, Lara C, Betev L, et al (2018) Arhuaco: Deep Learning and Isolation Based Security for Distributed High-Throughput Computing

Classification and Intelligent Mining of Anomalies in Industrial IoT

Nafiseh Sharghivand and Farnaz Derakhshan

1 Introduction

Today, Internet of Things (IoT) is being broadly used across various industries including manufacturing, energy, transportation, logistics, etc. It is often assumed that IIoT devices have continuous access to the Internet or other internal networks in their environment [1]. However, despite all the benefits that network accessibility and connectivity brings, it poses new security challenges to the system [2, 3]. Specifically, the Internet connectivity and data sharing between different IIoT devices increase the risk of various cyber-attacks, aimed at stealing or altering confidential or sensitive data.

In spite of the aforementioned security risks in IIoT systems, most of the machinery and equipment in modern industrial plants are not designed to be securely connected, making them more vulnerable to cyber-attacks [4]. This can in turn lead to a series of major problems from an individual machine breakdown to the shutdown of the entire production, or even loss of lives at the extreme point [5–11].

However, it should be noted that cyber-attacks are not the only origin of data corruption. In other words, data trustworthiness in IIoT can also be threatened by other reasons such as any hardware or software problems [12], without any motivations for a deliberate damage. Furthermore, the large-scale generated data by IIoT and the high dynamicity and heterogeneity of the industry environments make IIoT systems even more vulnerable to corrupted data [13].

Despite to the all above challenges, there is a general rule which is always exploited to improve the trustworthiness of the collected data. This rule is to exclude the corrupted data that do not exhibit a data pattern similar to the expected normal

N. Sharghivand (✉) · F. Derakhshan
Computer Engineering Department, Faculty of Electrical and Computer Engineering, University of Tabriz, Tabriz, Iran
e-mail: n.shaghivand@tabrizu.ac.ir

H. Karimipour, F. Derakhshan (eds.), *AI-Enabled Threat Detection and Security Analysis for Industrial IoT*, https://doi.org/10.1007/978-3-030-76613-9_9

behavior. As mentioned earlier, the corrupted data, which are also referred as data anomalies, may be a result of a hardware malfunction, a software problem, or even a malicious cyber-attack [14]. No matter which reason has caused anomalies, the corrupted data should be identified timely, before any critical loss or damage occurs [15].

Therefore, efficient anomaly detection schemes are needed to ensure the reliability of the collected data and to improve the efficiency of the IIoT. However, conventional security solutions do not meet industry standards and requirements and thus novel approaches need to be devised [16].

In this respect, a set of different classification and intelligent mining solutions have been proposed for the problem of anomaly detection in IIoT in recent years.

In this chapter, we aim to clarify the main challenges in designing efficient anomaly detection solutions in industrial IoT environments. Furthermore, we review the existing studies in the literature highlighting their major features. We also discuss the remaining open problems in the field that need to be addressed in future researches.

The rest of the chapter is organized as follows. In Sect. 2, we provide some preliminaries including anomaly detection definition and its challenges in IIoT. Next, in Sect. 3 we review the proposed intelligent anomaly detection approaches in the literature. We provide a discussion over these studies in Sect. 4. Furthermore, we highlight the open problems in anomaly detection in IIoT to shed light for future researches in this field in Sect. 5. Finally, in Sect. 6 we conclude the chapter.

2 Anomaly Detection and Its Challenges in IIoT

In this section, we provide some preliminary concepts about anomaly detection in IIoT. We first provide a general definition of anomaly detection. We then discuss the existing challenges in anomaly detection, and particularly in IIoT.

In general, anomaly detection is described as the process of recognizing patterns in the data that exhibits a behavior different from the one expected. Such non-conforming patterns are usually referred to as anomalies or outliers [17]. Figure 1 illustrates a two-dimensional dataset in which the observations with a normal behavior are shown with blue marks, whereas anomalies demonstrating a very different behavior are shown with blue marks.

The main goal of anomaly detection is to declare any observations outside the normal regions as anomalies. However, there are several general challenges for an efficient anomaly detection system in all application domains described as follows [17, 18]:

- The difficulty of defining normal regions
- Variation of normal behavior over time
- The difference of anomaly notion in various application domains
- Lack of sufficient training/valuation datasets
- Presence of noise in the dataset

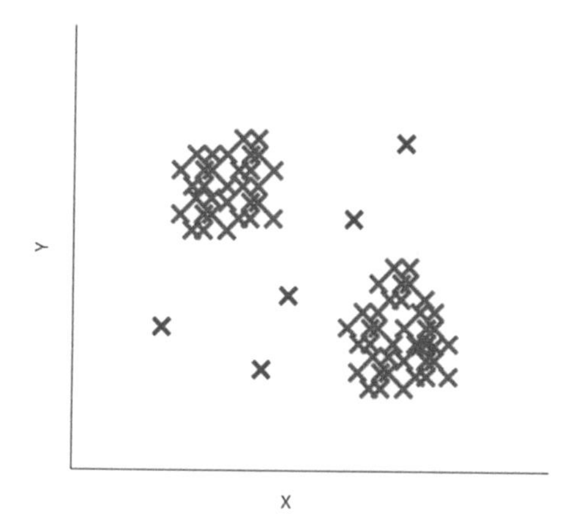

Fig. 1 A simple demonstration of anomalies in a 2-dimensional dataset

An efficient anomaly detection system is thereby the one that can improve the accuracy of detection while lowering the false alert rates.

When anomaly detection is applied in the context of IIoT, its main goal is to detect any kind of anomalies to discover any faults, malfunctions, or cyber-attacks [19]. However, several other challenges are specifically associated with the IIoT domain which we discuss in the following.

The first challenge is the time efficiency of the anomaly detection system which highly matters in IIoT. Therefore, time constraints should be considered in the whole process. In the first step, it should be noted that data collections and evaluations must be computed in an online fashion, using the latest data from IIoT devices. Next, requirements for a long series of past data should be taken into account, depending on the nature of the application of the collected data. Finally, the fast declaration of results (i.e. the anomalousness or trustworthiness of data) must be highly considered to make quick responses to the cause of the anomaly possible, before any critical loss or damage happens.

The second challenge is where the anomaly detection system must be deployed. This matters in terms of both computational and communication resources, and also security issues. Since, on one hand, anomaly detection systems often require both powerful computational resources and high bandwidth communication links. On the other hand, the anomaly detection systems usually access a set of sensitive data collected from different IIoT devices and thereby they should be able to guarantee the security requirements.

3 Literature Review for Anomaly Detection in IIoT

In this section, we review the proposed solutions in the literature for the problem of anomaly detection in IIoT and discuss their main features.

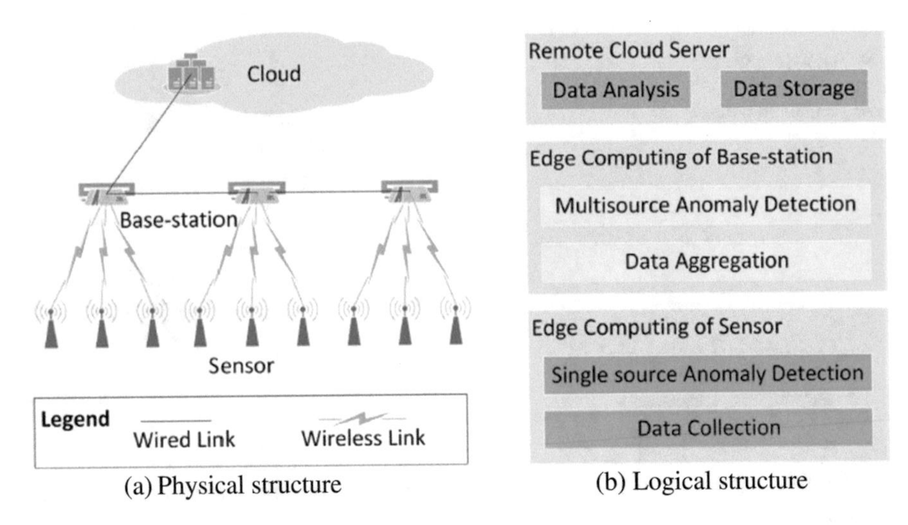

Fig. 2 Proposed hierarchical edge computing model in [20]. (**a**) Physical structure, (**b**) Logical structure

Peng et al. [20] have addressed the early anomaly detection problem in underground mining environments to improve safety. They propose a multi-source multi-dimensional data anomaly detection method based on hierarchical edge computing, which enables multi-source data anomaly detection at collection end (sensors) and sinks end (base-stations).

More specifically, first they propose a hierarchical edge computing model to realize load balance and low-latency data processing at the sensor and base-station ends. This model has been shown in Fig. 2.

As it can be seen, the physical structure (Fig. 2a) consists of three major parts including remote cloud server, base station and sensor. Also, according to Fig. 2b, the logical model consists of two edge computing units which are the base station edge and the sensor edge. The base stations have more powerful hardware infrastructure compared to the sensors. Hence, they are mainly responsible to execute the multi-source data anomaly detection algorithm, while the sensor nodes should execute the single source data anomaly detection algorithm.

Tthe proposed anomaly detection system works as follows. First, each sensor periodically collects environmental state data and then performs single source data anomaly detection. The proposed algorithm considers the temporal correlation of monitoring data in the anomaly detection process. Then, it sends the original data along with the detection results to the corresponding base station via a wireless link. Once the data is received by the base station, it performs multi-source heterogeneous data anomaly detection. It combines the received single-source data anomaly detection result with other detection results obtained by other sensors. Indeed, it considers the temporal and spatial correlation properties of multi-source data. The final result is then sent to the remote cloud with the original data through a wired link. Moreover, when an anomalous behavior is detected, the system will start an

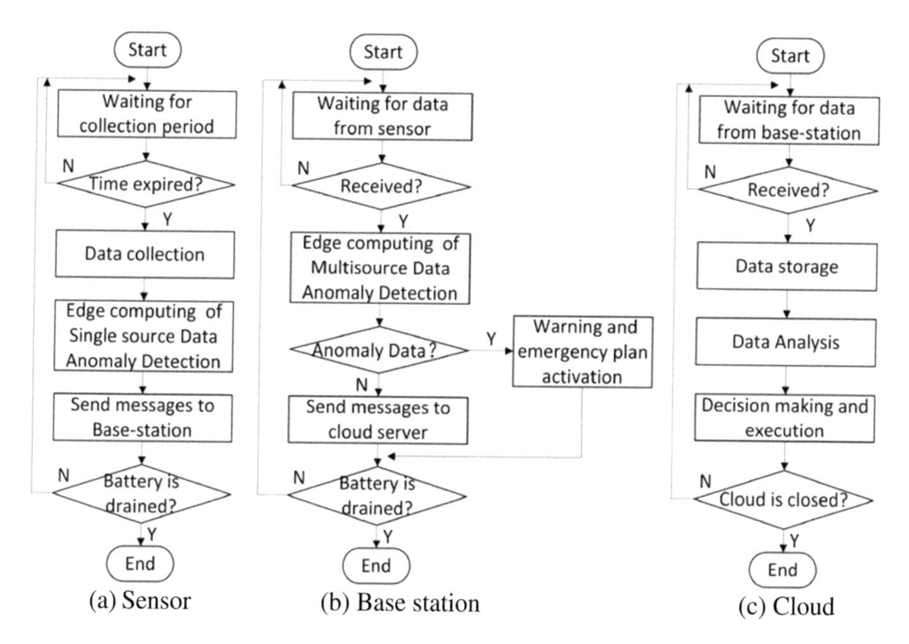

Fig. 3 Flow chart of data anomaly detection over different nodes in the proposed hierarchical edge computing model in [20]. (**a**) Sensor, (**b**) Base station, (**c**) Cloud

emergency warning and treatment plan according to the safety prevention and early warning level in underground mining.

Finally, at the highest level, the received data is stored in the database of the cloud platform. Then, the decision center uses data mining and other intelligent algorithms for analyzing the data and making decision.

Figure 3 illustrates the Flow chart of data anomaly detection over different nodes in the proposed hierarchical edge computing model.

Yang et al. [21] propose a secure and efficient distributed k-nearest neighbors classification algorithm (SEED-kNN) that can be implemented in the IIoT anomaly detection, while supporting large-scale data classification on distributed servers.

As shown in Fig. 4, they assume a system model which consists of three entities, namely, the control center, the cloud and the devices. The control center is not only responsible for managing directing, or regulating the behavior of devices, it is also in charge of running machine learning algorithms on the dataset in cloud to discover the added-values for automatic control and industrial process monitoring. The generated data by devices is pre-processed to provide the training samples and then maintained on the cloud infrastructure which includes multiple distributed servers. Indeed, each server maintains a different part of training samples.

However, the data exchanged between devices and servers can be vulnerable to a variety of cyber-attacks, such as eavesdropping, or intentional or unintentional data expose by the cloud. Therefore efficient mechanisms are required to ensure security.

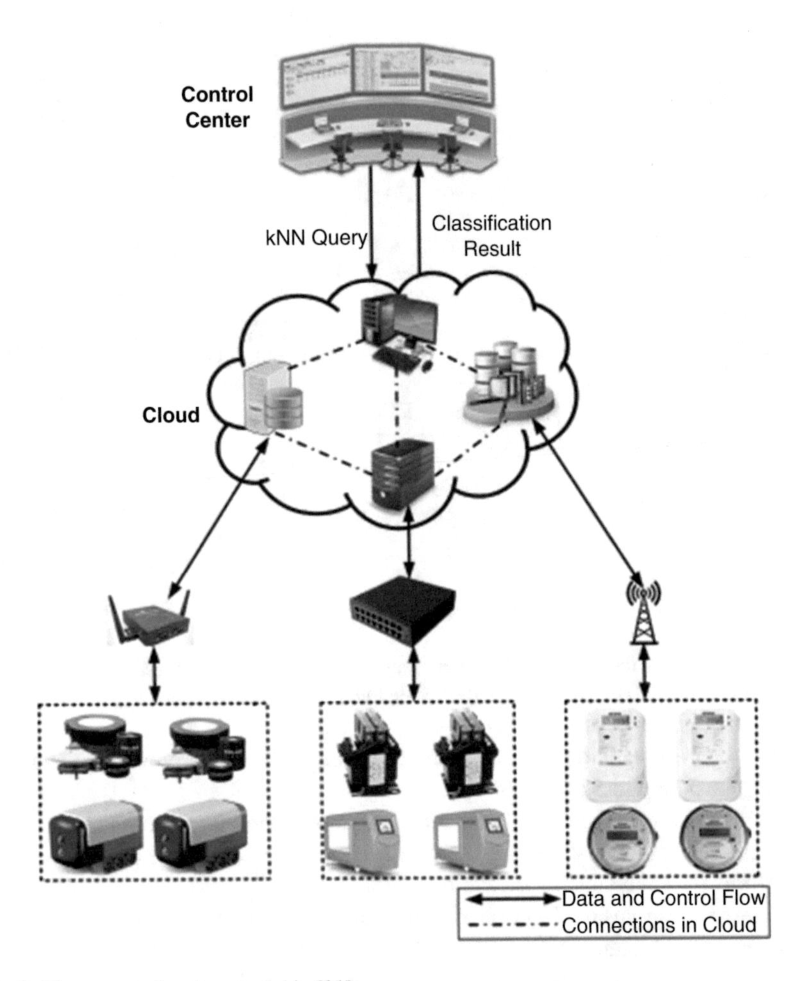

Fig. 4 The proposed system model in [21]

Hence, in order to preserve the security of training samples in cloud against data leakage, and also to prevent the control information exposure, the authors first design a secure and efficient vector homomorphic encryption (SE-VHE) scheme. The SE-VHE scheme is designed by constructing a key-switching matrix and a noise matrix for data encryption. Then, the SEED-kNN is proposed based on the designed SE-VHE to provide a secure and efficient kNN classification over the encrypted training samples.

Moreover, since the data are separately maintained on multiple servers, the Map/Reduce architecture is integrated to achieve the parallel and distributed data classification. Indeed, the encrypted query for classification which has been issued by the control center is split and mapped to all the distributed servers. Then, the classification results from servers are gathered and the final class label is returned to the control center.

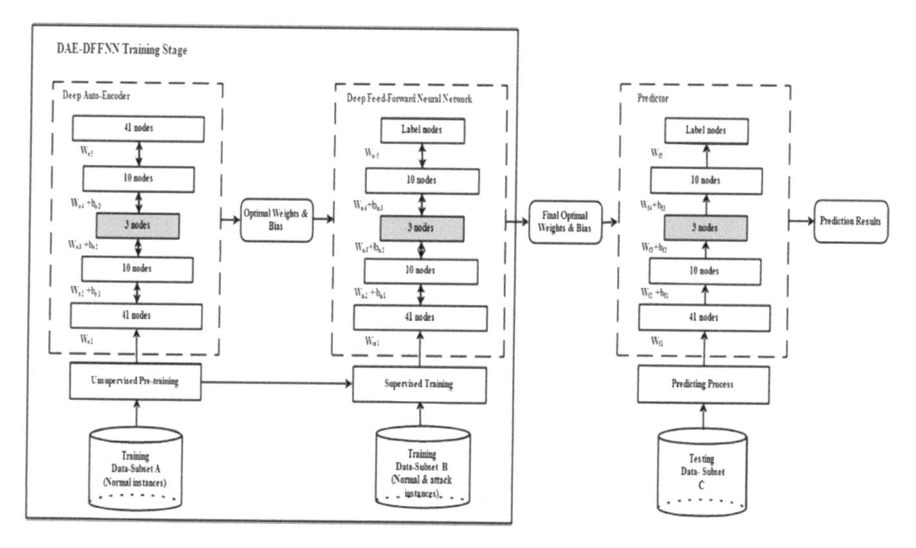

Fig. 5 Proposed architecture of DAE-DFFNN model based ADS for IICs in [22]

Muna et al. [22] propose an anomaly detection technique for Internet Industrial Control Systems (IICSs) based on deep learning models which are trained and validated using information collected from TCP/IP packets. In the training phase, a consecutive training process executed using an unsupervised Deep Auto-Encoder (DAE) algorithm to learn normal network behaviors and produce the optimal parameters (i.e., weights and biases). These parameters are then used as an initialization stage for the training of a supervised Deep FeedForward Neural Network (DFFNN) to classify network observations. In the testing phase, the DFFNN is used to discover attacks.

Figure 5 shows the overall structure of the proposed anomaly detection system. As it can be seen, only an unlabeled normal training dataset is used to train DAE to learn and discover the most important feature representations for normal behavior. Then, the trained model is used as the starting point for training the DFFNN using the labeled training dataset. In the testing phase, the new dataset sample is tested based on the final constructed network model.

Genge et al. [12] propose an anomaly detection approach in the context of aging IIoT. Indeed, a major novelty of this work in the field of anomaly detection systems for IIoT is the adding of aging parameter to the anomaly detection process.

In the proposed approach it is assumed that the IIoT's life cycle is split into distinct ages, while each age defines an operational time interval. Then, principle component analysis is used to create a model for the normal process behavior for each age. The proposed approach employs the correlation among process variables to detect stealthy cyber-attacks. It is based on Hotelling's T^2 statistics and the univariate cumulative sum. Another novel feature of their approach is the detection of attempts to alter the dataset in each age. Moreover, the leveraging of multivariate

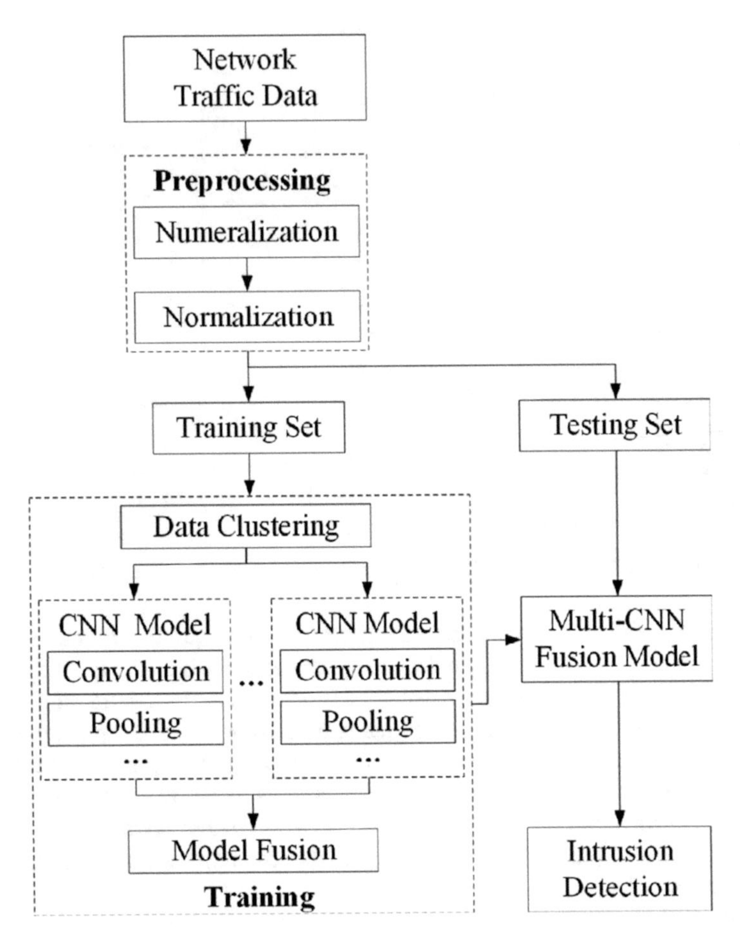

Fig. 6 Block diagram of the proposed intrusion detection system in [23]

process analysis enables the proposed anomaly detection system to detect stealthy attacks that cause minor process deviations by manipulating legitimate sensor data.

Li et al. [23] first propose a method for processing one-dimensional weakly correlated feature data. They apply this processing method on the benchmark NSL-KDD dataset provided by [24]. Then, they propose a deep learning approach for intrusion detection based on multi-convolutional neural network (multi-CNN) fusion algorithm. The authors believe that the processed data have a better training result for deep learning.

Figure 6 shows the diagram of their proposed intrusion detection system. In the first step the input dataset is preprocessed which involves numeralization and normalization. Numeralization is applied so that the one-dimensional feature data is converted into a grayscale image. However, normalization is performed to remove large numerical differences in the records by moving them within the range of [0,

1]. This will speed up the convergence speed of the model. The obtained dataset is then divided into a training set and a test set.

In the training phase, first data clustering is performed to improve the adaptability of the obtained model. According to their proposed approach, a data with m features should be divided into n parts according to prior knowledge or common clustering methods, where m > n. Then, the different parts of the data are processed separately. Hence, with respect to the existing correlation between features of their adopted dataset, they have divided the feature data into four parts, which are the basic features, the content features, the time-based network traffic statistics features, and the host-based network traffic statistics features. Then, the input data is converted into the form of images in order to better exploit the advantages of convolutional neural networks. Next, the same CNN structure is used for each part of the dataset. Finally, model fusion is performed to obtain the prediction result.

Yan et al. [25] propose a new hinge classification algorithm based on mini-batch gradient descent with an adaptive learning rate and momentum (HCA-MBGDALRM).

The most common method used for optimizing the hinge classification algorithm is the stochastic gradient descent. However, one of the major issues of this method is that it reduces the gradient descent only when the sample point maximizes the loss function. Also, the hinge classification training method is unstable and vulnerable to noise. Hence, the authors propose HCA-MBGDALRM to address the aforementioned shortcomings.

The algorithm significantly improves the performance of deep network training compared with traditional neural networks, decision trees, and logistic regression in terms of scale and speed. Indeed, the proposed parallel framework for HCA-MBGDALRM divides and executes program tasks on multiple microprocessors, accelerating the processing speed of very large traffic datasets.

HCA-MBGDALRM has been implemented using the parameter server architecture which enables distributed machine learning. In this architecture, data and workload are allocated to client nodes, while the global variables are retained by the server nodes.

In addition, the authors solve the data skew problem in the shuffle phase The proposed HCA-MBGDALRM method has been theoretically analyzed which shows that it can converge to the globally optimal solution effectively.

Demertzis et al. [4] propose an anomaly detection framework based on Deep Learning network architecture [26]. In this respect, they develop an innovative blockchain security architecture that aims to ensure secure network communication between the IIoT devices based on deep learning smart contracts. Indeed, a type of blockchain communication is considered in which smart contracts programmatically implement a bilateral traffic control agreement. This way, they are capable of detecting anomalies based on a trained deep autoencoder neural network. The implementation of the proposed approach was in fact based on the unary classification philosophy in which a deep autoencoder was trained using a dataset of normal IIoT behavior.

The proposed architecture provides a secure distribution platform for the associated transactions, without any intervention of a central authority. It can be also considered as a decentralized, reliable, peer-to-peer network architecture for device communication in order to improve security and functionality in industrial applications.

The presented architecture consists of three layers including, Authorization, Syndication, and Overlay layers [27]. These layers are shortly described as following.

The Authorization layer provides levels of access by expressing security policies, using entities, namespaces, resources, and delegations of trust. The Syndication layer provides publish/subscribe functions to system resources. The subscribe permission allows an entity to receive information from the published resource. The publish permission allows an entity to publish information and interact with the resources. This layer is directly related to the Authorization layer. Finally, the Overlay layer is responsible to form an overlay network over the existing physical network. In other words, it forms the communication network between the IoT devices.

However, the proposed approach exhibits several disadvantages. First, it assumes that the data is easily accessible. Second, the proposed system is not scalable as it is not applicable for very large data sets (terabytes).

Liu et al. [28] propose a new anomaly detection framework for sensing time-series data in IIoT. The proposed model enables on-device deep anomaly detection using Federated Learning (FL). In this model, a cloud aggregator and edge devices train a deep anomaly detection model by using a given training algorithm (e.g., LSTM) for anomaly detection.

More precisely, the edge devices train a shared global model on their own device using their own local dataset (i.e., sensing time series data from IIoT nodes). Then, they send their updated models (i.e., gradients) to the cloud aggregator. All the received models are then used by the cloud aggregator to obtain a new global model. In the end, the cloud aggregator send the new global model to all edge devices to achieve accurate and timely anomaly detection. Figure 7 illustrates the above steps.

It should be noted that local on-device training in their proposed model helps to preserve the privacy of edge devices, while solving the problem of data islands. Moreover, the proposed Attention Mechanism-based Convolutional Neural Network-Long Short Term Memory (AMCNN-LSTM) for anomaly detection avoids communication overhead during model training. The AMCNN-LSTM model uses attention mechanism-based CNN units to extract important fine-grained features of historical observation sensing time-series data. This way, memory loss and gradient dispersion problems are prevented which are common problems in encoder-decoder models such as LSTM model. Furthermore, this model uses LSTM modules for timeseries prediction. Finally, they propose a gradient compression mechanism based on Top-k selection to further improve the communication efficiency of the proposed framework.

Garmaroodi et al. [29] propose an anomaly detection system for a real-world dataset collected from SinaDarou Labs which is an industrial pharmaceutical

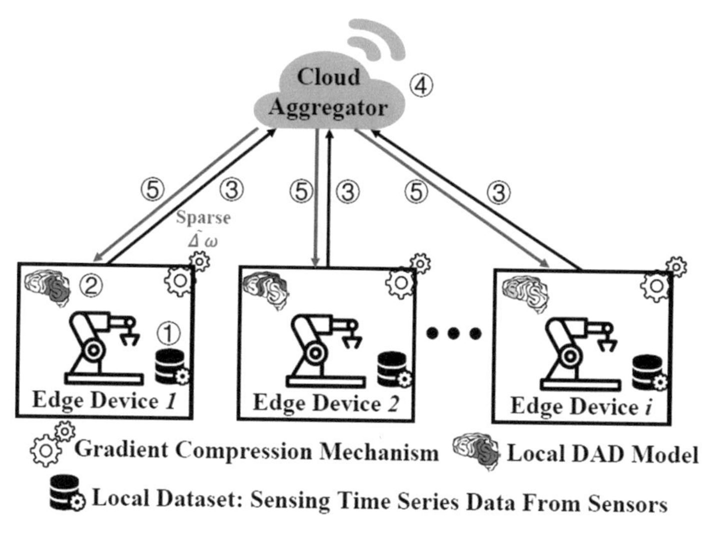

Fig. 7 The overview of proposed anomaly detection framework in [28]

company. They specifically address anomaly detection for CHRIST Osmotron water purifier.

In this respect, they first collect a dataset of normal and faulty operation samples over a two-week time interval. Given the data, they propose two anomaly detection approaches to detect system faults. The first one is based on a supervised learning model (Fig. 8a). However, due to the lack of enough faults data, the second model is based on normal system identification which models the system components by artificial neural networks (Fig. 8b).

Wu et al. [30] propose an anomaly detection method in IIoT, which is a synergy of the Long Short-Term Memory Neural Network (LSTM-NN) and the Gaussian Bayes model. A major employed idea in their work is that the time-dependency is closely related to the outlier detection of IIoT data. Because, any anomaly occurrence is not only related to the current state, but also related to the past states. Therefore, they propose a stacked LSTM model to deal with time series data with different types of time-dependency.

The proposed LSTM-NN builds a model on normal time series and then detects anomalies by utilizing the predictive error for the Gaussian Naive Bayes model [31]. This way, it exploits the advantages of both LSTM and Gaussian Naive Bayes models, which are LSTM's good prediction performance, and the excellent classification performance of the Gaussian Naïve Bayes model through the predictive error.

Figure 9 shows the overview of their proposed anomaly detection framework for IIoT time series data. As it can be seen, in the first step initial data processing is performed via data cleaning, data down-sample, and data normalization. Then, the pre-processed data is divided into training sets, validation sets and test sets. The training and validation sets contain only the normal data, while the test set contain both types of data. The training set is used to optimize and construct the stacked

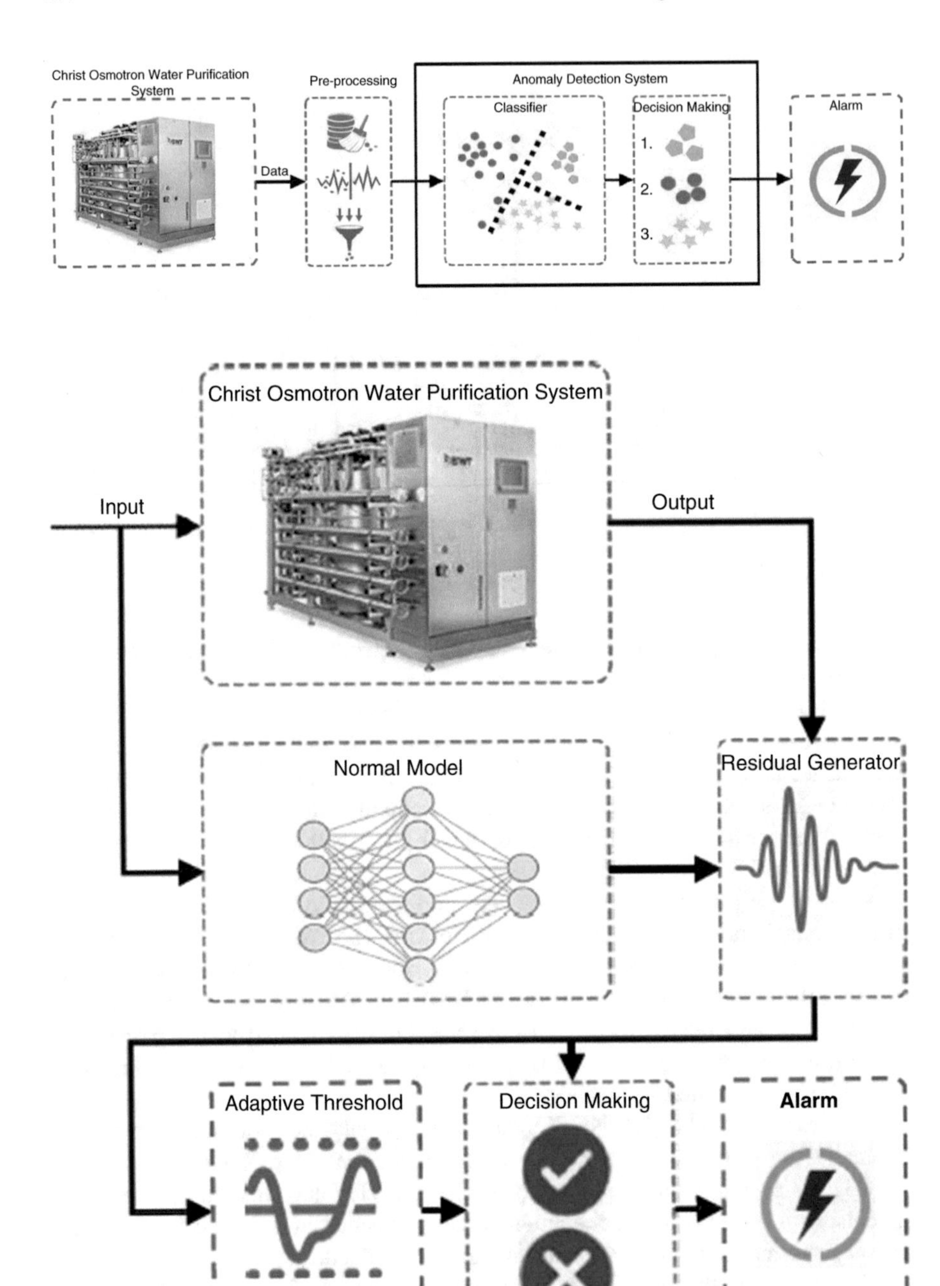

Fig. 8 The proposed anomaly detection approaches in [29]. (**a**) An anomaly detection model based on supervised learning in which abnormal/faulty classes need to be known beforehand [29]. (**b**) An anomaly detection model based on normal system identification in which fault samples are scarce [29]

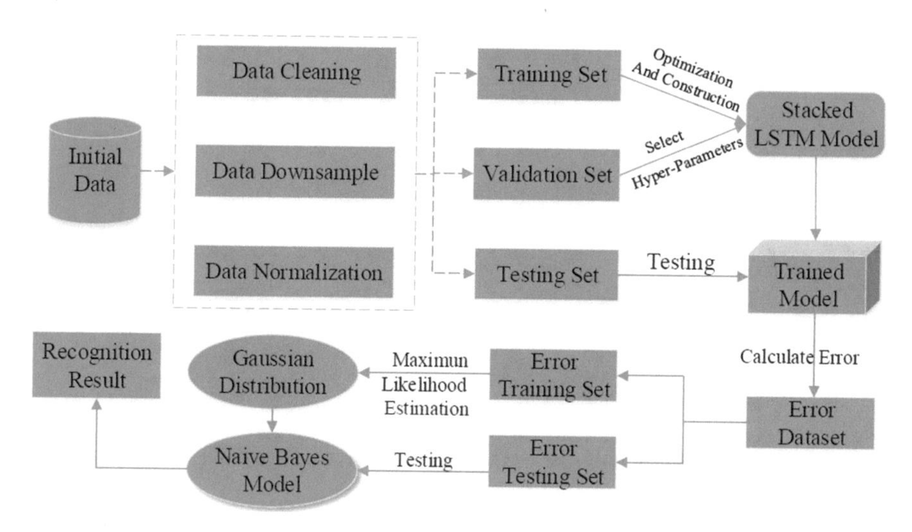

Fig. 9 Overview of the proposes anomaly detection framework for industrial IoT time series data in [30]

LSTM model. The validation set is used to select hyper-parameters. Finally, the test set is used to obtain error data sets which are also split into two sets of error training and error test. The error training set is used to make the maximum likelihood estimation in order to obtain the parameters of the Gauss distribution. These parameters are then used by the Naive Bayes model to build a Gaussian Naive Bayes model. Once the error test sets are imported into this Gaussian Naive Bayes model, the classification results are achieved.

Zolanvari et al. [32], study the applicability of ML-based anomaly detection systems to improve the security of the IIoT systems. In this respect, they first describe the four most popular IIoT protocols, along with their main communication network vulnerabilities. Then, they use a real-world testbed to deploy backdoor, command injection, and SQL injection attacks against the system and then show how an ML-based anomaly detection system can be effectively used to detect them.

Finally, a test methodology has been proposed in [33] for the comparison of cloud and edge-based implementation of deep learning algorithms for anomaly detection in IIoT. Since, deep learning algorithms often demand high computational and communication resources, raising serious questions on the system scalability.

In this regard, they use a real-world platform to study the tradeoff between scalability, communication delay, and bandwidth usage when using a full-cloud architecture and the edge-cloud architecture. They assume three possible architectures with respect to the above scenario considering the production Machine, the IIoT Edge Computer, and the Cloud App. In the edge-cloud architecture, the deep learning algorithm is run by the Edge Computer (Fig. 10a); In the full-cloud architecture, an Edge Computer is used only as a local gateway for data aggregation and thus the deep learning algorithm is executed in Cloud (Fig. 10b); Finally, in the full-cloud architecture the production Machine is directly connected to the Cloud (Fig. 10c).

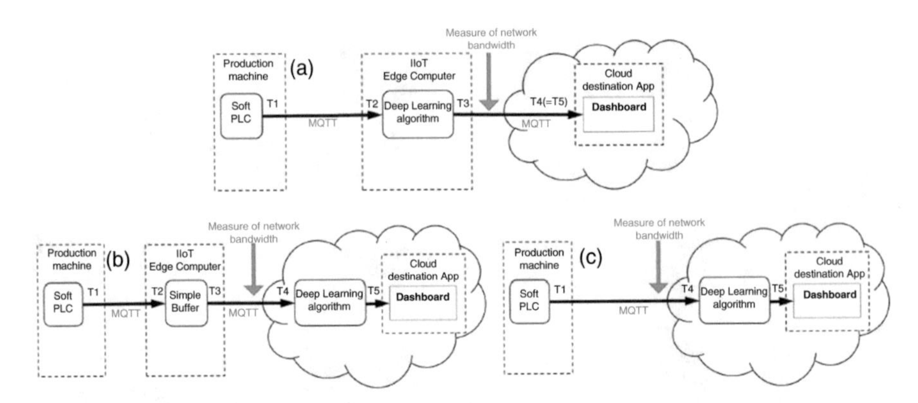

Fig. 10 Proposed experimental setup for the measurement of the performance metrics in [33], where the anomaly detection can be carried out in either the Edge Computer (**a**), or in the Cloud (**b**, **c**)

According to their obtained results, the complexity of the algorithm plays an important role in the decision about which architecture is most suitable. However, the full-cloud architecture can outperform the edge-cloud architecture when Cloud computation power is scaled.

4 Discussion

The proposed solutions in the literature for anomaly detection can be generally categorized based on their employed model, which may be parametric (e.g., distribution functions), or non-parametric (e.g., machine learning techniques). The non-parametric models can be also further categorized based on their requirement for prior knowledge (i.e., supervised and unsupervised learning) [34].

Most of the proposed models use machine learning-based approaches as they are more consistent with the dynamic nature of the IIoT environments. Many of these studies employ classification based models, supervised or semi-supervised learning techniques, which have expensive training times, but their testing time is much faster due to the existence of a pre-trained model.

Since in many application domains data acquisition for training and testing is a costly and time-consuming process, hence several works have employed unsupervised learning techniques. However, these models are less robust in handling noisy data and thus require prior assumptions on the anomaly distribution.

5 Open Challenges and Future Research Directions

In this section, we discuss the open challenges in the field to shed light on future research works.

5.1 *Lack of Training Data Sets*

One of the main challenges in IIoT environments is the difficulty of gathering sufficient training data for the anomaly detection system. This is specifically more challenging about anomalous samples so that a balanced training dataset can be provided. Because supervised learning approaches often show significant performance degradation for datasets with imbalanced classes. Hence, new studies are required for efficient training of the supervised and semi-supervised anomaly detection models concerning the aforementioned challenges.

5.2 *Real-Time Anomaly Detection*

As mentioned in Sect. 2, in many IIoT environments the real-time or near real-time detection of anomalies is crucial. If it takes too long to detect a malfunction in the system or a cyber-attack, critical losses or damages may happen. Hence, more studies are required in all aspects of data acquisition and evaluation for timely anomaly detection and declaration.

5.3 *Adaptive Learning*

In many cases, the normal system behavior may change over time. Hence, while offline approaches may be applicable in the initial steps, adaptive approaches are required to be developed to improve anomaly detection models over time to adapt to new changes in the data without requiring extensive retraining of the system.

5.4 *Resource and Energy Constraints*

Anomaly detection models often require both high computational and communication resources, raising serious questions on the system scalability because of major resource and energy constraints of IIoT devices.

Data elaboration close to the end IIoT devices (e.g. using on-site computing resources or edge computing) can reduce data transfer and thereby improve the time efficiency, however it increases imposed costs. In contrast, offloading anomaly computations to a distant cloud can decrease the costs, while deteriorating the system performance due to high data transfer delays. Therefore, a major challenge is where the anomaly detection system should be implemented regarding the performance and cost preferences, and the resource and energy constraints of IIoT devices.

5.5 *Privacy and Security Concerns*

Anomaly detection systems often access a set of sensitive data collected from different IIoT devices. Furthermore, the collected data may contain the user's private data, which arises new security concerns for user privacy. For example, a heart abnormal pulse detection model may reveal the patient's heart disease history [35, 36]. Hence, the anomaly detection system must be implemented by a trusted party in a secure place to prevent any data abuse or privacy leakage.

6 Conclusion

In this chapter, we discussed the necessity of anomaly detection in IIoT environments and the existing challenges in the field. We demonstrated that conventional anomaly detection approaches are not suitable for IIoT environments and novel solutions are required for the unique features of IIoT environments. We then reviewed existing studies in the literature highlighting their main features and discussing the overall pros and cons of the proposed solutions. Finally, we discussed the remaining open challenges in the field that demand further research.

References

1. S. Yousefi, F. Derakhshan, and H. Karimipour, "Applications of big data analytics and machine learning in the internet of things," in *Handbook of Big Data Privacy*: Springer, 2020, pp. 77–108.
2. T. A. Ahanger and A. Aljumah, "Internet of Things: A comprehensive study of security issues and defense mechanisms," *IEEE Access,* vol. 7, pp. 11020–11028, 2018.
3. H. HaddadPajouh, A. Dehghantanha, R. M. Parizi, M. Aledhari, and H. Karimipour, "A survey on internet of things security: Requirements, challenges, and solutions," *Internet of Things,* p. 100129, 2019.
4. K. Demertzis, L. Iliadis, N. Tziritas, and P. Kikiras, "Anomaly detection via blockchained deep learning smart contracts in industry 4.0," *Neural Computing and Applications,* vol. 32, no. 23, pp. 17361–17378, 2020.

5. N. Woolf. "DDos Attack That Disrupted Internet was Largest of Its Kind in History, Experts Say." https://www.theguardian.com/technology/2016/oct/26/ddos-attack-dyn-mirai-botnet (accessed Dec. 2020).
6. T. M. Chen and S. Abu-Nimeh, "Lessons from stuxnet," *Computer,* vol. 44, no. 4, pp. 91–93, 2011.
7. S. Karnouskos, "Stuxnet worm impact on industrial cyber-physical system security," in *IECON 2011-37th Annual Conference of the IEEE Industrial Electronics Society*, 2011: IEEE, pp. 4490–4494.
8. C. Garlati. "Owlet Baby Wi-Fi Monitor Worst IoT Security of 2016." https://www.informationsecuritybuzz.com/expert-comments/owlet-baby-wi-fi-monitor-worst-iot-security-2016/ (accessed Dec. 2020).
9. G. Liang, S. R. Weller, J. Zhao, F. Luo, and Z. Y. Dong, "The 2015 ukraine blackout: Implications for false data injection attacks," *IEEE Transactions on Power Systems,* vol. 32, no. 4, pp. 3317–3318, 2016.
10. M. Begli, F. Derakhshan, and H. Karimipour, "A layered intrusion detection system for critical infrastructure using machine learning," in *2019 IEEE 7th International Conference on Smart Energy Grid Engineering (SEGE)*, 2019: IEEE, pp. 120–124.
11. J. Sakhnini, H. Karimipour, A. Dehghantanha, R. M. Parizi, and G. Srivastava, "Security aspects of Internet of Things aided smart grids: A bibliometric survey," *Internet of things,* p. 100111, 2019.
12. B. Genge, P. Haller, and C. Enăchescu, "Anomaly Detection in Aging Industrial Internet of Things," *IEEE Access,* vol. 7, pp. 74217–74230, 2019.
13. S. M. Tahsien, H. Karimipour, and P. Spachos, "Machine learning based solutions for security of Internet of Things (IoT): A survey," *Journal of Network and Computer Applications,* vol. 161, p. 102630, 2020.
14. A. Al-Abassi, H. Karimipour, A. Dehghantanha, and R. M. Parizi, "An ensemble deep learning-based cyber-attack detection in industrial control system," *IEEE Access,* vol. 8, pp. 83965–83973, 2020.
15. H. Karimipour and V. Dinavahi, "Robust massively parallel dynamic state estimation of power systems against cyber-attack," *IEEE Access,* vol. 6, pp. 2984–2995, 2017.
16. S. Mohammadi, H. Mirvaziri, M. Ghazizadeh-Ahsaee, and H. Karimipour, "Cyber intrusion detection by combined feature selection algorithm," *Journal of information security and applications,* vol. 44, pp. 80–88, 2019.
17. H. Karimipour, S. Geris, A. Dehghantanha and H. Leung, "Intelligent Anomaly Detection for Large-scale Smart Grids," 2019 IEEE Canadian Conference of Electrical and Computer Engineering (CCECE), Edmonton, AB, Canada, 2019, pp. 1–4, doi: https://doi.org/10.1109/CCECE.2019.8861995.
18. A. Cook, G. Mısırlı, and Z. Fan, "Anomaly detection for IoT time-series data: A survey," *IEEE Internet of Things Journal*, 2019.
19. H. Karimipour and H. Leung, "Relaxation-based anomaly detection in cyber-physical systems using ensemble kalman filter," *IET Cyber-Physical Systems: Theory & Applications,* vol. 5, no. 1, pp. 49–58, 2020.
20. Y. Peng, A. Tan, J. Wu, and Y. Bi, "Hierarchical edge computing: A novel multi-source multi-dimensional data anomaly detection scheme for industrial Internet of Things," *IEEE Access,* vol. 7, pp. 111257–111270, 2019.
21. H. Yang, S. Liang, J. Ni, H. Li, and X. Shen, "Secure and Efficient kNN Classification for Industrial Internet of Things," *IEEE Internet of Things Journal*, 2020.
22. A.-H. Muna, N. Moustafa, and E. Sitnikova, "Identification of malicious activities in industrial internet of things based on deep learning models," *Journal of Information Security and Applications,* vol. 41, pp. 1–11, 2018.
23. Y. Li *et al.*, "Robust detection for network intrusion of industrial IoT based on multi-CNN fusion," *Measurement,* vol. 154, p. 107450, 2020.

24. M. Tavallaee, E. Bagheri, W. Lu, and A. A. Ghorbani, "A detailed analysis of the KDD CUP 99 data set," in *2009 IEEE symposium on computational intelligence for security and defense applications*, 2009: IEEE, pp. 1–6.
25. X. Yan, Y. Xu, X. Xing, B. Cui, Z. Guo, and T. Guo, "Trustworthy network anomaly detection based on an adaptive learning rate and momentum in IIoT," *IEEE Transactions on Industrial Informatics,* vol. 16, no. 9, pp. 6182–6192, 2020.
26. M. Dixit, A. Tiwari, H. Pathak, and R. Astya, "An overview of deep learning architectures, libraries and its applications areas," in *2018 International Conference on Advances in Computing, Communication Control and Networking (ICACCCN)*, 2018: IEEE, pp. 293–297.
27. M. P. Andersen, J. Kolb, K. Chen, G. Fierro, D. E. Culler, and R. A. Popa, "Wave: A decentralized authorization system for iot via blockchain smart contracts," *University of California at Berkeley, Tech. Rep*, 2017.
28. Y. Liu *et al.*, "Deep Anomaly Detection for Time-series Data in Industrial IoT: A Communication-Efficient On-device Federated Learning Approach," *IEEE Internet of Things Journal*, 2020.
29. M. S. S. Garmaroodi, F. Farivar, M. S. Haghighi, M. A. Shoorehdeli, and A. Jolfaei, "Detection of Anomalies and Faults in Industrial IoT Systems by Data Mining: Study of CHRIST Osmotron Water Purification System," *arXiv preprint arXiv:2009.03645,* 2020.
30. D. Wu, Z. Jiang, X. Xie, X. Wei, W. Yu, and R. Li, "LSTM learning with Bayesian and Gaussian processing for anomaly detection in industrial IoT," *IEEE Transactions on Industrial Informatics,* vol. 16, no. 8, pp. 5244–5253, 2019.
31. F. V. Jensen, *An introduction to Bayesian networks*. UCL Press London, 1996.
32. M. Zolanvari, M. A. Teixeira, L. Gupta, K. M. Khan, and R. Jain, "Machine learning-based network vulnerability analysis of industrial Internet of Things," *IEEE Internet of Things Journal,* vol. 6, no. 4, pp. 6822–6834, 2019.
33. P. Ferrari *et al.*, "Performance evaluation of full-cloud and edge-cloud architectures for Industrial IoT anomaly detection based on deep learning," in *2019 II Workshop on Metrology for Industry 4.0 and IoT (MetroInd4. 0&IoT)*, 2019: IEEE, pp. 420–425.
34. A. Al-Abassi, J. Sakhnini and H. Karimipour, "Unsupervised Stacked Autoencoders for Anomaly Detection on Smart Cyber-physical Grids," 2020 IEEE International Conference on Systems, Man, and Cybernetics (SMC), Toronto, ON, 2020, pp. 3123–3129, doi: https://doi.org/10.1109/SMC42975.2020.9283064.
35. E. Lundin and E. Jonsson, "Anomaly-based intrusion detection: privacy concerns and other problems," *Computer networks,* vol. 34, no. 4, pp. 623–640, 2000.
36. I. Butun, B. Kantarci, and M. Erol-Kantarci, "Anomaly detection and privacy preservation in cloud-centric Internet of Things," in *2015 IEEE International Conference on Communication Workshop (ICCW)*, 2015: IEEE, pp. 2610–2615.

A Snapshot Ensemble Deep Neural Network Model for Attack Detection in Industrial Internet of Things

Hossein Mohammadi Rouzbahani, Amir Hossein Bahrami, and Hadis Karimipour

1 Introduction

In the past few years, as a result of developments in the field of electronics and improvements of wireless systems, the term Internet of Things (IoT) emerged. The opportunity to connect devices together and share information and data while performing their individual tasks without being bound to locations and physical equipment [1].

Industrial Internet of Things (IIoT) is a new application of the Internet of Things (IoT) in the industrial sector. The IIoT enables an enterprise to perform operations in an efficient way while maintaining quality and validation [2]. IIoT makes monitoring and maintenance tasks more convenient, which will be discussed under the category of smart manufacturing systems [3]. By integrating Cyber-Physical Systems CPS, a smart manufacturing execution system can be created such that, documents all data obtained from production and performs decision-making based on predictions on the data for better and optimized future steps [4]. IoT has been progressively used in different sectors of the industry and created a new revolution, IIoT or Industry 4.0 [5, 6], which improves the efficiency, security and productivity in the industry [7–10]. Based on the environment and the purpose of its application, IIoT can have different architectures, but generally, it can be described in a four-layered architecture, as can be observed in Fig. 1.

The physical layer, consists of all physical elements such as actuators, sensors, machines, etc. The network layer consists of communication networks and

H. M. Rouzbahani (✉) · H. Karimipour
Department of Electrical and Software Engineering, University of Calgary, Calgary, AB, Canada
e-mail: hmoham15@uoguelph.ca; hadis.karimipour@ucalgary.ca

A. H. Bahrami
Peter the Great St. Petersburg Polytechnic University, St. Petersburg, Russia

H. Karimipour, F. Derakhshan (eds.), *AI-Enabled Threat Detection and Security Analysis for Industrial IoT*, https://doi.org/10.1007/978-3-030-76613-9_10

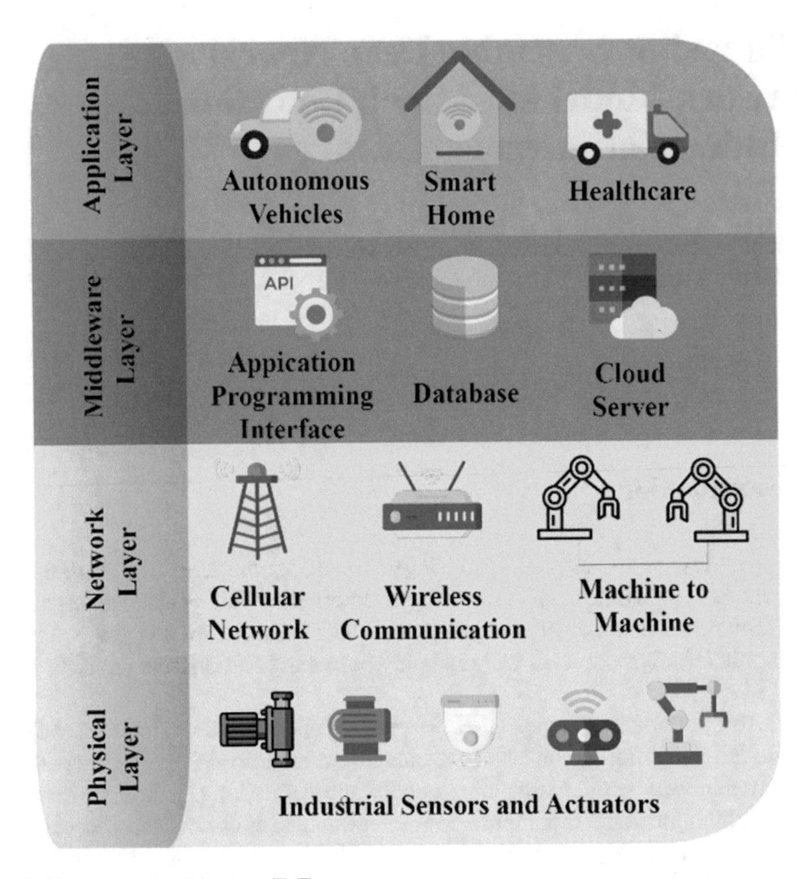

Fig. 1 Four layered architecture IIoT

protocols. The middleware layer makes the communication between the second layer and the first layer possible. It consists of an application programming interface (API), database, cloud server, etc. The fourth layer, the application layer, describes the application of the IIoT. Some instances of the applications are autonomous vehicles, smart home, healthcare, etc. [11–15]. for instance, Rouzbahani et al. proposed an Incentive-based Demand Response Optimization (IDRO) model in order to efficiently schedule household appliances for minimum usage during peak hours [16, 17], which demonstrates noticeable improvements in power factor and cost-saving during peak hours for individual households.

While IIoT is an excellent solution to facilitate industrial processes, it creates new challenges with its application. As the devices start to operate simultaneously, they generate valuable data for online monitoring and control of the system, which can also be used by attacker to manipulate the system performace [18, 19]. There are several attacks that can be performed in IIoT, one of such is cyber-attacks, and this type of attack has other variations itself, such as Denial of Service (DoS), Datatype Probing (DP), Scan, and etc.

In order to perform data processing and analysis, Machine Learning (ML) is preferred over the traditional methods due to the huge quantity of data that is being generated throughout the operations. ML is considered to be a useful paradigm for detecting security threats [20]. Apruzzese et al. [21] conducted an investigation on the effectiveness of ML for cyber threat detection to find and address the limitation of it in such tasks. Lee et al. [22] conducted the same topic of research, but the focus was on the reduction of error in the solution. These are a few instances to show the effectiveness of ML in the area of detection and classification of cyber-security threats.

In this paper, we proposed a Snapshot Ensemble Deep Neural Network (SEDNN) for cyber-attack detection. The model has high accuracy in the detection of cyber threats. It is worth noticing that the classification of the attacks was not considered in this paper. In sect. II, some previous work on the same area of study will be presented, section III will be devoted to the methodology, in sect. IV the results will be presented and in sect. V conclusion and future steps will be discussed.

2 Previous Works in IIoT Security

As the system becomes more complex and data quantity becomes enormous [23–25], the computation and control become more challenging, resulting in traditional methods not to perform as expected because of latency and long response time [26, 27]. ML algorithms improve industrial processes' security and reliability and are rapidly used to detect and address security threats in IIoT [28, 29]. Previous studies in the area of ML application in IIoT security show promising results in using ML algorithms for addressing cyber threats in IIoT.

Rouzbahani, Karimipour and Lei [30] proposed an Ensemble Deep Convolutional Neural Network (EDCNN) model for electricity theft detection in smart grids. In this study, they used a dataset consisting of the daily consumption of 42,372 users. They used an unbalanced dataset in which 8% of customers were attackers, and the rest were normal users. They compared the results with other models and concluded that EDCNN could detect electricity theft in smart grids with an accuracy of 0.981, which indicates that the model is precise.

Farahnakian and Heikkonen [31] approached intrusion detection by presenting a Deep Auto-Encoder (DAE) based system. They used the model on the KDD-CUP'99 dataset and achieved an accuracy of 94.71% for attack detection, which then they concluded that their approach obtained better results as opposed to other deep learning-based approaches. Moukhafi et al. [32] chose a novel hybrid genetic algorithm and support vector machine with the particle swarm optimization feature selection approach for detecting Denial of Service (DoS) attack detection, which they implemented on KDD 99 dataset and obtained an accuracy of 96.38%. Rouzbahani et al. [33] presented research on using ML algorithms for the classification of False Data Injection (FDI) attacks in CPS.

Vajayanand et al. [34] proposed a support vector machine (SVM)-based model, and by doing so, they improved the classification algorithm. They used the ADFA-LD dataset for the implementation of their model and obtained an accuracy of 94.51%. In the research of Khalvati et al. [35], they proposed the SVM and Bayesian model to successfully classify IoT attacks. They conducted research with their proposed model on KDD CUP 99 dataset and achieved an accuracy of 91.50%. Li et al. [36] proposed a bidirectional long and short-term memory network with a multi-feature layer (B-MLSTM) on the classical IIoT datasets: CTU-13 [37], Gas-Water [38], and AWID [39] in order to detect low-frequency and multi-stage attacks in IIoT. After the implementation of the model, an accuracy of 95.01% on the CTU-13 and 97.58% on AWID was obtained. Rouzbahani et al. [40] conducted research and performed cyber-attack detection in smart cyber-physical grids by using different ML algorithms, which resulted in a great performance for Random forest K-Nearest Neighbor (KNN).

Overall, investigations show that ML can efficiently and precisely detect security threats in IIoT. What is worth noticing is that the datasets in these studies are classical datasets that are available on the internet and are considered to be outdated. We are obligated to use new datasets because of the modern security requirements of IIoT. This paper proposes a modern ML model that will be implemented on newer datasets and will also address the compatibility of the model with resource-constrained devices.

3 Methodology

In this section, a brief description of the dataset has been presented. The section will then continue with a description of the preprocessing of the dataset, the proposed model, and evaluation parameters that were considered to evaluate the model's performance.

3.1 Dataset

The dataset used in this paper is an open-source dataset obtained from Kaggle [41]. It was provided by Pahl et al. [42]. This dataset contains communications between different IoT nodes, sensors and applications. In this dataset, multiple attacks were performed on the IIoT applications, for example, "spying", "wrong setup" and etc., which resulted in an anomaly in some of the 357,952 data samples [43, 44]. This paper tried to address the cyber-attack performed on the data. Classification of the attacks will be discussed in another paper.

3.2 Preprocessing of Data

In order to obtain acceptable results from ML models, a comprehensive dataset is the main requirement. Most of the time in data mining is devoted to data processing [45], and the most essential problem in data processing is missing values, which can be caused by various reasons such as power outage, sensor damage or cyber-attacks [46].

In this dataset, there are missing values. Deleting them can result in losing valuable data on other columns. Therefore, the missing values need to be replaced. Figure 2 shows a diagram of the algorithm for attack detection. The processing of replacing the missing values is as follows:

3.2.1 Features

First, we need to select the features that we want to create our model based on. Table 1 shows the features that were selected. It demonstrates which methods were considered in order to encode the features as well.

3.2.2 Replacing Missing/NaN Values

Backward Difference Encoding: this coding system is one of the coding systems of categorical encoding. When a regression is performed on a set of variables with K categories, these variables will enter the regression as a sequence of K-1 dummy variables. The regression coefficient of these K-1 variables corresponds to linear hypotheses on the cell means.

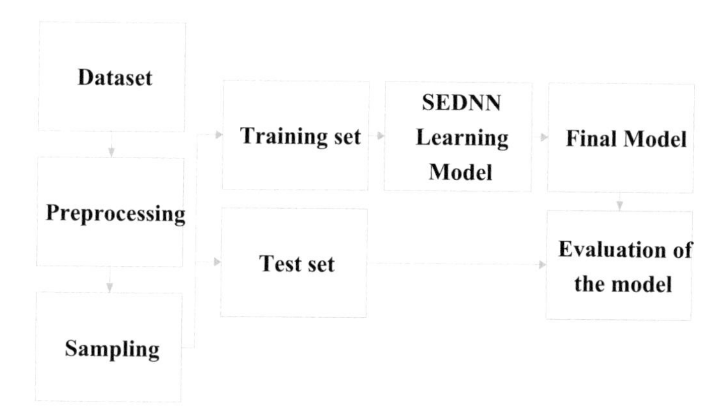

Fig. 2 Diagram of the attack detection algorithm

Table 1 Methods for feature encoding

Feature	Method
Source ID	Label encoding
Source type	Label encoding
Source location	Label encoding
Destination service type	Label encoding
Destination location	Label encoding
Accessed node type	Label encoding
Operation	Label encoding
Source address	Backward difference encoding
Destination service address	Backward difference encoding
Accessed node address	Backward difference encoding

Table 2 Replacing missing values

Variable	Assigned value
Blank	−2
False	−1
True	0.1
20	20
None	0

In this coding system, the mean of the dependent variable for one level of the categorical variable is compared to the mean of the dependent variable for the prior adjacent level.

Label Encoding: in this encoding, a number will be assigned to each variable. The model should be able to understand the difference between "blank," "False," and "None" variables. Therefore we cannot assign 0 to all of them. Table 2 demonstrates the values which were assigned to each variable.

3.3 Snapshot Ensemble Deep Neural Network

In this paper, a Snapshot Ensemble Deep Neural Network (SEDNN) was proposed in order to detect cyber-attacks on the dataset. The disadvantage of an ordinary Ensemble Deep Neural Network (EDNN) is a high computational cost, so that with ordinary hardware, the time of the training and testing will be high. In order to overcome this problem, this paper approached this problem with an SEDNN model [47]. The difference between and ordinary EDNN and SEDNN is that every time the SEDNN reaches a local minimum, it will save the model's weights and biases and continues to do so until the model finds the optimal minimum, resulting in a set of neural networks with low errors. After this process, the model will ensemble all models in this set and obtains the perfect model. The algorithm uses Gradient Descent in order to find the minimum in each step. Two types of activation functions

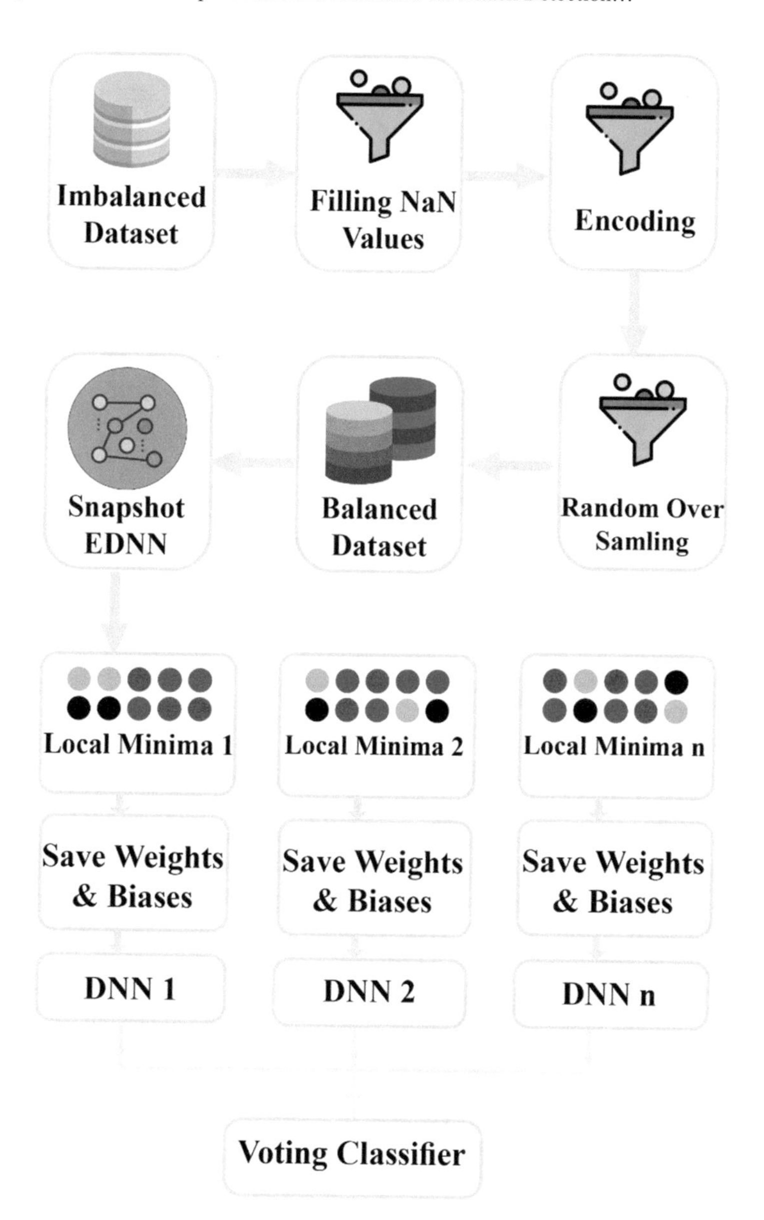

Fig. 3 Architecture of The Proposed Algorithm

were considered for the DNN layers, for the first three layers, a "Relu" activation function was assigned, and for the last layer, a "Sigmoid" function was considered to conduct a binary classification in this paper. As an output, each of the test set data will be given a label of 0 (Normal) or 1 (Attack). Figure 3 shows a visualization of the proposed algorithm, and the architecture of the DNNs can be observed in Fig. 4.

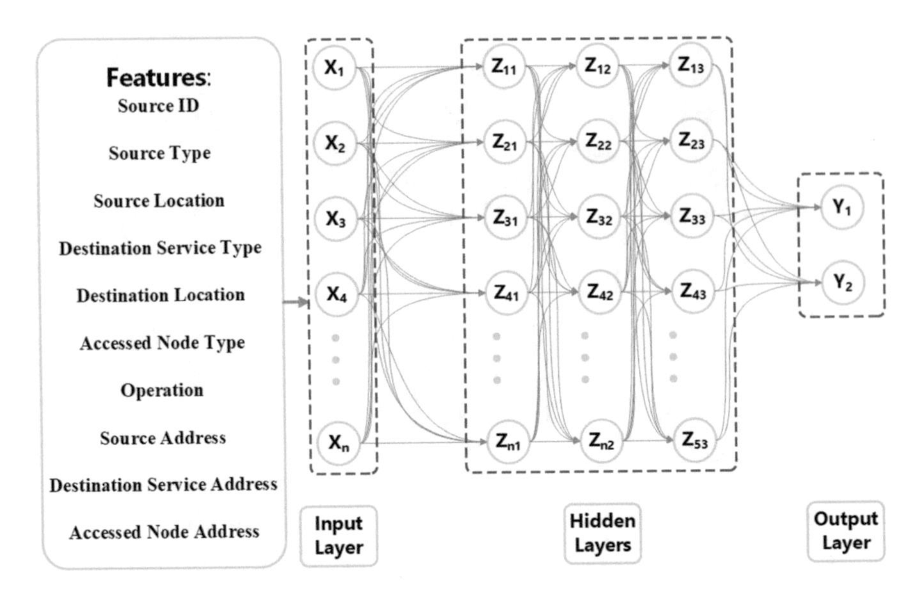

Fig. 4 Deep Neural Network Architecture

3.4 Evaluation Parameters

In order to evaluate a ML model, there are some parameters that can be used. In this section, these parameters will be briefly explained. There are some terms used in the calculation of the evaluation parameters that need to be defined.

True positive is the resulting term where the model correctly predicted the positive class. True negative is the resulting term where the model correctly predicted the negative class. False-positive is the resulting term where the model incorrectly predicted the positive class. False-negative is the resulting term where the model incorrectly predicted the negative class.

Accuracy is the most common measure for evaluating the ML model, and it is defined as the ratio of correctly predicted results to the total predicted results. It may be implied that the higher the accuracy, the more precise model. This is not true in all possible cases. This assumption is only correct when there are symmetric datasets where false positives and false negatives are almost the same. Therefore, we have to look for other parameters to evaluate our model more accurately. The mathematical formula for accuracy calculation is described in Eq. 1.

$$Accuracy = \frac{T_{Pos} + T_{Neg}}{T_{Pos} + T_{Neg} + F_{Pos} + F_{Neg}} \tag{1}$$

Precision is the ratio of true positives to all optimistic predictions. The formula for precision calculation is described in Eq. 2. High precision will result in low false-positive rate.

$$Precision = \frac{T_{Pos}}{T_{Pos} + F_{Pos}} \quad (2)$$

The recall is the ratio between true positive to all predictions (true positive and false negative) of the same class. The formula for recall calculation is described in Eq. 3.

$$Recall = \frac{T_{Pos}}{T_{Pos} + F_{Neg}} \quad (3)$$

F1-Score is the weighted average of Precision and Recall. Therefore, it takes false positives and false negatives into account. The formula for F1-score calculation is described in Eq. 4.

$$F1 - Score = \frac{2 \times (Precision + Recall)}{Precision + Recall} \quad (4)$$

4 Implementation and Results

In this section, hardware and software equipment will be discussed. The section will continue to present the results in detail.

4.1 *Software and Hardware*

The proposed model has been tested using Python 3.7.4 on a system with an Intel Core i7-97580H CPU, 16.0 GB of RAM, and the model's design is structured based on TensorFlow. In order to analyze the performance of the model, we need to obtain the confusion matrix, which will offer us true positive, false positive, true negative and false negative.

4.2 *Results*

The general form of a confusion matrix can be observed in Table 3.

In this research, different classifiers have been tested on the dataset in order to compare the results and accuracy percentage. In Table 4, the confusion matrix of the proposed model can be observed; moreover, Table 5 presents the proposed model's performance with evaluation parameters.

Table 3 Confusion matrix

Actual/Detected	Normal	Attacker
Normal	T_{Pos}	F_{Neg}
Attacker	F_{Pos}	T_{Neg}

Table 4 Confusion matrix of the proposed model

Actual/Detected	Normal	Attacker
Normal	49,478	7123
Attacker	3285	50,613

Table 5 Result comparison of different classifiers

Classifier	Accuracy	Precision	Recall	F1-Score
SEDNN	0.9058	0.8742	0.9377	0.9048

As it can be implied from Tables 4 and 5, the model presents promising results. Obtained accuracy of 90.58% and F1-Score of 90.48% show the great performance of SEDNN in detecting cyber-attacks in IIoT applications. Figures 5 and 6 show the accuracy and loss rate of the model.

In Fig. 5, the accuracy is not stable, and this is caused by changes of DNN between each time it reaches a local minimum, the algorithm uses a new DNN with new weights and biases. Overall, the test set's accuracy is higher than the train set, which shows the model's outstanding performance.

In Fig. 6, we can observe the loss diagram of the train and test set. The nose in the test diagram was caused by utilizing multiple DNN in between each local minimum, as was described before. It can be observed that overall, the loss of the test set is lower than the train set, which shows the model is performing great.

5 Conclusion and Future Work

In this paper, a SEDNN model was proposed for cyber-attack detection in industrial IoT systems. As the model searches for a global minimum, upon finding every local minimum, it will save the weights and biases of that particular DNN (Snapshots), and when it reaches the global minima, it generates the best possible model from the set of DNNs, instead of training and testing different models on the entire dataset. The proposed model has a high accuracy of 90.58%, demonstrating the model's excellent performance in cyber-attack detection. The model was tested on an open-source dataset, DS2OS, which showed promising results. The dataset consists of communication between different IoT nodes such as sensors and actuators. In the future steps, more real-time experiments and ìnvestigations can be conducted with the proposed model to test the model on real IIoT systems; furthermore classification of the attacks with the proposed model will be conducted in future researches.

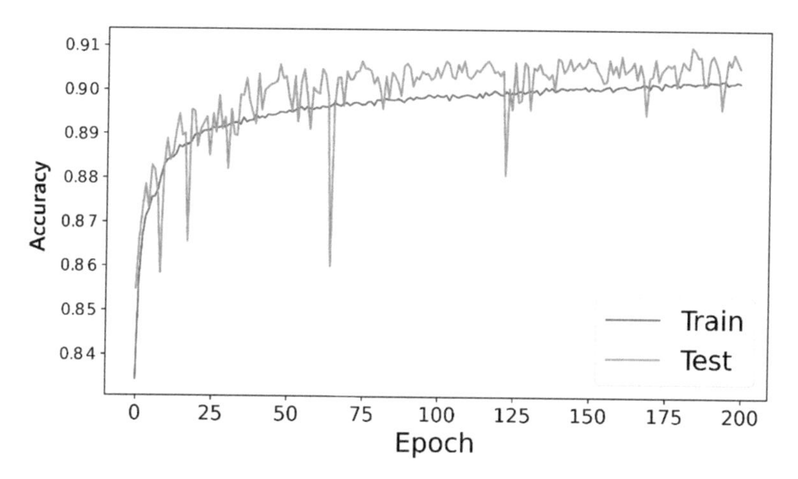

Fig. 5 SEDNN accuracy rate

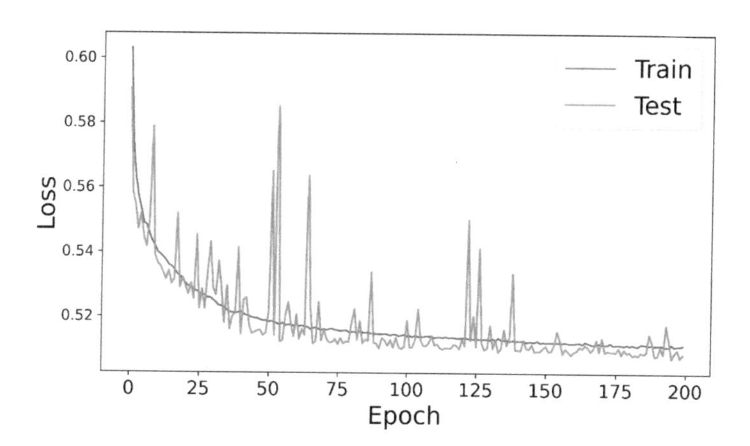

Fig. 6 SEDNN loss rate

References

1. E. Borgia, "The Internet of Things vision: Key features, applications and open issues," Comput. Commun., vol. 54, pp. 1–31, 2014, https://doi.org/10.1016/j.comcom.2014.09.008.
2. A. Alabasi, H. Karimipour, A. Dehghantanha, "An Ensemble Deep Learning-based Cyber-Attack Detection in Industrial Control System", IEEE Access, vol. 8, pp. 83965-83973, April. 2020. doi: https://doi.org/10.1109/ACCESS.2020.2992249.
3. J. Lin, W. Yu, N. Zhang, X. Yang, H. Zhang, and W. Zhao, "A Survey on Internet of Things: Architecture, Enabling Technologies, Security and Privacy, and Applications," IEEE Internet Things J., vol. 4, no. 5, pp. 1125–1142, 2017, doi: https://doi.org/10.1109/JIOT.2017.2683200.
4. V. V Potekhin, A. H. Bahrami, and B. Katalinič, "Developing manufacturing execution system with predictive analysis," IOP Conf. Ser. Mater. Sci. Eng., vol. 966, p. 12117, 2020, doi: https://doi.org/10.1088/1757-899x/966/1/012117.

5. J. Sakhnini, H. Karimipour, A. Dehghantanha, R. Parizi, G. Srivastava "Security Aspects of Internet of Things Aided Smart Grids: a Bibliometric Survey", Internet of Things Journal—Elsevier, pp. 1–15, Sept. 2019. https://doi.org/10.1016/j.iot.2019.100111
6. F. Anwaar, N. Iltaf, H. Afzal, and R. Nawaz, "HRS-CE: A hybrid framework to integrate content embeddings in recommender systems for cold start items," J. Comput. Sci., vol. 29, pp. 9–18, 2018, doi: https://doi.org/10.1016/j.jocs.2018.09.008.
7. S. M. Tahsien, H. Karimipour, P. Spachos, "Machine Learning Based Solutions for Security of Internet of Things (IoT): A Survey", Journal of Network and Computer Applications—Elsevier, vol. 161, pp. 1–18, April. 2020. https://doi.org/10.1016/j.jnca.2020.102630
8. A. Yazdinejad, R. M. Parizi, A. Dehghantanha, H. Karimipour, G. Srivastava, and M. Aledhari, "Enabling Drones in the Internet of Things with Decentralized Blockchain-based Security," IEEE Internet of Things Journal, 2020.
9. E. Nowroozi, A. Dehghantanha, R. M. Parizi, and K.-K. R. Choo, "A survey of machine learning techniques in adversarial image forensics," *Computers & Security,* vol. 100, p. 102092, 2021.
10. A. Yazdinejad, G. Srivastava, R. M. Parizi, A. Dehghantanha, H. Karimipour, and S. R. Karizno, "SLPoW: Secure and Low Latency Proof of Work Protocol for Blockchain in Green IoT Networks," in 2020 IEEE 91st Vehicular Technology Conference (VTC2020-Spring), 2020, pp. 1–5: IEEE.
11. M. Aledhari, R. M. Parizi, A. Dehghantanha and K. R. Choo, "A Hybrid RSA Algorithm in Support of IoT Greenhouse Applications," *2019 IEEE International Conference on Industrial Internet (ICII)*, Orlando, FL, USA, 2019, pp. 233–240, https://doi.org/10.1109/ICII.2019.00049
12. A. Yazdinejadna, R. M. Parizi, A. Dehghantanha, and M. S. Khan, "A kangaroo-based intrusion detection system on software-defined networks," *Computer Networks,* vol. 184, p. 107688, 2021.
13. A. Yazdinejad, R. M. Parizi, A. Dehghantanha, and K.-K. R. Choo, "Blockchain-enabled authentication handover with efficient privacy protection in SDN-based 5G networks," IEEE Transactions on Network Science and Engineering, 2019.
14. M. Aledhari, R. Razzak, R. M. Parizi and F. Saeed, "Federated Learning: A Survey on Enabling Technologies, Protocols, and Applications," in IEEE Access, vol. 8, pp. 140699–140725, 2020, doi: https://doi.org/10.1109/ACCESS.2020.3013541
15. A. Yazdinejad, G. Srivastava, R. M. Parizi, A. Dehghantanha, K.-K. R. Choo, and M. Aledhari, "Decentralized authentication of distributed patients in hospital networks using blockchain," IEEE journal of biomedical and health informatics, vol. 24, no. 8, pp. 2146–2156, 2020.
16. H. M. Ruzbahani and H. Karimipour, "Optimal incentive-based demand response management of smart households," in 2018 IEEE/IAS 54th Industrial and Commercial Power Systems Technical Conference (I&CPS), 2018, pp. 1–7, https://doi.org/10.1109/ICPS.2018.8369971.
17. H. M. Ruzbahani, A. Rahimnejad, and H. Karimipour, "Smart Households Demand Response Management with Micro Grid," in 2019 IEEE Power & Energy Society Innovative Smart Grid Technologies Conference (ISGT), 2019, pp. 1–5, https://doi.org/10.1109/ISGT.2019.8791595.
18. E. Modiri, A. Azmoodeh, A. Dehghantanha, H. Karimipour, "Fuzzy Pattern Tree for Edge Malware Detection and Categorization in IoT", Journal of Systems Architecture, vol. 9, pp. 1–7, Jan. 2018. https://doi.org/10.1016/j.sysarc.2019.01.017
19. Z. Ling, J. Luo, Y. Xu, C. Gao, K. Wu, and X. Fu, "Security Vulnerabilities of Internet of Things: A Case Study of the Smart Plug System," IEEE Internet Things J., vol. 4, no. 6, pp. 1899–1909, 2017, doi: https://doi.org/10.1109/JIOT.2017.2707465.
20. M. Hazrati, H. Karimipour, A. N. Jahromi, A. Dehghantanha, "Ensemble sparse representation-based cyber threat hunting for security of smart cities", Computer & Electrical Engineering Journal- Elsevier (IF: 2.6), Dec. 2020. https://doi.org/10.1016/j.compeleceng.2020.106825
21. G. Apruzzese, M. Colajanni, L. Ferretti, A. Guido, and M. Marchetti, "On the effectiveness of machine and deep learning for cyber security," in 2018 10th International Conference on Cyber Conflict (CyCon), 2018, pp. 371–390, https://doi.org/10.23919/CYCON.2018.8405026.

22. J. Lee, J. Kim, I. Kim, and K. Han, "Cyber Threat Detection Based on Artificial Neural Networks Using Event Profiles," IEEE Access, vol. 7, pp. 165607–165626, 2019, doi: https://doi.org/10.1109/ACCESS.2019.2953095.
23. A. Yazdinejad, H. HaddadPajouh, A. Dehghantanha, R. M. Parizi, G. Srivastava, and M.-Y. Chen, "Cryptocurrency malware hunting: A deep recurrent neural network approach," Applied Soft Computing, vol. 96, p. 106630, 2020.
24. A. Namavar Jahromi et al., "An improved two-hidden-layer extreme learning machine for malware hunting," Computers & Security, vol. 89, p. 101655, 2020.
25. A. Yazdinejad, R. M. Parizi, A. Dehghantanha, G. Srivastava, S. Mohan, and A. M. Rababah, "Cost optimization of secure routing with untrusted devices in software defined networking," Journal of Parallel and distributed Computing, vol. 143, pp. 36–46, 2020
26. J. Chen et al., "Collaborative Trust Blockchain Based Unbiased Control Transfer Mechanism for Industrial Automation," IEEE Trans. Ind. Appl., vol. 56, no. 4, pp. 4478–4488, 2020, doi: https://doi.org/10.1109/TIA.2019.2959550.
27. W. G. Hatcher and W. Yu, "A Survey of Deep Learning: Platforms, Applications and Emerging Research Trends," IEEE Access, vol. 6, pp. 24411–24432, 2018, doi: https://doi.org/10.1109/ACCESS.2018.2830661.
28. M. Lavassani, S. Forsström, U. Jennehag, and T. Zhang, "Combining Fog Computing with Sensor Mote Machine Learning for Industrial IoT," Sensors, vol. 18, p. 1532, May 2018, doi: https://doi.org/10.3390/s18051532.
29. A. N. Jahromi, A. Dehghantanha, R. Choo, H. Karimipour, R. Parizi, "An Improved Two-Hidden-Layer Extreme Learning Machine for Malware Hunting", Computer and Security, vol. 89, pp. 1–11, Sept. 2019. https://doi.org/10.1016/j.cose.2019.101655
30. H. M. Rouzbahani, H. Karimipour, and L. Lei, "An Ensemble Deep Convolutional Neural Network Model for Electricity Theft Detection in Smart Grids," in 2020 IEEE International Conference on Systems, Man, and Cybernetics (SMC), 2020, pp. 3637–3642, https://doi.org/10.1109/SMC42975.2020.9282837.
31. F. Farahnakian and J. Heikkonen, "A deep auto-encoder based approach for intrusion detection system," in 2018 20th International Conference on Advanced Communication Technology (ICACT), 2018, pp. 178–183, https://doi.org/10.23919/ICACT.2018.8323688.
32. Mehdi Moukhafi, Khalid El Yassini, and Seddik Bri, "A novel hybrid GA and SVM with PSO feature selection for intrusion detection system," I. J. Adv. Sci. Res. Eng. (ISSN 2454—8006), vol. 4, no. 5 SE-, pp. 129–134, May 2018, https://doi.org/10.31695/IJASRE.2018.32724.
33. H. Mohammadi Rouzbahani, H. Karimipour, A. Rahimnejad, A. Dehghantanha, and G. Srivastava, "Anomaly Detection in Cyber-Physical Systems Using Machine Learning BT—Handbook of Big Data Privacy," K.-K. R. Choo and A. Dehghantanha, Eds. Cham: Springer International Publishing, 2020, pp. 219–235.
34. R. Vijayanand, D. Devaraj, and B. Kannapiran, "A novel intrusion detection system for wireless mesh network with hybrid feature selection technique based on GA and MI," J. Intell. Fuzzy Syst., vol. 34, pp. 1243–1250, 2018, doi: https://doi.org/10.3233/JIFS-169421.
35. L. Khalvati, M. Keshtgary, and N. Rikhtegar, "Intrusion Detection based on a Novel Hybrid Learning Approach," J. AI Data Min., vol. 6, no. 1, pp. 157–162, 2018, doi: https://doi.org/10.22044/jadm.2017.979.
36. X. Li, M. Xu, P. Vijayakumar, N. Kumar, and X. Liu, "Detection of Low-Frequency and Multi-Stage Attacks in Industrial Internet of Things," IEEE Trans. Veh. Technol., vol. 69, no. 8, pp. 8820–8831, 2020, doi: https://doi.org/10.1109/TVT.2020.2995133.
37. P. Bereziński, B. Jasiul, and M. Szpyrka, "An Entropy-Based Network Anomaly Detection Method," Entropy, vol. 17, no. 4, pp. 2367–2408, Apr. 2015, doi: https://doi.org/10.3390/e17042367.
38. T. Morris and W. Gao, "Industrial Control System Traffic Data Sets for Intrusion Detection Research BT—Critical Infrastructure Protection VIII," 2014, pp. 65–78.

39. M. Tavallaee, E. Bagheri, W. Lu, and A. A. Ghorbani, "A detailed analysis of the KDD CUP 99 data set," in 2009 IEEE Symposium on Computational Intelligence for Security and Defense Applications, 2009, pp. 1–6, https://doi.org/10.1109/CISDA.2009.5356528.
40. H. M. Rouzbahani, Z. Faraji, M. Amiri-Zarandi, and H. Karimipour, "AI-Enabled Security Monitoring in Smart Cyber Physical Grids", Security of Cyber-Physical Systems: Vulnerability and Impact, Eds. Cham: Springer International Publishing, 2020, pp. 145–167.
41. M.-O. P. and F.-X. Aubet., "Ds2Os Traffic Traces IoT Traffic Traces Gathered in a The Ds2Os IoT Environment. [Online].," 2018. https://www.kaggle.com/francoisxa/ds2ostraffictraces.
42. M. Pahl and F. Aubet, "All Eyes on You: Distributed Multi-Dimensional IoT Microservice Anomaly Detection," in 2018 14th International Conference on Network and Service Management (CNSM), 2018, pp. 72–80, [Online]. Available: http://doi.ieeecomputersociety.org/.
43. M. Hasan, M. M. Islam, M. I. I. Zarif, and M. M. A. Hashem, "Attack and anomaly detection in IoT sensors in IoT sites using machine learning approaches," Internet of Things, vol. 7, p. 100059, 2019, doi: https://doi.org/10.1016/j.iot.2019.100059.
44. O. Brun, Y. Yin, and E. Gelenbe, "Deep Learning with Dense Random Neural Network for Detecting Attacks against IoT-connected Home Environments," Procedia Comput. Sci., vol. 134, pp. 458–463, 2018, doi: https://doi.org/10.1016/j.procs.2018.07.183.
45. F. Honghai, C. Guoshun, Y. Cheng, Y. Bingru, and C. Yumei, "A SVM Regression Based Approach to Filling in Missing Values BT—Knowledge-Based Intelligent Information and Engineering Systems," 2005, pp. 581–587.
46. Sh. Yousefi, F. Derakhshan, H. Karimipour, "Applications of Big Data Analytics and Machine Learning in the Internet of Things", Handbook of Big Data and Privacy, Springer Books, pp. 1–32, Feb. 2020. https://doi.org/10.1007/978-3-030-38557-6_5
47. G. Huang, Y. Li, G. Pleiss, Z. Liu, J. Hopcroft, and K. Weinberger, "Snapshot Ensembles: Train 1, get M for free," Mar. 2017.

Privacy Preserving Federated Learning Solution for Security of Industrial Cyber Physical Systems

Seyed Hossein Majidi and Hadi Asharioun

1 Introduction

Traditionally, monitoring and controlling physical processes were performed using embedded computers via a feedback loop [1]. However, the advancement of society is directed at using interconnected devices to improve daily life. One of the major factors that literally revolutionized economic activities and urban infrastructure was the emergence of Information and Communication Technologies (ICTs). This dramatic change motivates researchers to investigate the topic of integrating ICTs in urban development projects such as smart grids and smart cities.

Today's industry and society seek to utilize digital infrastructure for regulatory and entrepreneurial purposes and embrace ICT in their development strategies [2]. This integration of computation technologies with traditional embedded physical systems leads to a new type of system, called Cyber-Physical System (CPS). CPSs increase the physical system's efficiency and enhance its monitoring and control process [3–7].

The major components of CPS are sensors, actuators, and controllers. The cooperation of these components helps to decrease operational costs and optimize the data preprocessing. The role of sensors and actuators is communicating information between the network and physical components. Also, controllers send commands to different parts of the system.

Most CPSs are not centralized, and they consist of many distributed physical systems. Cyber systems cooperate with this structure to control all physical systems. CPSs include several domains such as smart manufacturing, healthcare, smart buildings and infrastructures, smart cities, wearable devices, smart grids, etc. [8]. For example, in healthcare systems, CPS enables real-time observation of patients'

S. H. Majidi (✉) · H. Asharioun
School of Engineering, Shahid Beheshti University, Tehran, Iran
e-mail: se.majidi@mail.sbu.ac.ir; asharioun@sbu.ac.ir

H. Karimipour, F. Derakhshan (eds.), *AI-Enabled Threat Detection and Security Analysis for Industrial IoT*, https://doi.org/10.1007/978-3-030-76613-9_11

conditions and reduces costs. In transportation, management and scheduling become efficient since vehicles and passengers can be located easily. One of the notable examples in this field is European Rail Traffic Management System (ERTMS), which manages train scheduling by using communications among Internet of Things (IoT) integrated vehicles and devices [9].

The main goals of CPSs can be summarized as the combination of three technologies, namely computation, communication, and control. In fact, this integration distinguishes CPSs from traditional control systems. Improper control of CPSs can cause its failure, which can cause adverse effects to machines and humans involved in the operation [10]. There are many constraints in the physical aspect of CPS. Because of these constraints and ICT limitations, the network's data is without proper security protection. CPS systems' high dependency on cyber-based technologies caused them to face various new vulnerabilities like cyber-attacks [4, 11]. Therefore, it is critical to consider the security of CPSs [12–14]. However, every user involved in CPSs wants to be free of any potential threat and vulnerability, which is practically impossible for real-world systems [15].

This paper discusses the CPS security and different threats that endanger CPSs. Next, some major attacks in different domains of CPSs are pointed out. After introducing these attacks, privacy issues and some of its countermeasures are presented. Finally, Federated Learning (FL) is introduced as a solution for privacy using machine learning models.

2 Cyber-Physical System (CPS) Security

Nowadays, CPSs cover many critical domains of our life. Because of the physical and cyber technology aspects of CPSs, those critical domains utilize these aspects as well [14]. Because of each of these aspects' intrinsic vulnerabilities, critical domains will expose many cyber threats that can lead to catastrophic consequences. As an example, many devices are connected to a network in a hospital. If adversaries try to compromise these devices, it can have some destructive impacts [16]. Moreover, it is not just about healthcare. Other domains might also have the same devastating results. The next example is concerning smart grid infrastructure. Faulty measurements resulting from compromised smart meters lead control centers to make wrong decisions, resulting in wrong energy generation and distribution and inducing a blackout or physical damage to critical infrastructure [17]. Thus, the safety and security of CPS are vital. A secured and functioning CPS is defined by satisfying the following categories:

- *Confidentiality:* some information needs to be private and should not be available by unauthorized individuals. Confidentiality is a subcomponent of privacy, and it aims to protect personal data. For example, in healthcare systems, patients' personal information needs to be transmitted confidentially between doctors and medical devices. So, it should be encrypted.

- *Integrity:* data needs to be kept complete and accurate in the whole process. This is called integrity that data cannot be modified in an unauthorized manner [3].
- *Availability:* all components and systems in a CPS need to be always available and ready for service. This availability should be in both aspects of physical control and communication channels.
- *Authenticity:* It means all transactions and communications need to be real and genuine. Authenticity means you make sure that you are really communicating with one whom you want to [18, 19].

As mentioned before, the inadequacy of security in CPSs can be catastrophic depending on the application. Furthermore, each CPS application's security violations could lead to service loss to consumers and financial losses to utility companies. Base on [20] there are five types of threats including: (1) criminal; (2) financial; (3) political; (4) privacy; and (5) physical [20].

- *Criminal Threats*: Attackers who know the system configuration or are partially familiar with the system can exploit wireless capabilities to remotely control an industrial control system application and possibly disrupt its operations. Also, thieves who try to rob a house can infer private information from communications between the smart meter and utility company to perform a successful robbery [13].
- *Financial Threats:* A customer who tricks the utility company by tampering with smart meters to reduce the electricity bill is an example of this threat [21]. Also, when utility companies collect customers' privacy information for analyzing their electricity usage to infer habits and types of house appliances to sell such information for advertisement purposes, it results in privacy violation [22].
- *Political Threats:* In case of political issues, a country's policy may lead to a cyberwar against another country and gain remote access to the smart grid's infrastructure. Therefore, they can make large-scale blackouts, disturbances, or financial losses [23].
- *Privacy Threats:* Unauthorized access to private data has always been one of the motivations for attackers. For example, in healthcare infrastructures, medical devices communicate with other parties, such as hospitals. Hence, a large amount of private data is stored in various locations, and it may tempt attackers and result in privacy invasion [24].
- *Physical Threats:* For example, in smart grids, attackers may sabotage components that are physically exposed across the power grid to cause service disruption or even potentially blackouts.

2.1 Major Attacks on Cyber-Physical Systems (CPS)

To ensure the security of CPSs, it is required to identify different types of threats and vulnerabilities that cause systems to face danger. There are many types of attacks that exploit CPS's vulnerabilities. These attacks can be divided into two categories: passive attacks and active attacks [25]. The goal of passive attacks is to be stealthy and undetected over time. These attacks can intercept sensitive data without causing any destruction. In contrast, active attacks' goal is to cause direct damage or take control of a system [26]. In general, CPS can endure any of the following attacks [27]:

- *Compromised-key attacks:* The most important thing in security is the "Key." Attackers who have the key can access any desired information. For example, an attacker can compromise the security key with the sensor node's access having a pre-shared key [19].
- *False Data Injection Attacks (FDIAs):* In this type of attack, attackers try to add malicious data to the measurements of the system in order to mislead the state estimation and control the performance of the system without being detected by any of the existing techniques for bad measurement detection.
- *Replay Attacks:* Adversaries in this type of attack try to resend a packet. Basically, they repeat or even delay a valid data transmission.
- *Eavesdropping:* An eavesdropping attack, also known as a sniffing or snooping attack, is a theft of information transmitted over a network. In an eavesdropping attack, the attacker passively listens to network communications to access private information, such as node identification numbers, routing updates, or application sensitive data. The attacker can use this private information to compromise nodes in the network, disrupt routing, or degrade application performance. CPS is mostly affected by eavesdropping by traffic analysis and interrupting the data in the sensors and monitoring [28].
- *Man-in the middle attacks:* In these types of attacks, the attacker secretly relays and possibly alters the communications between two parties who believe that they are directly communicating with each other. False messages are sent to the operator, and they form a false negative or a false positive. This is an active attack, and many common attacks like modification and replay attacks come in these domains.
- *Denial of service (DoS):* DoS attacks can be categorized into three subclasses: permanent DoS, distributed DoS, and reflected attacks [29]. Adversaries can try to exploit some of the unpatched vulnerabilities in order to install the firmware. This kind of attack is called permanent Dos. However, if an attacker tries to perform an attack with multiple systems simultaneously against a single system and block its bandwidth, it is called a distributed attack [30]. Some attackers tend to send some forged requests to many systems with IP address set to target the victim. This will lead to a flooded response in the target system. This attack is called reflected attacks [31].
- *Spoofing:* The attacker pretends to be a legal part of a system. Authors in [32] named three types of spoofing attacks: Global Positioning System (GPS) spoofing, Address Resolution Protocol (ARP) spoofing and Internet Protocol (IP) spoofing. In IP spoofing, the attacker attempts to change IP and pass through security systems with the changed IP. On the other hand, GPS spoofing spread some wrong signals stronger than the one received from satellite to deceive the

victim. ARP spoofing uses some fake ARP messages in order to link the attacker's Media Access Control (MAC) address with IP of the victim.

Different types of CPSs have different susceptibility to particular attacks. For example, in the healthcare domain, replay attacks and traffic analysis are more common. Other CPSs are also vulnerable to this attack, but some critical infrastructures have additional vulnerabilities with high potential impact. Therefore, the security of these systems like smart grids is a more elaborate task. Additionally, particular types of CPSs are susceptible to specific types of attacks specially designed for these systems. For example, FDIAs as stealthy attacks try to inject some malicious data into system measurements [33, 34]. This attack can potentially cause blackouts or serious infrastructural damage [35–37].

Some different CPS applications and their specific attacks were investigated in [15]. These attacks are categorized as cyber, physical, and cyber-physical attacks. This taxonomy is shown in Table 1.

2.2 *Privacy*

One of the most important issues in the deployment of CPSs is their privacy since any privacy breach can result in severe consequences [38–41]. Particularly, the complex architectures of CPSs, especially in smart grids and healthcare, arise new privacy issues. Thus, privacy risks are difficult to assess. It is also strenuous to trace,

Table 1 Different attacks in different domains of CPSs [15]

	Industrial control systems	Smart grids	Medical devices
Cyber	Communication protocols	DoS	Replay attacks
	Espionage	False data injection	Privacy invasion
	Unintentional attack	Customers' information	–
	Web-based attacks	Untargeted malware	–
Physical	Untargeted attacks	Natural and environmental incidents	Acquiring unique IDs
	–	Theft	–
	–	Car accidents	–
	–	Vandalism	–
	–	Terrorist attacks	–
Cyber-physical	Legacy communication channels	Cyber extortion	DoS attacks
	Disgruntled insiders	Blackouts	False data and unauthorized commands injection
	Modbus worm	–	Replay attacks
	Malware	–	–
	Web-based attacks	–	–

identify, examine, and eliminate privacy attacks that may target multiple components of CPSs such as real-time sensors, wearable health devices, industrial control systems, etc. [39–42].

Data centers that contain a lot of personal and private data and sensors are based on CPSs. For example, patients who use some wearable devices for medical cases share their real-time and personal data with doctors [43]. If a strong privacy preservation scheme is not considered in this communication, any attacker may hack this personal data and use illegal benefits, possibly blackmailing [44, 45].

Many cryptographic techniques are proposed in the literature by researchers to preserve data privacy [46–48]. However, most of them need a key to be kept by the user, which is computationally expensive. Additionally, it becomes more difficult to ensure privacy in a situation when public sharing of data is required. Another privacy scheme proposed in the literature is anonymization like k-anonymity [49]. However, these anonymization strategies do not guarantee a complete level of protection from adversaries because the chances of re-identification increase if the size of attributes in the dataset increases [50].

One of the other privacy schemes proposed by researchers is differential privacy introduced to overcome the aforementioned privacy issues [51]. This technique protects statistical and real-time data by adding a desirable amount of noise and maintaining a healthy trade-off between privacy and accuracy. However, this approach suffers from dimensionality. Also, selecting a desirable trade-off between privacy and accuracy is difficult.

3 Federated Learning (FL)

Several directions have been taken to detect and defend against cyber-attacks in CPSs. While these directions differ massively, the two main themes are model-based detection algorithms and data-driven detection algorithms. In order to implement model-based algorithms, we need to have a system model and system parameters. Also, they are computationally expensive, and they are not scalable for large scale systems. Due to these disadvantages, data-driven techniques are becoming more popular in recent studies. Among all the data-driven approaches, machine learning-based algorithms achieved more attention in the literature. With the advent of information and communications technology, it has become technically easier to collect a large amount of data. These data are spread across different edge devices owned by different individuals or organizations [51, 52]. All collected data is valuable as it contains insights into various application domains. In the traditional centric approach, data collected is uploaded and processed centrally in a server or data center [51, 52].

In the last few years, some legislative attempted to develop new laws about how data privacy should be preserved due to privacy concerns. The General Data Protection Regulation (GDPR) by European Union (EU) is an example of these regulations in 2018. The California Consumer Privacy (CCPA) in 2020 in the United

States, California, and China's Cyber Security Law and the General Provisions of Civil Law was implemented in 2017 and imposed strict controls on data collection and transactions [53]. Due to these regulations, it is infeasible to bring data from different organizations to a data center in a simple way [54, 55]. If privacy protection is ignored, many cyber threats and risks will threaten sensitive data. Some records represent privacy violations in recent years. One of them is Equifax, which was attacked in 2017 while having 17.9 million customers. Another record is Mariott, which was compromised in 2018 with 500 million customers. Also, eBay was attacked in 2014 with 145 million users [56].

Google proposed a privacy-preserving distributed machine learning framework, called FL, to train machine learning models without compromising privacy [57–60]. Inspired by this framework, different edge devices or clients can contribute to the global model training while keeping the training data locally.

Besides privacy and security concerns, centric approaches still involve some drawbacks [59–61]. They lead to long propagation delays and incur unacceptable latency for applications like self-driving car systems, in which real-time decisions have to be made. Another strong motivation for FL is that maximum computing power ability at the client can be used. On the client-side, communication is efficient because the only thing transmitted between clients and the server is computed result instead of raw data [62].

Multiple clients can collaboratively train a global model on their dataset in a FL framework while keeping data locally [53]. This framework has some rounds to do. At first, the server initializes a model and share it with other clients. Every client trains this shared model with their dataset, and only model updates are sent back to the server. Server after a weighted aggregation makes a single update to the global, thereby concluding the round. Figure 1 shows the structure of a centralized learning network and FL. As shown in Fig. 1a, the server collects all datasets from clients and then train the model. However, in Fig. 1b, which is representative of FL framework, just model updates are transmitted, and clients' dataset is kept locally.

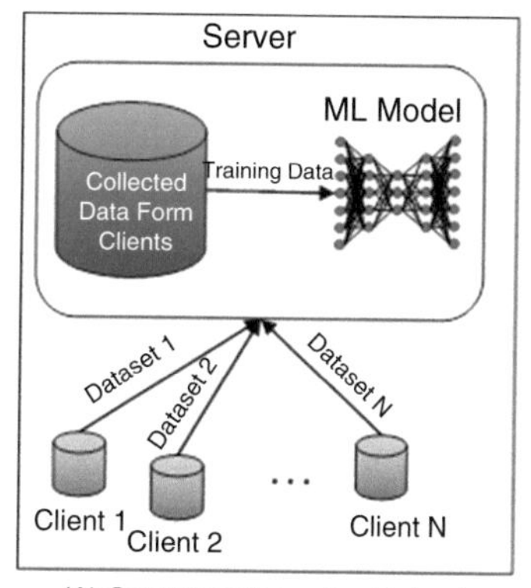

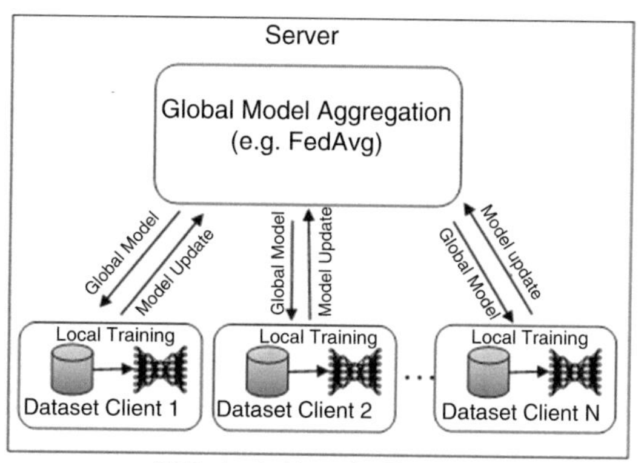

(B) Federated Learning Network

Fig. 1 Illustration of a centralized learning network and federated learning network

3.1 Architectures of Federated Learning (FL)

There are some architectures for FL that can be classified into Horizontal Federated Learning (HFL), Vertical Federated Learning (VFL), and Federated Transfer learning (FTL), according to how data's feature and sample spaces are partitioned among various parties.

In the FL network, wherever clients share the same features but have different data samples, it is called HFL as shown in Fig. 2. This architecture is also called sample-partitioned FL [56, 63]. In contrast, wherever clients share the same data samples but differ in data features, it is called VFL as shown in Fig. 3. This architecture is also called feature-partitioned FL. Finally, FTL is when neither is overlapping in data samples nor features.

Due to minimum accuracy loss in FTL, it is a common architecture in various domains like healthcare [64]. In FTL, encryption techniques are used to preserve privacy [65]. FTL is shown in Fig. 4.

Some other architectures are proposed in the literature for different domains. A study in [54] has surveyed other architectures like Multi-participant Multi-class Vertical Federated Learning (MMVFL) framework [66], Federated Learning Framework (FEDF), PerFit [67], and Federated-Autonomous Deep Learning (FADL) [68]. These papers have considered personalization and also other domains like medical devices and IoT etc.

3.2 Algorithms of FL

Algorithms towards FL are proposed in several papers in the literature. Authors in [69] introduced and compared several algorithms: Federated Averaging (FedAvg), Federated Stochastic Variance Reduced Gradient (FSVGR), and CO-OP.

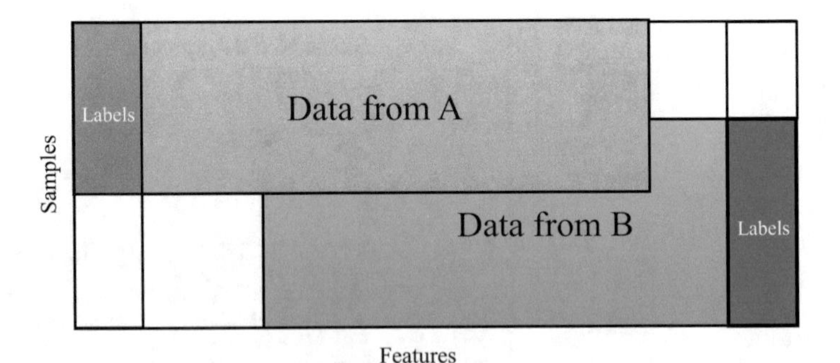

Fig. 2 Illustration of horizontal federated learning. Horizontal federated learning shares the same data features

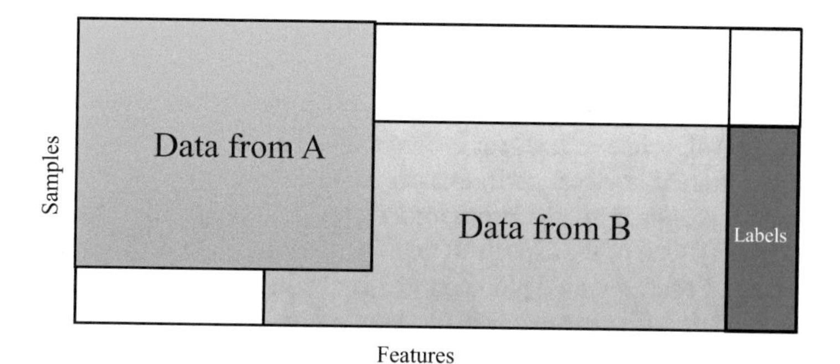

Fig. 3 Illustration of vertical federated learning. Vertical federated learning shares the same data samples

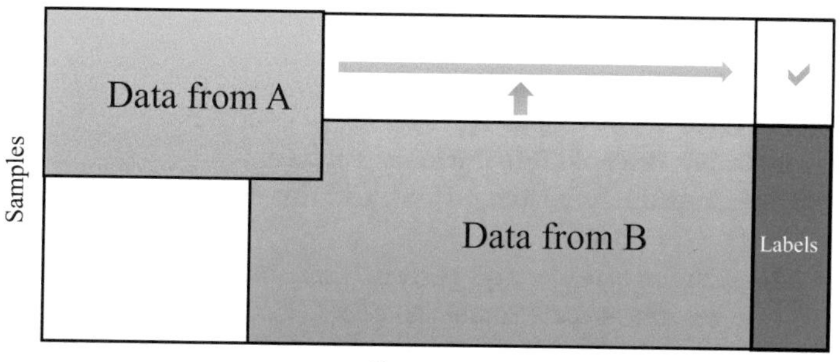

Fig. 4 Federated transfer learning

FedAvg is one of the most common algorithms for FL. In the FedAvg algorithm, clients train the generic neural network model using the gradient descent algorithm, and the trained weights are sent back to the server. The server then takes the average of all such updates to return the final weights.

Five parameters matter to FedAvg: the number of clients, batch sizes, number of epochs, learning rate, and decay. This is how The FedAvg algorithm works. First, a global model is built and is sent to several clients. Then clients train this shared model according to batch sizes and epochs of Stochastic Gradient Descent (SGD). After training the local model, the client transfers model updates to the server. The server aggregates all these updates by taking the average and makes a new global model. Despite all of the advantages, this algorithm still has challenges in tackling heterogeneity [63].

The FSVRG algorithm concerns the case in which some features are rarely represented in the dataset, and its main focus is sparse data. In this algorithm, there will be just one full calculation, and after that, lots of updates on each client will be available [63].

Sahu et al. proposed another algorithm in [70] called FedProx. FedProx is the same as FedAvg and can be viewed as a generalization and re-parametrization of

FedAvg. This algorithm is introduced to tackle heterogeneity. After training the local model and gaining local updates, a global update is produced by grouping all these local updates. FedProx is meant to be a modification of the original FedAvg algorithm. This algorithm allows for different amounts of work to be performed by considering different devices' performances at different iterations.

Wang et al. proposed an algorithm for FL called Federated Matched Averaging (FedMA) in [71]. One of the aspects of FedMA is communication. In this aspect, at the beginning of each round, a global model is obtained, and they can modify it as their local models with sizes equivalent to the original models. Accordingly, sizes can be smaller and therefore easier to manage.

3.3 Challenges and Vulnerabilities of FL

FL networks can be deployed flexibly in various industries and applications. Despite all promising FL results in various applications and domains, some challenges still make this framework susceptible to threats and prevent FL from being fully adopted. These challenges mostly stem from privacy, security, and in some cases the technical requirements.

The data for training is not always perfect. There may be some missing labels or imbalances in some attack and normal samples [71]. This will result in inaccurate models. Furthermore, communication, system heterogeneity, and privacy are other kinds of FL challenges. In many domains like medical, communication is essential for FL [72, 73]. There are lots of devices in FL, so that it can affect communication a lot. To solve this issue, a communication-efficiency model is needed [74]. This means that smaller messages or smaller model updates are sent over for the training process instead of just sending the entire training dataset [75]. Accordingly, two factors need to be considered:

- decreasing the iterations for communication
- decreasing each message size for each iteration.

In FL, privacy is often the biggest concern. Sensitive information can be extracted in the training phase while communicating model updates [76, 77]. However, there have been many papers in literature proposing new methods to tackle privacy, but they affect performance and accuracy. It is crucial to find a compromise considering this trade-off to solve privacy issues [60, 75]. Also, Wadu et al. in [78] mentioned that security and privacy are the main issues in FL. Consequently, proposing a model with high performance and privacy protection while avoiding computational burden is desirable [79, 80]. Since local models are continuously updating with new data, attackers may affect local datasets and results.

Therefore, in general, since FL relies on all clients' collaborative effort to train a machine learning model, the machine learning model will tamper even if only a few clients work abnormally. We can investigate FL vulnerabilities from two aspects based on [81]:

- *Malicious clients:* Since smart devices are getting more complicated, flaws are inescapable, and these devices can get compromised easily. Moreover, many FL network designs do not include any security scheme like an authentication mechanism, and thus they cannot prevent adversaries from setting up malicious clients to join FL networks.
- *Insecure Connections:* In an FL network, many connections might be insecure, but it is not easy to ensure their security and distinguish secure from insecure connections. Nowadays, many methods like encryption and verification are used to secure the connections, but they lead to other issues like additional overheads, and thus, they are not always preferred. Consequently, it is possible that some insecure connections exist, over which any of the global model sent by the server and local model updates sent by clients can be compromised or manipulated.

The global model aggregation can easily get tampered with, and the performance and accuracy be reduced by injecting poisoned model updates and exploiting the vulnerabilities of FL networks by an attacker. These kinds of attacks are called poisoning attacks. The purpose of poisoning attacks is to achieve a poor accuracy in the machine learning model by tampering with the global model aggregation of FL with poisoned model updates. According to where a poisoned model update comes from, poisoning attacks can be categorized as follows [58]:

- *Data Poisoning:* Data poisoning can be done by changing the training data in the target clients. For example, an attacker can change the signs of the labels in training data, such that the target clients train the local models using the poisoned data and generate false model updates. These attacks can be intentional or random. In an intentional attack, the attacker changes the signs of the specific labels. On the one hand, unintentional attacks aim at decreasing the prediction accuracy on all classes, and thus attackers can change the labels randomly [82–84]. On the other hand, the purpose of intentional attacks is to make the machine learning model achieve a poor accuracy on only certain classes, for which reason attackers only change the labels of the training data in the targeted classes [84, 85].
- *Model Poisoning:* To produce poisoned model updates, some rules need to be set instead of modifying training data. For example, Dong et al. [86] describe how model updates can be sampled from a Gaussian distribution. Additionally, adversaries can manipulate benign model updates into poisoned ones. Another way to make an aggregated model have lower accuracy is to design some clients to flip the sign of benign model updates proposed in [86–88]. As a matter of fact, all the aforementioned attacks are not intentional as defined before. Alternatively, intentional ones are considered in [84, 88, 89], where adversaries use a pre-designed compromised model to craft poisoned model updates. They try to use the compromised model instead of training machine learning models.

Another type of attack that FL is susceptible to is called a Free-Riding Attack, in which the adversaries want to leech benefits from the model while keeping themselves outside of the learning process. This will lead to using more and more resources in the training phase [58].

3.4 Countermeasures

There are some methods and countermeasures in the literature to decrease the risk of poisoning attacks on FL networks. Tan et al. in [81] defined these three categories as countermeasures:

- *Robust Aggregation:* In FL networks, clients send their model updates to the server then the server will aggregate by taking the average of them. This will let poisoned model updates affect the global model directly. Therefore, proposing a method that is robust to poisoned model updates is necessary. In literature, some techniques were proposed, namely: Component-Wise Median (COMED), Geometric Median (GEOMED) and Component-Wise Trimmed Median (COTMED). These methods are used for aggregation and can be used instead of average operation [83, 86]. Dong et.al proposed another technique called KRUM in [86]. This method tries to find the most representative model update, and then it will update the global model. In those studies, the most representative model can be, for example, the shortest Euclidean distances from others. In contrast, Sun et al. proposed a new method in [89] in which all of the model updates need to be in a limited norm. This will avoid overwhelming global model by just a few poisoned model updates to prevent global model. Also, Li et al. [87] introduces the RSA method. This method uses a robust stochastic aggregation.
- *Anomaly Detection:* The anomaly detection method is used to distinguish between benign and poisoned ones. Benign model updates and poisoned model updates are used for different intentions. Li and Munoz et.al in [82, 88] offer to calculate models' cosine similarities and map data into low-dimensional. To analyze model updates, the anomaly detection method is used to distinguish between benign and poisoned ones. Benign model updates and poisoned model updates are used for different intentions. Li and Munoz et.al in [82, 88] offer to calculate models' cosine similarities and map data into low-dimensional in order to analyze model updates.
- Another method proposed in the literature is using a combination of two types of methods to leverage the benefits of both of them. It means before aggregating the model updates, it is required to evaluate model updates. For example, Fang et al. in [82] improved the aforementioned algorithms, COMED, COTMED, and KRUM, with a preliminary evaluation procedure. This model will be discarded if accuracy and loss are unacceptable. In contrast, authors in, [84, 85] try to reweight model updates by evaluation results. In [84], a repeated median estimator is used to build an aggregated confidence record for each client. Base on confidence records, it is required to reweight the model updates for aggregation. In contrast, Fung et al. in [85] reweight model updates according to their cosine similarities. Authors in [81] proposed a new method called Verify Before Aggregate (VBA). This method enables the server to distinguish and identify poisoned model updates. Then Deep Reinforcement Learning (DRL) is applied to learn the clients' behaving patterns, which are typically determined by attackers and cannot be known to the server. Based on historical identification results

with the learned knowledge, DRL allows the server to actively select the clients that can provide benign model updates at low training fees.

References

1. H. Karimipour, P. Srikantha, J. Wei-Kocsis, "Security of Cyber-Physical Systems: Vulnerability and Impact", Springer Books, Aug. 2020. https://doi.org/10.1007/978-3-030-45541-5
2. Z. K. Aldein Mohammed and E. S. Ali Ahmed, "Internet of Things Applications, Challenges and Related Future Technologies," *World Sci. News*, vol. 67, no. 2, pp. 126–148, 2017.
3. H. Karimipour, V. Dinavahi, "Extended Kalman Filter Based Massively Parallel Dynamic State Estimation", IEEE Transaction in Smart Grid, vol. 6, no. 3, pp.1539-1549, May 2015. DOI: https://doi.org/10.1109/TSG.2014.2387169
4. H. Karimipour and V. Dinavahi, "On False Data Injection Attack Against Dynamic State Estimation on Smart Power Grids," in *2017 5th IEEE International Conference on Smart Energy Grid Engineering, SEGE 2017*, 2017.
5. A. Alabasi, H. Karimipour, A. Dehghantanha, "An Ensemble Deep Learning-based Cyber-Attack Detection in Industrial Control System", IEEE Access, vol. 8, pp. 83965-83973, April. 2020. doi: https://doi.org/10.1109/ACCESS.2020.2992249
6. A. Al-Abassi, J. Sakhnini and H. Karimipour, "Unsupervised Stacked Autoencoders for Anomaly Detection on Smart Cyber-physical Grids," 2020 IEEE International Conference on Systems, Man, and Cybernetics (SMC), Toronto, ON, 2020, pp. 3123-3129, doi: https://doi.org/10.1109/SMC42975.2020.9283064..
7. L. Monostori, "Cyber-physical production systems: Roots, expectations and R&D challenges," in *Procedia CIRP*, 2014.
8. H. Karimipour and H. Leung, "Relaxation-based anomaly detection in cyber-physical systems using ensemble kalman filter," *IET Cyber-Physical Syst. Theory Appl.*, 2020.
9. H. H. Pajouh, A. Dehghantanha, R. Parizi, H. Karimipour, "A Survey on Internet of Things Security: Requirements, Challenges, and Solutions", Internet of Things Journal, pp. 1–16, Oct. 2019. https://doi.org/10.1016/j.iot.2019.100129
10. B. McMillin and T. Roth, "Cyber-Physical Security and Privacy in the Electric Smart Grid," *Synth. Lect. Inf. Secur. Privacy, Trust*, vol. 9, no. 2, pp. 1–64, 2017.
11. H. Karimipour, A. Dehghantanha, R. M. Parizi, K. K. R. Choo, and H. Leung, "A Deep and Scalable Unsupervised Machine Learning System for Cyber-Attack Detection in Large-Scale Smart Grids," *IEEE Access*, 2019.
12. C. S. Wickramasinghe, D. L. Marino, K. Amarasinghe, and M. Manic, "Generalization of deep learning for cyber-physical system security: A survey," in *Proceedings: IECON 2018—44th Annual Conference of the IEEE Industrial Electronics Society*, 2018.
13. S. Sridhar, A. Hahn, and M. Govindarasu, "Cyber-physical system security for the electric power grid," *Proc. IEEE*, 2012.
14. R. Rajkumar, I. Lee, L. Sha, and J. Stankovic, "Cyber-physical systems: The next computing revolution," in *Proceedings—Design Automation Conference*, 2010.
15. A. Humayed, J. Lin, F. Li, and B. Luo, "Cyber-Physical Systems Security—A Survey," *IEEE Internet Things J.*, vol. 4, no. 6, pp. 1802–1831, 2017.
16. S. Mohammadi, H. Mirvaziri, M. G. Ahsaee, H. Karimipour, "Cyber Intrusion Detection by Combined Feature Selection Algorithm", Journal of Information Security & Applications—Elsevier (IF: 2.6), pp. 80-88, vol. 44, Feb. 2018. https://doi.org/10.1016/j.jisa.2018.11.007
17. Z. El Mrabet, N. Kaabouch, H. El Ghazi, and H. El Ghazi, "Cyber-security in smart grid: Survey and challenges," *Comput. Electr. Eng.*, 2018.
18. E. K. Wang, Y. Ye, X. Xu, S. M. Yiu, L. C. K. Hui, and K. P. Chow, "Security issues and challenges for cyber physical system," in *Proceedings—2010 IEEE/ACM International Conference*

on Green Computing and Communications, GreenCom 2010, 2010 IEEE/ACM International Conference on Cyber, Physical and Social Computing, CPSCom 2010, 2010.
19. Q. Shafi, "Cyber physical systems security: A brief survey," in *Proceedings—12th International Conference on Computational Science and Its Applications, ICCSA 2012*, 2012.
20. A. N. Jahromi, J. Sakhnini, H. Karimipour, A. Dehghantanha, "A Deep Unsupervised Representation Learning Approach for Effective Cyber-physical Attack Detection and Identification on Highly Imbalanced Data", 29th Annual International Conf. on Computer Science and Software Engineering, pp.1–10, Toronto, Canada, Nov. 2019. https://dl.acm.org/doi/10.5555/3370272.3370274#sec-terms
21. R. Anderson and S. Fuloria, "Who Controls the off Switch?," 2010.
22. R. Chow, E. Uzun, A. A. Cárdenas, Z. Song, and S. Lee, "Enhancing Cyber-Physical Security through Data Patterns," *Work. Found. Dependable Secur. Cyber-Physical Syst.*, 2011.
23. P. McDaniel and S. McLaughlin, "Security and privacy challenges in the smart grid," *IEEE Secur. Priv.*, 2009.
24. I. Lee *et al.*, "Challenges and research directions in medical cyber-physical systems," *Proc. IEEE*, 2012.
25. Y. Shoukry, P. Martin, P. Tabuada, and M. Srivastava, "Non-invasive spoofing attacks for anti-lock braking systems," in *Lecture Notes in Computer Science (including subseries Lecture Notes in Artificial Intelligence and Lecture Notes in Bioinformatics)*, 2013.
26. Y. Chen, S. Kar, and J. M. F. Moura, "Cyber-Physical Attacks with Control Objectives," *IEEE Trans. Automat. Contr.*, 2018.
27. J. Sakhnini, H. Karimipour, A. Dehghantanha, R. M. Parizi, and G. Srivastava, "Security Aspects of Internet of Things Aided Smart Grids: a bibliometric survey," *Internet of Things*, 2020.
28. D. Papp, Z. Ma, and L. Buttyan, "Embedded systems security: Threats, vulnerabilities, and attack taxonomy," in *2015 13th Annual Conference on Privacy, Security and Trust, PST 2015*, 2015.
29. P. G. Neumann, *Computer-related risks*. Addison-Wesley Professional, 1994.
30. O. Osanaiye, K. K. R. Choo, A. Dehghantanha, Z. Xu, and M. Dlodlo, "Ensemble-based multi-filter feature selection method for DDoS detection in cloud computing," *arXiv*. 2018.
31. S. M. Tahsien, H. Karimipour, and P. Spachos, "Machine Learning Based Solutions for Security of Internet of Things (IoT): A survey," *J. Netw. Comput. Appl.*, vol. 161, no. February, 2020.
32. P. Jokar, N. Arianpoo, and V. C. M. Leung, "Spoofing detection in IEEE 802.15.4 networks based on received signal strength," *Ad Hoc Networks*, 2013.
33. H. Karimipour, V. Dinavahi, "Robust Massively Parallel Dynamic State Estimation of Power Systems Against Cyber-Attack", IEEE Access, vol. 6, pp. 2984–2995, Dec. 2017. DOI: https://doi.org/10.1109/ACCESS.2017.2786584
34. J. Tian, B. Wang, X. Li, and J. Wei, "Data-Driven and Low-Sparsity False Data Injection Attacks in Smart Grid," *Secur. Commun. Networks*, vol. 2018, 2018.
35. F. Ghalavand, B. M. Alizadeh, H. Karimipour, H. Gaber, "Micro Grid Islanding Detection Based on Mathematical Morphology", Journal of Energies, vol. 11, no. 10, pp. 456-477, Sept. 2018. DOI: https://doi.org/10.3390/en11102696
36. A. Yazdinejad, R. M. Parizi, A. Dehghantanha, H. Karimipour, G. Srivastava, and M. Aledhari, "Enabling Drones in the Internet of Things with Decentralized Blockchain-based Security," IEEE Internet of Things Journal, 2020.
37. E. Nowroozi, A. Dehghantanha, R. M. Parizi, and K.-K. R. Choo, "A survey of machine learning techniques in adversarial image forensics," *Computers & Security,* vol. 100, p. 102092, 2021.
38. A. Yazdinejad, G. Srivastava, R. M. Parizi, A. Dehghantanha, H. Karimipour, and S. R. Karizno, "SLPoW: Secure and Low Latency Proof of Work Protocol for Blockchain in Green IoT Networks," in 2020 IEEE 91st Vehicular Technology Conference (VTC2020-Spring), 2020, pp. 1–5: IEEE.
39. M. Aledhari, R. M. Parizi, A. Dehghantanha and K. R. Choo, "A Hybrid RSA Algorithm in Support of IoT Greenhouse Applications," *2019 IEEE International Conference on*

Industrial Internet (ICII), Orlando, FL, USA, 2019, pp. 233–240, doi: https://doi.org/10.1109/ICII.2019.00049
40. A. Yazdinejad, R. M. Parizi, A. Dehghantanha, G. Srivastava, S. Mohan, and A. M. Rababah, "Cost optimization of secure routing with untrusted devices in software defined networking," *Journal of Parallel and Distributed Computing,* vol. 143, pp. 36–46, 2020
41. A. Yazdinejad, R. M. Parizi, A. Dehghantanha, and K.-K. R. Choo, "Blockchain-enabled authentication handover with efficient privacy protection in SDN-based 5G networks," *IEEE Transactions on Network Science and Engineering*, 2019.
42. Q. Xu, P. Ren, H. Song, and Q. Du, "Security-Aware Waveforms for Enhancing Wireless Communications Privacy in Cyber-Physical Systems via Multipath Receptions," *IEEE Internet Things J.*, 2017.
43. S. S. Gowtham, M. and Ahila, "Privacy Enhanced Data Communication Protocol for Wireless Body Area Network for Wireless Body Area Network," 2017.
44. M. Li, W. Lou, and K. Ren, "Data security and privacy in wireless body area networks," *IEEE Wirel. Commun.*, 2010.
45. Z. Wang, H. Chen, Q. Cao, H. Qi, Z. Wang, and Q. Wang, "Achieving location error tolerant barrier coverage for wireless sensor networks," *Comput. Networks*, 2017.
46. L. Chen *et al.*, "Robustness, Security and Privacy in Location-Based Services for Future IoT: A Survey," *IEEE Access*, 2017.
47. K. Muhammad, R. Hamza, J. Ahmad, J. Lloret, H. Wang, and S. W. Baik, "Secure surveillance framework for IoT systems using probabilistic image encryption," *IEEE Trans. Ind. Informatics*, 2018.
48. W. Meng, E. W. Tischhauser, Q. Wang, Y. Wang, and J. Han, "When intrusion detection meets blockchain technology: A review," *IEEE Access*, 2018.
49. L. Sweeney, "k-anonymity: A model for protecting privacy," *Int. J. Uncertainty, Fuzziness Knowlege-Based Syst.*, 2002.
50. T. Wang, Z. Zheng, M. H. Rehmani, S. Yao, and Z. Huo, "Privacy preservation in big data from the communication perspective—A survey," *IEEE Commun. Surv. Tutorials*, 2019.
51. C. Dwork, "A firm foundation for private data analysis," *Communications of the ACM*. 2011.
52. Z. Du, C. Wu, T. Yoshinaga, K. L. A. Yau, Y. Ji, and J. Li, "Federated Learning for Vehicular Internet of Things: Recent Advances and Open Issues," *IEEE Comput. Graph. Appl.*, 2020.
53. Intel AI, "Federated Learning," pp. 1–7, 19AD.
54. R. Kanagavelu *et al.*, "Two-Phase Multi-Party Computation Enabled Privacy-Preserving Federated Learning," in *Proceedings—20th IEEE/ACM International Symposium on Cluster, Cloud and Internet Computing, CCGRID 2020*, 2020.
55. W. Y. B. Lim *et al.*, "Federated Learning in Mobile Edge Networks: A Comprehensive Survey," *IEEE Commun. Surv. Tutorials*, 2020.
56. S. A. Rahman, H. Tout, H. Ould-Slimane, A. Mourad, C. Talhi, and M. Guizani, "A Survey on Federated Learning: The Journey from Centralized to Distributed On-Site Learning and Beyond," *IEEE Internet Things J.*, 2020.
57. V. Mothukuri, R. M. Parizi, S. Pouriyeh, Y. Huang, A. Dehghantanha, and G. Srivastava, "A survey on security and privacy of federated learning," Future Generation Computer Systems, vol. 115, pp. 619-640, 2021.
58. A. Yazdinejadna, R. M. Parizi, A. Dehghantanha, and M. S. Khan, "A kangaroo-based intrusion detection system on software-defined networks," Computer Networks, vol. 184, p. 107688, 2021.
59. R. M. Parizi, S. Homayoun, A. Yazdinejad, A. Dehghantanha, and K.-K. R. Choo, "Integrating privacy enhancing techniques into blockchains using sidechains," in 2019 IEEE Canadian Conference of Electrical and Computer Engineering (CCECE), 2019, pp. 1–4: IEEE.
60. KPMG LLP, "Federated Learning: Strategies for Improving Communication Efficiency," *Iclr*. 2018.
61. Y. Liu *et al.*, "Deep Anomaly Detection for Time-series Data in Industrial IoT: A Communication-Efficient On-device Federated Learning Approach," *arXiv*. 2020.

62. V. Kulkarni, M. Kulkarni, and A. Pant, "Survey of personalization techniques for federated learning," in *Proceedings of the World Conference on Smart Trends in Systems, Security and Sustainability, WS4 2020*, 2020.
63. M. Aledhari, R. Razzak, R. M. Parizi, and F. Saeed, "Federated Learning: A Survey on Enabling Technologies, Protocols, and Applications," *IEEE Access*. 2020.
64. Q. Jing, W. Wang, J. Zhang, H. Tian, and K. Chen, "Quantifying the performance of federated transfer learning," *arXiv*. 2019.
65. S. Caldas, V. Smith, and A. Talwalkar, "Federated Kernelized Multi-Task Learning," *Conf. Syst. Mach. Learn.*, 2018.
66. S. Feng and H. Yu, "Multi-participant multi-class vertical federated learning," *arXiv*. 2020.
67. Q. Wu, K. He, and X. Chen, "Personalized Federated Learning for Intelligent IoT Applications: A Cloud-Edge based Framework," *IEEE Comput. Graph. Appl.*, 2020.
68. D. Liu, T. Miller, R. Sayeed, and K. D. Mandl, "FADL:Federated-Autonomous Deep Learning for Distributed Electronic Health Record," *arXiv*. 2018.
69. A. Nilsson, S. Smith, G. Ulm, E. Gustavsson, and M. Jirstrand, "A performance evaluation of federated learning algorithms," in *DIDL 2018—Proceedings of the 2nd Workshop on Distributed Infrastructures for Deep Learning, Part of Middleware 2018*, 2018.
70. V. S. Li, Tian, Anit Kumar Sahu, Maziar Sanjabi, Manzil Zaheer, Ameet Talwalkar, "On the convergence of federated optimization in heterogeneous networks," 2018.
71. Y. Zhao, M. Li, L. Lai, N. Suda, D. Civin, and V. Chandra, "Federated learning with non-iid data," *arXiv*. 2018.
72. S. Samarakoon, M. Bennis, W. Saad, and M. Debbah, "Federated Learning for Ultra-Reliable Low-Latency V2V Communications," in *2018 IEEE Global Communications Conference, GLOBECOM 2018—Proceedings*, 2018.
73. L. Wang, W. Wang, and B. Li, "CMFL: Mitigating communication overhead for federated learning," in *Proceedings—International Conference on Distributed Computing Systems*, 2019.
74. F. Chen, M. Luo, Z. Dong, Z. Li, and X. He, "Federated meta-learning with fast convergence and efficient communication," *arXiv*. 2018.
75. T. Li, A. K. Sahu, A. Talwalkar, and V. Smith, "Federated Learning: Challenges, Methods, and Future Directions," *IEEE Signal Process. Mag.*, 2020.
76. B. Liu, L. Wang, M. Liu, and C. Z. Xu, "Federated imitation learning: A privacy considered imitation learning framework for cloud robotic systems with heterogeneous sensor data," *arXiv*. 2019.
77. B. S. Ciftler, A. Albaseer, N. Lasla, and M. Abdallah, "Federated Learning for RSS Fingerprint-based Localization: A Privacy-Preserving Crowdsourcing Method," in *2020 International Wireless Communications and Mobile Computing, IWCMC 2020*, 2020.
78. M. M. Wadu, S. Samarakoon, and M. Bennis, "Federated learning under channel uncertainty: Joint client scheduling and resource allocation," *arXiv*. 2020.
79. F. Ang, L. Chen, N. Zhao, Y. Chen, W. Wang, and F. R. Yu, "Robust Federated Learning with Noisy Communication," *IEEE Trans. Commun.*, 2020.
80. X. Yao, C. Huang, and L. Sun, "Two-Stream Federated Learning: Reduce the Communication Costs," in *VCIP 2018—IEEE International Conference on Visual Communications and Image Processing*, 2018.
81. J. Tan, Y. C. Liang, N. C. Luong, and D. Niyato, "Toward smart security enhancement of federated learning networks," *arXiv*. 2020.
82. L. Munoz-González, K. T. Co, and E. C. Lupu, "Byzantine-robust federated learning through adaptive model averaging," *arXiv*. 2019.
83. C. Xie, O. Koyejo, and I. Gupta, "SLSGD: Secure and Efficient Distributed On-device Machine Learning," in *Lecture Notes in Computer Science (including subseries Lecture Notes in Artificial Intelligence and Lecture Notes in Bioinformatics)*, 2020.
84. S. Fu, C. Xie, B. Li, and Q. Chen, "Attack-resistant federated learning with residual-based reweighting," *arXiv*. 2019.

85. C. Fung, C. J. M. Yoon, and I. Beschastnikh, "Mitigating sybils in federated learning poisoning," *arXiv*. 2018.
86. Y. Dong, J. Cheng, M. Jahangir Hossain, and V. C. M. Leung, "Secure distributed on-device learning networks with byzantine adversaries," *IEEE Netw.*, 2019.
87. L. Li, W. Xu, T. Chen, G. B. Giannakis, and Q. Ling, "RSA: Byzantine-robust stochastic aggregation methods for distributed learning from heterogeneous datasets," in *33rd AAAI Conference on Artificial Intelligence, AAAI 2019, 31st Innovative Applications of Artificial Intelligence Conference, IAAI 2019 and the 9th AAAI Symposium on Educational Advances in Artificial Intelligence, EAAI 2019*, 2019.
88. S. Li, Y. Cheng, W. Wang, Y. Liu, and T. Chen, "Learning to detect malicious clients for robust federated learning," *arXiv*. 2020.
89. Z. Sun, P. Kairouz, A. T. Suresh, and H. Brendan McMahan, "Can you really backdoor federated learning," *arXiv*. 2019.

A Multi-Stage Machine Learning Model for Security Analysis in Industrial Control System

Prabhat Semwal

1 Introduction

Industrial control systems (ICS) are cyber-physical systems, which incorporate sensors, computers, networks, communications, and other digital management components into critical infrastructure to manage or track facilities remotely and independently [1]. ICS plays an important role in the monitoring and control of critical infrastructures such as smart power grids, oil and gas, aerospace, and transportation [2, 3]. The inclusion of the Internet of Things (IoT) in ICSs increase the vulnerabilities of these systems towards cyber-attacks. While the security concerns of critical infrastructure facilities are already considered in the Information Technology (IT) community, limited efforts have been made to develop security solutions that are specific to ICSs [4–9].

ICS has unique performance and reliability requirements and also uses operating systems, applications, and procedures that may be considered uncommon by mainstream IT professionals. Such standards typically reflect the concept of quality and integrity, followed by confidentiality, which involves the management of procedures. If not properly enforced, this will pose a major risk to human health and safety, environmental damage, and serious financial problems, such as loss of production. Sensitive infrastructure (e.g., electrical power, transport) unavailability may have an economic effect well beyond immediate and physical harm structures [10, 11]. The local, regional, national, and perhaps global economies will suffer from these impacts.

Industrial processes require enormous supervision of machinery. As long as a system is operating, some knowledge is required to ensure the proper and effective operation of the device. If the people do this, the machine must be patrolled 24/7,

P. Semwal (✉)
School of Computer Science, University of Guelph, Guelph, ON, Canada

H. Karimipour, F. Derakhshan (eds.), *AI-Enabled Threat Detection and Security Analysis for Industrial IoT*, https://doi.org/10.1007/978-3-030-76613-9_12

with constant and tedious data measurements and equipment checks. It is not only a waste of technical skill but also expensive for a company to pay for the continuous monitoring of a machinery part by a person. The risk of human error is often added in a mechanism that includes inherent risks. It allows automation systems such as SCADA (Supervisory Controlling and Data Acquisition networks are generally referred to as these commercial command and control networks) and Automation Monitoring systems [12, 13].

The large, scattered complexes include modern manufacturing plants, such as refineries for gasoline, chemical plants, generation plants for electricity, and production plants. To ensure their proper service, plant managers need to track and control several different parts constantly. This remote access has been understood with the advancement of networking technologies. The previous networks of control were simply point-to-point networks to link a monitoring or control device to a remote sensor or actuator. Since then, these are built into complex networks that allow communication over a specific communication route between a central control unit and many remote devices. The nodes of these networks are typically built into computing devices, such as switches, actuators, and PLCs, for different purposes.

SCADA manages and controls systems or equipment and works from a single position with the supervisory staff. It consists of a main SCADA host computer, several distributed data units, including sensors, and PLCs that are programmable logic controllers.

With ICS and SCADA we have huge datasets that required low cost and high computational analysis techniques. Machine Learning is used as an alternate tool or an external strategy to analyze these datasets and protect against ransomware, botnets, and other threats [14]. Machine learning technologies are used by many applications to address network and security-related issues. Anomaly detection is used in numerous fields of machine learning, including intrusion detection and fraud detection. This can help predict traffic patterns and identify anomalies in the behavior of the network.

This chapter aims to provide ICS with an efficient method of detecting cyber-attacks. It is achieved by a mixture of the analysis of literature and the experiment. The main objective of this project is to examine how machine learning can assist in the cyber-attack detection in ICS, to analyze how the accuracy and efficiency of attack detection algorithms can be improved and to create a multi-stage attack detection model.

The remaining structure is as follows:

Section 2 contains context information on the Industrial control system datasets and Machine supervised learning algorithms. In Sect. 3 there is an examination of the research-related literature contained in this thesis. It begins by recognizing the kinds of literature attacks and threats. Section 4 demonstrates the proposed model and describe various sections of the multi-stage model and Sect. 5 describes the experimental methods used. It highlights the experimental process, explains data sets, and experimental processes. Section 6 reveals the outcomes of these research studies and evaluation methods. Section 7 concluded by summarizing its findings and proposes potential studies and changes that can be made on this research.

2 Background

This section provides the context required for an interpretation of the research conducted. The chapter starts with the description of the ICS and SCADA infrastructure, a short overview of datasets, and then addresses techniques of machine learning for the experiments (Fig. 1).

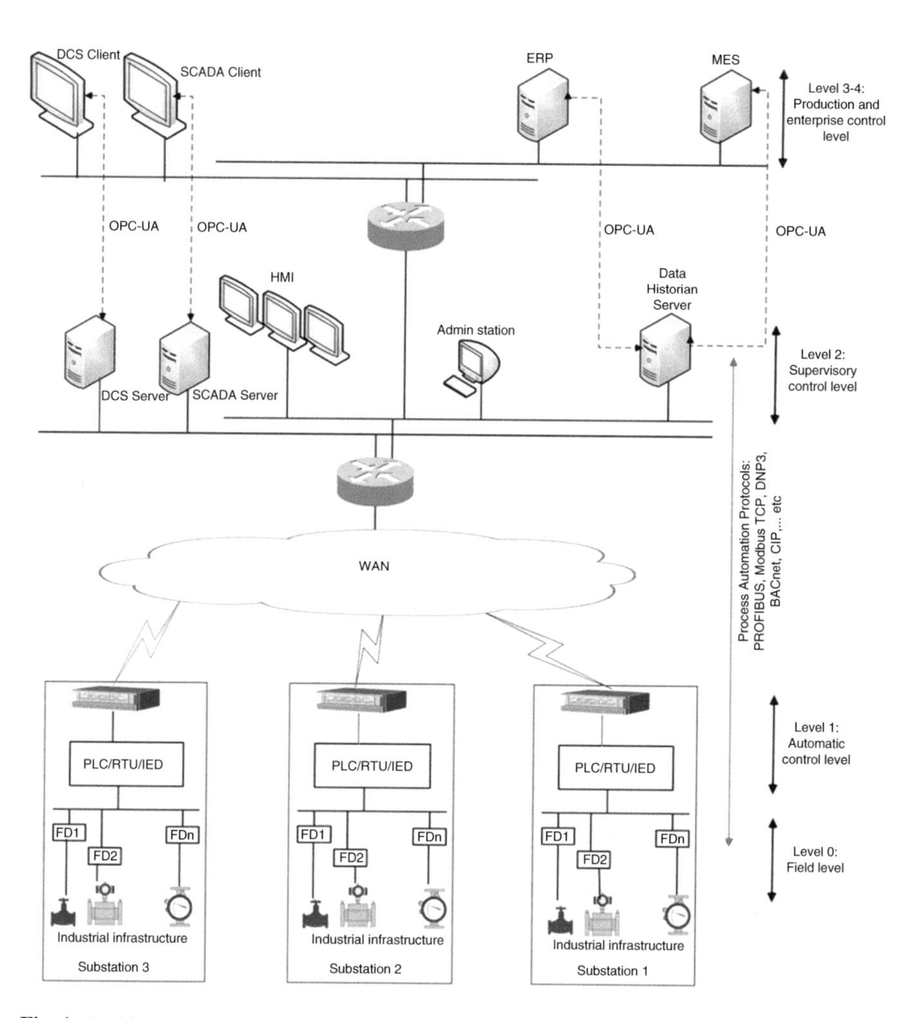

Fig. 1 Architecture of ICS

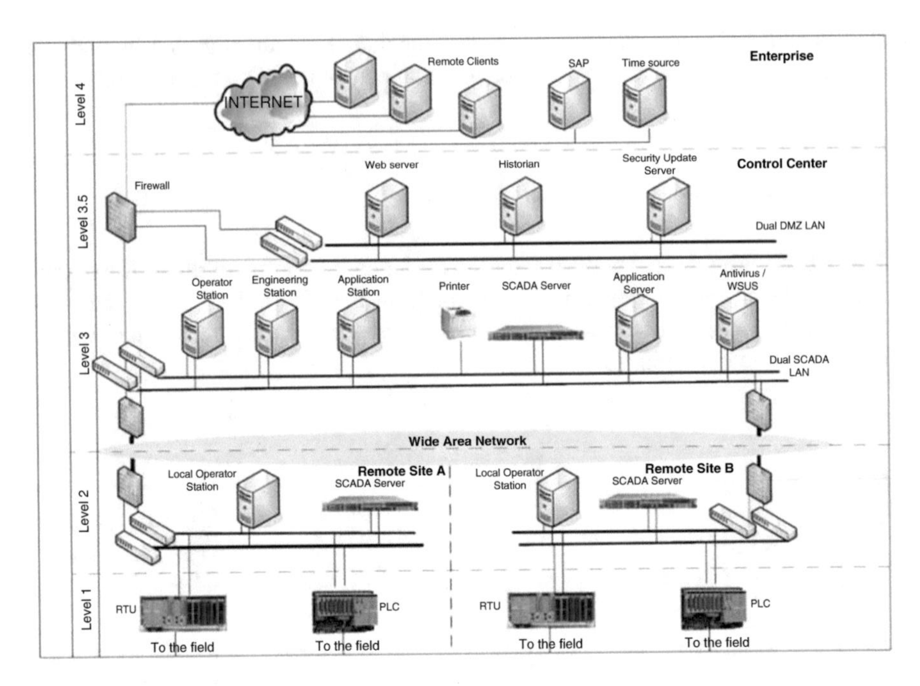

Fig. 2 The architecture of SCADA

As can be seen in Fig. 2, the architecture of ICS systems has been designed to provide tools for discovering, integrating, and responding to user requirements, and data products, software, and services (DDSS) [15]. Integrating IoT solutions with ICS enables the collection and analysis of a large data set across the entire industrial area. These ICS include SCADA and other device configurations, such as programmable logic controller (PLCs). SCADA systems typically manage distributed assets using centralized data storage and monitoring controls as shown in Fig. 2.

The SCADA System interacts with the AMI through two methods of communication, defining precise power usage and appropriately transmitting electricity. Such bidirectional connectivity is achieved by various size networks, including the Home Area Networks, which allow contact in household communications, the Neighborhood Area Networks, and the wide-area Networks, which connect all major sections such as plants, sub-stations, and operating centers. The network is safely shared between the residences, as well.

ICS and SCADA are used in industries such as electric, water and wastewater, oil and natural gas, chemical, and pharmaceutical. Gas pipelines and water storage tanks are examples of these systems.

2.1 Gas Pipeline System

The gas pipeline system for data collection was supplied by the in-house SCADA laboratory of the Mississippi state university. The system comprises three main components: sensors and actuators, a network of communications and monitoring [7].

At a lower level, two actuators with a pressure sensor are included in the gas pipeline. To control the physical operation of the system, the actuators, a pump, and a solenoid are used to sustain the tension generated by the sensors. There are three major modes of gas pipeline system: automated, manual, off. Once the system is automatically configured, the supervisory controls have two mechanisms for controlling the load. The first scheme is the pump mode, which activates and deactivates the pump to retain pressure at the specified point in the tubing. This scheme has been developed to model a continuous machine load. The second scheme is solenoid mode, where the solenoid-controlled relief valve is opened and closed for pressure management. Both the pump and the solenoid modes have a PID-Function control mechanism. In machine mode, the operator can also be operated manually for the pump and the solenoid to be operated manually.

2.2 Water Tank Storage System

The completed water tank system consists of a water tank, a water pump, a source reservoir, a solenoid valve, two float sensors, two switches mounted in a switch housing unit, and a relay in a NEMA 12 compliant container. SCAD systems have elements that are monitored and controlled. The water level of this system inside the tank is monitored, and the ability to drain or fill the tank is regulated by automation based on my water level.

2.3 Machine Learning Algorithm

2.3.1 Decision Tree

A decision tree is a guided, inverted machine learning (ML) framework, the algorithm in which the internal (non-leaf) node is a (predictor) element, the branch of the test outcome is a branch between the node and each leaf node is an entity (response variable) entity. The Decision Tree is a Machine Learning(ML) algorithm that is managed by an inverted tree structure, wherein each internal (non-page) Node represents a function (predictor variable), and the node branches represent the test result [8].

$$E(S)=\sum_{i=1}^{c}-p_i p_i \quad (1)$$

Here 'pi' is the probability of class 'i', Entropy is computed as the proportion of class 'i' in the set.

2.3.2 Random Forest

Random forests are an ensemble learning framework that works by constructing a range of decision trees through preparation and providing the class that predicts class or middle trees. Random forest models are like a decision tree, but random forest models generate several decision trees, not just one. The maximum depth was set at 2 in our random forest model.

Averaging the predictions from all the individual regression trees (Eq. 2).

$$f=\frac{1}{B}\sum_{b=1}^{B}f_b(x') \quad (2)$$

2.3.3 K-Nearest Neighbors

K-nearest neighbors (KNN) is a supervised classification problem-solving learning algorithm. The findings of the KNN primarily rely on three elements: distancing metric used to determine the closest neighbor, the K-nearest classification distance law, and the number of neighbors in the current categorized study. The high-performance KNN cyber-attack detector is primarily used as an identification algorithm.

Euclidean Distance formula (Eq. 3).

$$d(p,q)=\sqrt{\sum_{i=1}^{n}(q_{i-}p_i)^2} \quad (3)$$

2.3.4 Logistic Regression

Logistic regression applies if the dependent target is categorical. The logistic regression is like every regression analysis a statistical model. n Logistic regression is used to explain the relationship between one and one or more independent nominal, common, and interval or relation-level variables. Logistic regression is frequently referred to as log-linear or maximum entropy classification. This states that a categorical dependent variable can be expected from a certain set of independent variables. The algorithm for logistic regression uses a linear decision area and, therefore, cannot solve non-linear issues.

$$l = \frac{p}{1-p} = \beta_0 + \beta_1 x_1 + \beta_2 x_2 \quad (4)$$

Linear relationship between the predictor variables and the log-odds of the event (Eq. 4).

2.3.5 Multi-Layer Perceptron Algorithm

More than one linear layer (combinations of neurons) may exist in the Multilayer perceptron, as we can see in Fig. 3. If we take the simple example of a network with three layers, the input layer is the first layer, and the output layer last is the hidden layer. Our data is added to the input layer, and the output is taken from the output layer. We can increase as much as we like the amount of the secret layer to complex the model according to our work [9].

Feed-Forward Network, the neural network architecture that is most distinctive. Its objective is some function f) (approximate. Given the classification y = f f(x) mapping the input x to output y, by defining the mapping, y = f(x; f) and by knowing the best parameterš ā for that classification, for instance, the MLP can consider the best approach to that classification. The MLP networks are made up of many chained functions. F(x) = f(3)(f(2).(f(1)(x))) will be a network of three functions or layers. Each of these layers consists of units that transform a linear sum of inputs in an affiliate way. -- layer has y = f(WxT + b) representations. Where f is the activation function, W is the layer parameter or weights; x is the vector of entry, which may be the value of the previous layer, and b is the vector of inclination. The layers of an MLP are several fully connected layers since each device in the previous layer

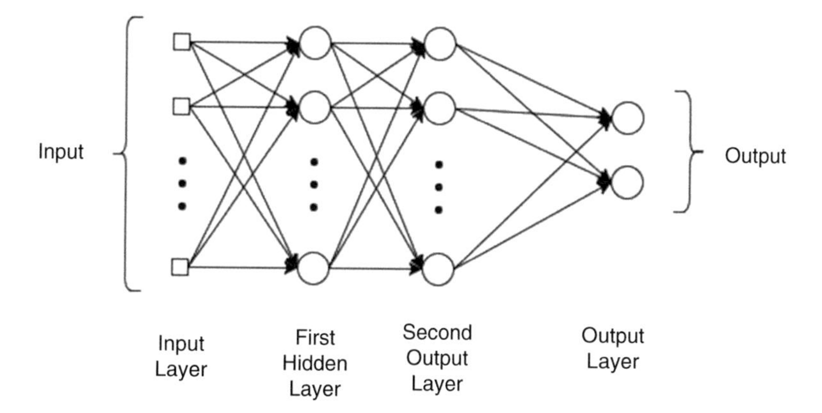

Fig. 3 Multi-layer perceptron algorithm layer representation

is connected to all the devices. The parameter of each unit in a completely connected layer is distinct from other units in the layer, which ensures that each unit has a special set of weights.

3 Literature Review

Many computer scientists in earlier studies have suggested various solutions to cyber threat problems hunting with various machine-learning techniques. This section addresses the types of cyber-challenges faced by smart cyber-physical grids, as well as the strategies outlined in the literature to counter these challenges. Introducing an appropriate approach to this issue includes an in-depth analysis of the relevant works.

3.1 Types of Cyber Attacks

Computer Network Attack (CNA) or Cyber Attack is the intentional exploitation of computer systems, networks, and technology-dependent enterprises. In this section, we will discuss different types of Cyberattacks based on our dataset.

Naïve Malicious Response Injection Naive Malicious Response Injection (NMRI) attacks lack sophistication [10, 11]. The NMRI attacks use injecting response packets into the network, but it lacks monitoring and control knowledge about the operation. Invalid payloads can be sent by NMRI attacks. For example, an attacker might be aware that a series of reconciliation attacks have been carried out to learn system addresses, function codes, and memory maps, but they lack details of what is the monitoring process or details about the valid data content for each server item. In that event, an attacker may make an injecting attack with a payload of all zeroes, negative numbers, very high numbers, or any other contents that are likely to be invalid. Otherwise, NMRI attacks could be based on limited information about the process [12]. For instance, an attacker can understand process data such as process limitations or valid contents for each server item but cannot perform more advanced attacks. For instance, an assailant may cause an alarm. The first NMRI attack is Naïve Read Payload Size. The attack on the Naïve Read Payload Size is based on the knowledge of the network protocol. The reading coil, discrete inputs, record holders, and input registration queries for MODBUS include a quantity field for specifying the number of objects the server will return. An NMRI attack can produce malicious reactions, including the correct number of returned objects, all or zeroes. The right number of products with random content can also be retrieved by the NMRI. In reading coils and discreet input situations, random input is especially important since, for each coil or discrete input, such returned values must be limited to 0x00 and 0xFF.

Complex Malicious Response Injection Complex Malicious Response Injection (CMRI) provides a degree of complexity above the NMRI [10, 11]. CMRI needs an understanding of the attacking cyber-physical system. CMRI attacks attempt to cover up the real state of the controlled physical process to affect the cyber-physical system control feedback loop [12].

Malicious State Command Injection Altered Actuator State is an MSCI attack scenario that changes system actuator states one time [10–12]. For the gas pipeline system, Altered Actuator State attacks include command injections that turn the pump on or off and command injections that open or close the relief valve. For the water storage tank system, an Altered Actuator State attack was implemented to turn the pump on or off.

Malicious Parameter Command Injection Altered Control SetPoint is an MPCI attack which changes device set points. Setpoints are typically used to provide variable control over a system [10, 11]. For example, the water storage tank system uses an ON/OFF control scheme to keep the amount of liquid in a tank between a low set point and a high set point. A level sensor continuously monitors liquid level as a percentage of tank full and turns a pump on an off to add liquid to the tank. A MODBUS write register command was used to change both the high and low setpoints [12]. This attack also alters alarm values stored in PLC registers to disable alarms by changing setpoints liquid level alarms to values in line with the altered high and low setpoints.

Malicious Function Code Injection This attack is part of the subclasses of the Command Injection Injection Control. The server will not transmit the data to the network anymore because of this attack. Normally, polling techniques such as HMI are used for most industrial control systems [12].

Denial of Service Denial of Service (DOS) attacks against industrial control system attempt to stop the proper functioning of some portion of the cyber-physical system to disable the entire system [10, 11] effectively. As such, DOS attacks may target the cyber system or the physical system. DOS attacks against the cyber system target communication links or attempt to disable programs running on system endpoints that control the system, log data, and govern communications. DOS attacks against the physical system vary from the manual opening or closing of valves and switches to the destruction of portions of the physical process, which prevents operation. This work concentrates on DOS attacks against the communication system.

Reconnaissance Reconnaissance attacks gather control system network information, map the network architecture, and identify the device characteristics such as manufacturer, model number, supported network protocols, system address, and system memory map [10, 12]. The device identification attack allows an attacker to learn a discovered device's vendor name, product code, major and minor revision, et cetera. The points scan allows the attacker to build a device memory map.

3.2 Detection of Cyber Attacks

Numerous reports have addressed the traffic-based intrusion detection network in [16], the proposed method the decision tree is used as the classifier. The suggested approach was applied to large datasets of KDD Cup 99 in addition to evaluation with an accuracy of 95.03%. Similarly, Ghaeini et al. [17] applying this approach to the SWaT data collection used in our study. In [18] an optimized SVM solution with a mixture of features of two machine learning strategies demonstrated a low false-positive rate. In [13], an ensemble approach is tested on several malware samples, including Windows, Ransomware, the Internet of Things (IoT). An estimate of the accuracy of 99.65 percent using an IoT-specific data collection. Further, [19] proposed a deep learning-based approach for developing an efficient and flexible Network Intrusion Detection System. Most research on ICS intrusion detection uses unsupervised (learning from unlabeled real data) [20] and semi-supervised approaches. In other research [21], 1D CNNs can be effectively used for detecting cyber-attacks in complex multivariate ICS data. Recent studies [22] shows architecture involving ensemble and base classifiers for intrusion detection model results in significantly larger improvement of prediction accuracy than the base classifiers. In the paper [23], The experimental results revealed that the hybrid approach had a significant effect on the minimization of the computational and time complexity with low false-negative rates. In 2017, [24] Wathiq Laftah Al-Yaseen and Zulaiha AliOthman proposed a multi-level hybrid intrusion detection model that uses a support vector machine and extreme learning machine to improve the efficiency of detecting known and unknown attacks. In [25] a major step forward in protecting the entire network by using fuzzy and fast fuzzy patterns or malware detection methods in which malicious activities are detected by nodes.

In another paper [26], The model was tested with a data set that contains valid and anomalous data. Further, in most cases, confidences measured by the Naive Bayes classifier is quite significant, with several predictions above 90%. Another paper, [27], uses the Random Forest Algorithm for the identification of cyber threats and achieves a significant 94% accuracy. A Machine Learning sequence technique (ML) to identify suspect incidents or attacks that may alter CPS behavior [4]. This approach not only recognizes the cyber-attack on a physical process layer but also recognizes a certain type of attack.

Furthermore, Moshe Kravchik introduced a model based on a 1D convolutional neural network that detected 31 out of 36 cyber-attacks successfully [28]. Its superior operating time and performance have been shown ICS cyber-attack detection great promise. In [29] proposed a deep automated, unsupervised representation method for detecting the status of the power system using the stacked Auto Encoder model. In which an unsupervised ensemble model evaluates a highly imbalanced dataset and attack detection with a low amount of false alarms (FP) besides too high accuracy (f-measure in multiclass classification).

Similarly, in [30] with C4.5 and iSVM as base classifications, developed a hybrid C4.5-iSVM- model and ensemble approach. The empirical results show that C4.5 provides better or equal accuracy to classes of standard and test and iSVM provides better accuracy to classes of standard and DOS. In comparison to individual classification accuracy, the hybrid C4.5-iSVM classification improves accuracy for R2L and U2R classes. Another paper [31] focused on the selection of functions to detect distributed cyber-attacks early and carried out the early detection of distributed attacks by detecting C&C communication because this communication is at its preparation phase.

The list of open ports in a system provides an attacker with very critical information. There are many tools for identifying open ports like antivirus and IDS. In [4, 32], the port scan attempt was based on a new CICIDS2017 data set using deep learning and support vector machine (SVM) algorithms, and 97.80% and 69.69% accuracy were achieved respectively. On the other hand, Denial-of-service attacks can prohibit computers or network infrastructure is used and can be difficult to identify and counteract as, to a large degree, they exploit the facilities involved according to their intended purposes. A support Vector Machine based machine learning model is used to learn a pattern and to select suitable detection algorithm features to minimize computational costs and look only for the patterns that are most relevant to your detection. The experimental results show that 99.19% of LDoS attacks can be detected by the proposed method with the time complexity of O(n log n) in the best cases [17, 33].

Many IDSs have been developed to detect network attacks, but the problems which often arise in the IDS are either false positive or false negative issues. This study [34] explores how the precision of threat identification is achieved by various approaches of machine learning with the IDS. In this analysis, an IDS prototype is anticipated. The IDS system is built with a mixture of machine-learning approaches to improve precision in detecting multiple attacks. In [35] it has been shown that the use of machine learning techniques increases the efficiency in state estimation and a promising approach to the problems of Bad Data Detection (BDD).

In [36] a supervised model of machine learning to evaluate the detection of attacks is proposed. In [6], the authors used unsupervised algorithms to detect attacks on physical dataset sensors. Another paper [37] focused on ML-based approaches to attack detection for industrial networks.

3.3 Summary

This section shows the types of cyber-attacks studied in previous researches and identifies the most common attacks and a literature survey designed to identify the types of existing cyber threats and methods of detecting them.

4 Proposed Models

The method followed to implement the multi-stage model is defined in this section. In the first phase, the dataset was analyzed to identify the cyber threat. The feature extraction and selection were carried out on the initial dataset during the pre-processing period. In the second phase, our machine learning models were analyzed based on train and test datasets as shown in Fig. 4 and Table 1. In our experiment, we tested the efficiency of the supervised machine learning algorithm in cyber-attack detection with a multi-layer perceptron algorithm.

4.1 Dataset Processing

The first feature contains the slave device's station address. The address of the station is a special 8bit value allocated to each master and slave unit. The address is used to mark the slave that the master sends commands and the responding slave. The Modbus protocol is designed to accept all master transactions from all slave computers. To determine whether the message is intended for oneself or another slave computer, the slave device must search the station address area. This feature is used to improve system scan attack detection, which transmits commands to all possible station addresses

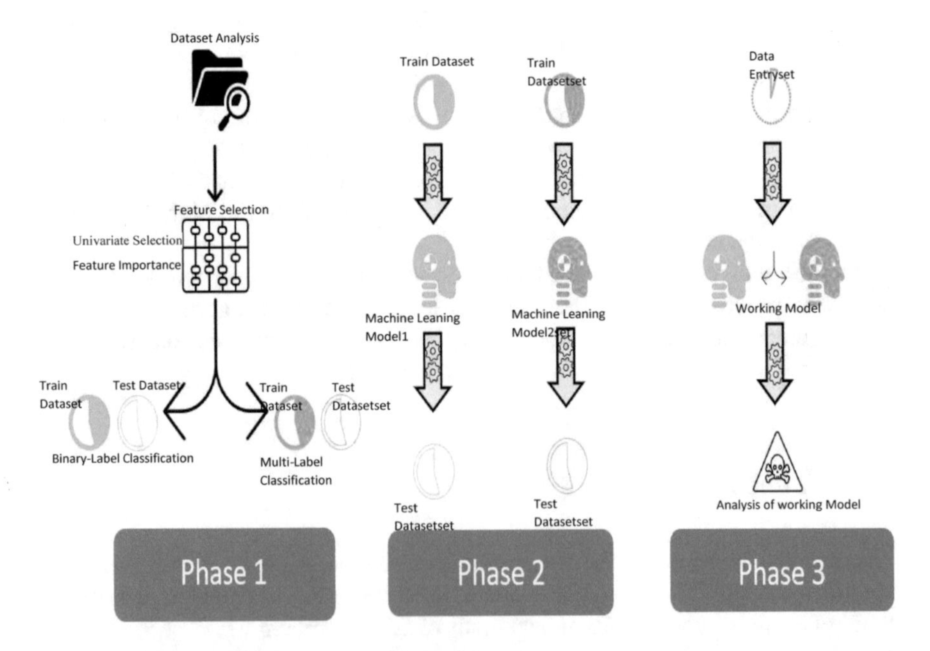

Fig. 4 Proposed Model: Phase1 (Data Processing), Phase2 (Training and Testing Model), Phase3 (Working Model Analysis)

Table 1 Dataset features description

Features	
Address	Control scheme
Function	Pump
Length	Solenoid
Setpoint	Pressure measu.
Gain	CRC rate
Reset rate	Command response
Headband	Time
Cycle time	Binary result
Rate	Categorized result
System mode	Specific result

to determine what addresses are active. The second attribute contains the code of the method. The main function codes in the gas pipeline can be read (0x03) and write (0x16) but there are 256 valid function codes. Certain function codes, such as the '0x08' function, can be used for malicious purposes. The function code '0x08' is commonly used for diagnostic purposes but can be used to force a slave device to listen only. A denial of service utilizing a legitimate feature code will result in a similar attack. This functionality can be used by IDSs to identify irregular function codes. The third function includes the length of the Modbus frame. The length of the Modbus frame for each command or response query is set in a similar way to the function code. A set of writing and reading commands is used for repeated block writing and block readings from registers in the gas pipeline system. Frames of a given duration are detected as anomalous when attacks are observed [38].

The fourth function contains the set-point value that governs gas pipeline pressure. If the mode is set to 'automatic,' the fixed-point function is used. The principle of the slave ladder is built to preserve the value set by either opening a valve or flipping on and off the generator. The setpoint feature dramatically affects the physical structure and is a specific point for an attacker to be malicious. The following five features reflect the importance of the PID controller. Benefit, retrieval rate, dead unit, cycle time, and rate are all values that the PID controller can change. Based on these five conditions, an error is measured, and the PID controller can be triggered, turned, and of the pump to open and shut the relief valve to mitigate the error. The tenth feature includes the attribute that governs the service cycle of the system [36].

4.2 Machine Learning Model

The model implemented in this paper is based on an approach of train and test model. This section will provide an overview of our multistage model and how it helps in attack detection in ICS (Industrial Control System). In our multi-stage model, we have two different algorithms running.

The first is an algorithm of a decision tree that uses algorithms to divide a node into two or more sub-nodes. Sub nodes are generated to increase the homogeneity of subsequent sub-nodes. In other words, concerning the target value, we can assume that the node's purity decreases. The decision tree separates the nodes on all the variables available and chooses the division, resulting in the most uniform sub-nodes [6].

The second algorithm consists of at least three-node layers: the input layer, the hidden layer, and an output layer. Multi-layer sensor Each node is a neuron that uses a nonlinear activation function except for the input nodes. For training, MLP is using a supervised learning method known as backpropagation [37].

4.3 Summary

In this section, the proposed frameworks used to complete each contribution are outlined and explained. The chapter discusses the supervised machine learning algorithm and multi-layer perceptron model implemented to meet the second contribution of the thesis.

5 Methodology

5.1 Datasets Collection Methodology

For generating the dataset, a new approach has been employed for relaxation and data processing. The first step to update the dataset was to parameterize and change the order of the attacks. The implementation was achieved by cutting out and executing all attacks manually. The aim of the midway is for all types of attacks to be included. Such definitions fell in line with the assaults on the gas and water storage facilities, but are also more affected. In the following Table 2, the types of attacks are shown in the dataset [36].

Table 2 Attack Categorization

Type of Attacks	Abbreviation	Threat Type
Normal	Normal(0)	N/A
Naïve malicious response injection	NMRI(1)	Modification/fabrication
Complex malicious response injection	CMRI(2)	Modification/fabrication
Malicious state command injection	MSCI(3)	Modification/fabrication
Malicious parameter command injection	MPCI(4)	Modification/fabrication
Malicious function code injection	MFCI(5)	Modification/fabrication
Denial of service	DoS(6)	Interruption
Reconnaissance	Recon(7)	Interception

In Figs. 5 and 6, we have the dataset evaluation based on attack categories. As we can see that dataset is highly unbalanced with respect and this may cause unbiased learning for our models. In the water tank storage dataset, we have 4132 data entries for reconnaissance but for Denial of Service and Malicious Function Code Injection we have 135 and 155 data entries respectively.

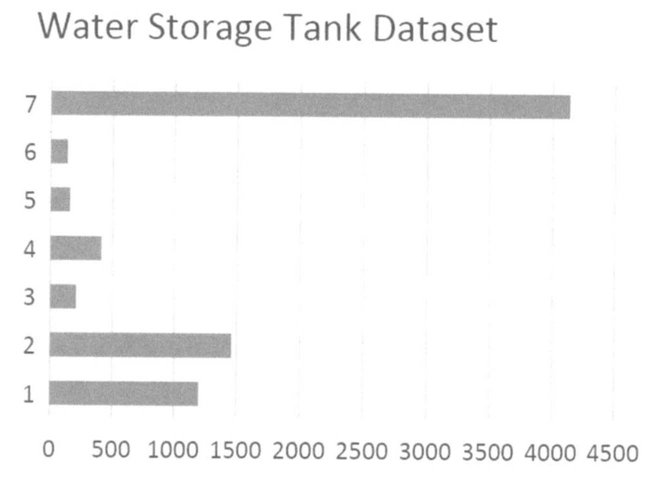

Fig. 5 Data sample based on Attack categories

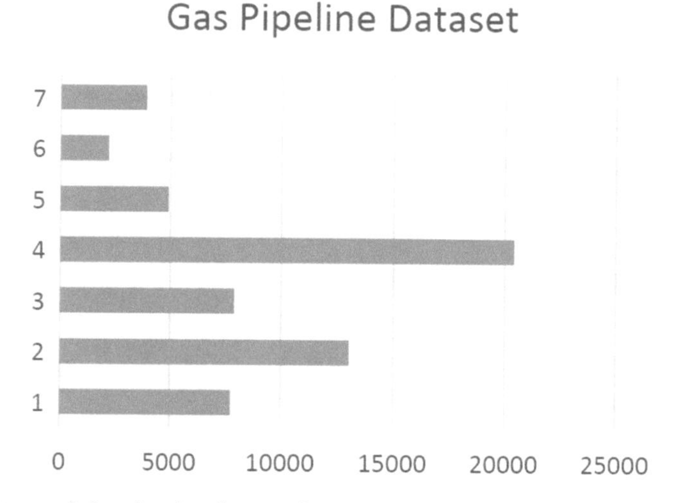

Fig. 6 Data sample based on Attack categories

5.2 Feature Selection Methodology

The method is chosen to minimize duplication, improve consistency, and decrease training time. The procedure is a practical alternative. We used additional tree classifiers for the collection of features-incredibly random trees that selected features based on their score using an ensemble learning technique [38].

We performed the function extraction process to reduce the overall dimensionality of the data set. A combination of the Univariate Choice and the Importance of Features results in Figs. 7 and 8 show the best feature to contribute to the target variable. Extraction of the most popular and highly trained apps.

Univariate Selection Statistical checks should be used to select the most closely linked characteristics to the performance component. The scikit-learn library provides the SelectKBest class to be used to select a certain set of features through a series of various statistical checks. As figure reveals. To pick 10 of the best features in both data sets, we used the chi-squared (chi2) statistic method to classify non-negative features [39].

Feature Importance Use the value of the model feature to obtain the feature of each feature of your dataset. For each attribute of your results, the more significant or more appropriate the value for your performance variable is. The higher the score, the more significant. App value is that we use Extra Tree Classification to select the top 10 data set functions from our built-in class for Tree-Based Classifiers [39].

Fig. 7 WTS dataset Features

Specs	Score
measurements	1.487742e+21
resp length	4.534864e+05
response memory	4.370406e+05
response memory count	1.875711e+04
resp write fun	1.875711e+04
response address	1.312998e+04
LL	1.251467e+04
sub function	3.927969e+03
crc rate	3.738359e+03
H	2.797450e+03

Fig. 8 GP dataset Features

Specs	Score
pressure measurements	9.631416e+41
crc rate	1.843978e+07
fucntion	3.252275e+06
gain	1.687004e+06
length	3.133059e+05
setpoint	2.486708e+05
cycle time	1.777755e+04
system mode	1.651371e+04
command response	1.142457e+04
control scheme	1.072158e+04

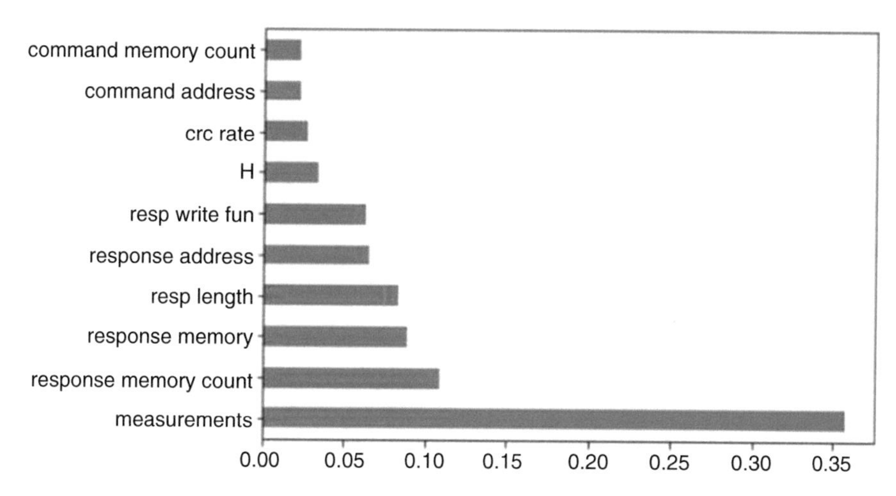

Fig. 9 WTS dataset Features

5.3 *Machine Learning Classifiers*

In the first stage of feature selection, feature extracted from the dataset based on the 'Target' attribute of the dataset. The target attribute is the classification of the dataset as attack and normal. After the selection of features, the dataset is then divided into two different sets, i.e., train and test. These two datasets are now helping to evaluate our model based on several computational matrices. First, with the training dataset, we train our model with a decision tree algorithm. Our decision tree model was trained with the processed dataset. The decision tree model simply designs an inverted tree structure on the base of a trained dataset and then classify a sample by tracing the down the designed tree.

After training the decision tree model with features extracted in the first step in the next step, all the features are again evaluated based on 'Attack Categories'. Attack categories classify dataset in multiple categories such as Naïve Malicious Response Injection, Complex Malicious Response Injection, Malicious State, Command Injection, Malicious Parameter Command Injection, Malicious Function, Code Injection, Denial of Service, Reconnaissance. Now in the next stage, we need to train our multi-layer perceptron model. After Training both models, we evaluated both models based on the test dataset.

5.4 *Summary*

In this section, the methods of this research are outlined. First, the general methodology of research progress is presented. Then a framework for data collection is discussed. This is followed by a thorough explanation of the multi-stage model.

6 Results and Discussion

In this section, the results achieved utilizing machine learning techniques in cyber attack detection have been highlighted, and the different measures used to evaluate the performance of basic and multi-stage machine learning algorithms are described. This model incorporates techniques of generalization for improved scalability and performance with imbalanced data.

6.1 Model Performance

The confusion matrix can be used to calculate the efficiency of the implemented machine learning system by comparing the actual and expected outcomes for a model: True Positive (TP), True Negative (TN), and False Negative (FP) (Table 3). The different values in the learning algorithm will be used to determine the various steps. In our evaluation methodology, we used four commonly used metrics, namely Accuracy (ACC), True Positive Rating (TPR), FPR, ROC, and AUC [40].

Table 3 Description of evaluation metric used for comparative analysis

Evaluation Metric	Description
TPR	$\frac{TP}{TP+FN}$
FPR	$\frac{FP}{FP+TN}$
Accuracy	$\frac{TP+TN}{TP+TN+FP+FN}$

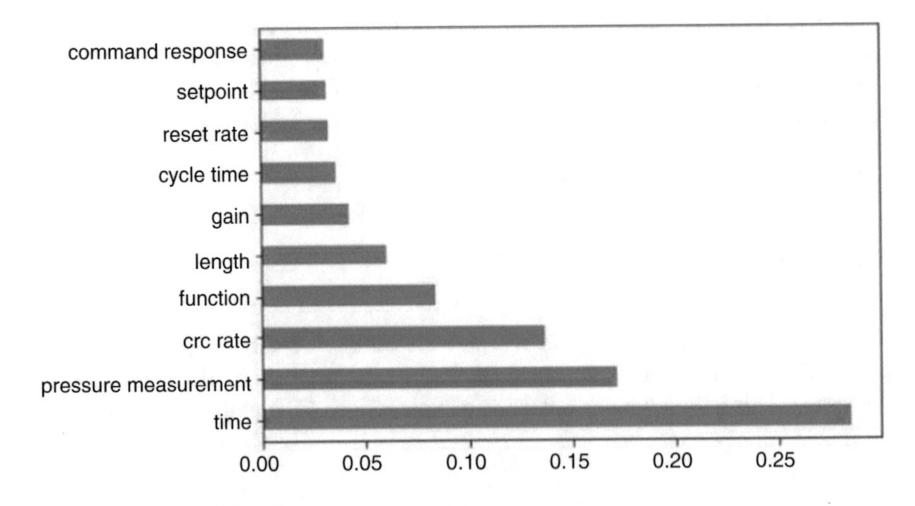

Fig. 10 GP dataset Features

We have used commonly used evaluation metrics to estimate and compare the performance of models. Table 3 describes the evaluation metrics used for benchmarking.

Both models were tested using the processed water storage tank and gas pipeline dataset with the total samples of 27,200 and 274,269 with selected features, respectively. On the processed dataset for binary classification of cyber-attack, the KNN, the decision tree, and the random forest model have been trained, and the results observed based on assessment metrics (Table 3) are shown in Fig. 11, similarly, for multi-label classification of cyber-attack categories, the logistic regression and multi-layer perceptron model has been trained and the results observed based on assessment metrics (Table 3) are shown in Fig. 12.

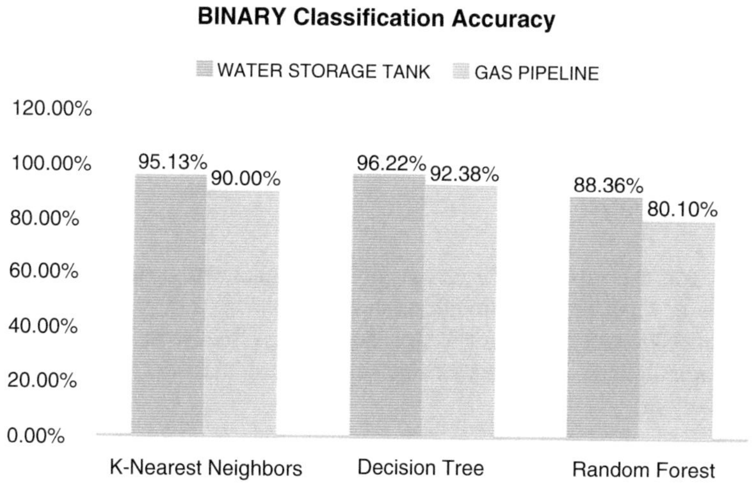

BINARY Classification Accuracy

WATER STORAGE TANK GAS PIPELINE

120.00%
100.00%
80.00%
60.00%
40.00%
20.00%
0.00%

95.13% 90.00%
96.22% 92.38%
88.36% 80.10%

K-Nearest Neighbors Decision Tree Random Forest

Fig. 11 Analysis of K-NN, Decision Tree, and Random Forest

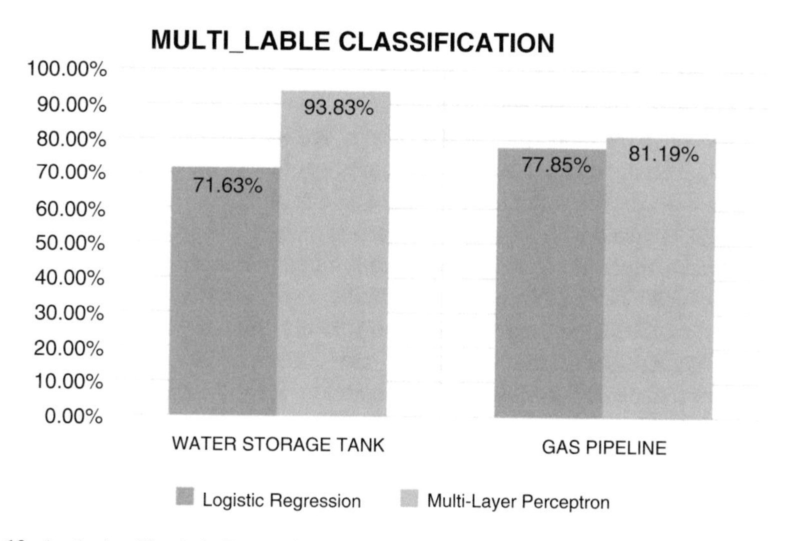

Fig. 12 Analysis of Logistic Regression and Multi-layer perceptron

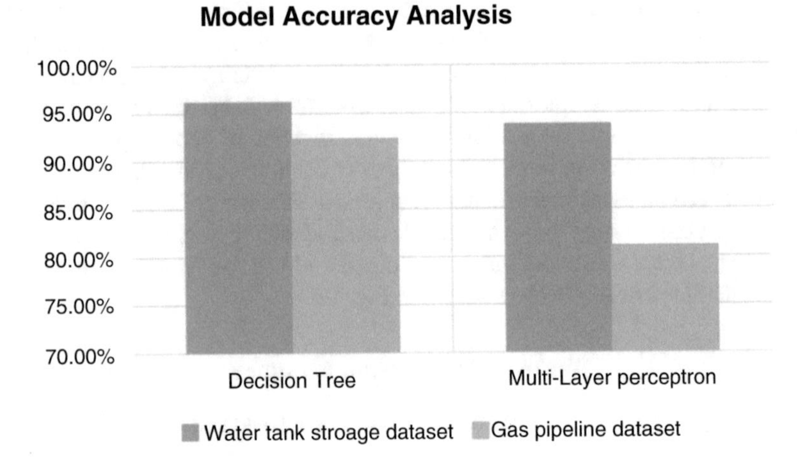

Fig. 13 Working Model Analysis

6.2 Summary

In this section, the experimental process of cyber-attack was demonstrated through two stages of experiments. Through the results demonstrated in these two experiments, it was shown that the multi-stage model is an effective technique of enhancing the computational efficiency of machine learning classifiers in the context of attack detection.

7 Conclusions

With increased network connectivity, SCADA systems are becoming more vulnerable to outsiders' threats. The need for IDS research into an industrial control system is increasing. As it has been demonstrated that machine learning can achieve attack detection in the industrial control system. We were able to successfully design multi-stage machine learning models to accurately classify the cyber-attack samples from the data set for the water storage tank and gas pipeline.

The results obtained with the critical evaluation metrics enabled us to perform effective comparative analysis and to propose the most appropriate algorithm. We achieved an overall accuracy of 96.22% and 93.83% for the water tank storage dataset with decision tree and multilayer perceptron model, respectively. Similarly, with the gas pipeline data set, we achieved an accuracy of 92.38% and 81.90%, as shown in Fig. 7.

We found a tradeoff depending on the dataset used for the evaluation. The multi-stage model shows a bias in performance based on the data provided to train our model. There are several challenges, such as defining an appropriate version of fairness [41–59]. Immediate future research is expected to improve the efficiency of the machine learning model in terms of data set independence.

References

1. Hadis Karimipour and Henry Leung. 2020. Relaxation-based anomaly detection in cyber-physical systems using ensemble kalman filter. IET Cyber-Physical Systems: Theory Applications 5, 1, 49–58. https://doi.org/10.1049/iet-cps.2019.0031
2. Hadis Karimipour, Ali Dehghantanha, Reza M. Parizi, Kim-Kwang Raymond Choo, and Henry Leung. 2019. A Deep and Scalable Unsupervised Machine Learning System for Cyber-Attack Detection in Large-Scale Smart Grids. IEEE Access 7, 80778–80788. https://doi.org/10.1109/ACCESS.2019.2920326
3. Hadis Karimipour and Venkata Dinavahi. Extended Kalman Filter-Based Parallel Dynamic State Estimation. IEEE Transactions on Smart Grid 6, 3 (May 2015), 1539–1549. https://doi.org/10.1109/TSG.2014.2387169
4. Moshe Kravchik and Asaf Shabtai. 2018. Detecting Cyber Attacks in Industrial Control Systems Using Convolutional Neural Networks. In Proceedings of the 2018 Workshop on Cyber-Physical Systems Security and PrivaCy (CPS-SPC '18), Association for Computing Machinery, Toronto, Canada, 72–83. https://doi.org/10.1145/3264888.3264896
5. Khurum Nazir Junejo and Jonathan Goh. 2016. Behaviour-Based Attack Detection and Classification in Cyber Physical Systems Using Machine Learning. In Proceedings of the 2nd ACM International Workshop on Cyber-Physical System Security—CPSS '16, ACM Press, Xi'an, China, 34–43. https://doi.org/10.1145/2899015.2899016
6. Jun Inoue, Yoriyuki Yamagata, Yuqi Chen, Christopher M. Poskitt, and Jun Sun. 2017. Anomaly Detection for a Water Treatment System Using Unsupervised Machine Learning. In 2017 IEEE International Conference on Data Mining Workshops (ICDMW), 1058–1065. https://doi.org/10.1109/ICDMW.2017.149
7. Abdulrahman Al-Abassi, Hadis Karimipour, Ali Dehghantanha, and Reza M. Parizi. 2020. An Ensemble Deep Learning-Based Cyber-Attack Detection in Industrial Control System. *IEEE Access* 8, (2020), 83965–83973. https://doi.org/10.1109/ACCESS.2020.2992249
8. Farnaz Seyyed Mozaffari, Hadis Karimipour, and Reza Parizi. 2020. Learning Based Anomaly Detection in Critical Cyber-Physical Systems. 107–130. https://doi.org/10.1007/978-3-030-45541-5_6
9. Jacob Sakhnini and Hadis Karimipour. 2020. AI and Security of Cyber Physical Systems: Opportunities and Challenges. In *Security of Cyber-Physical Systems: Vulnerability and Impact*, Hadis Karimipour, Pirathayini Srikantha, Hany Farag and Jin Wei-Kocsis (eds.). Springer International Publishing, Cham, 1–4. https://doi.org/10.1007/978-3-030-45541-5_1
10. Hadis Karimipour and Venkata Dinavahi. 2017. On false data injection attack against dynamic state estimation on smart power grids. In *2017 IEEE International Conference on Smart Energy Grid Engineering (SEGE)*, 388–393. https://doi.org/10.1109/SEGE.2017.8052831
11. Jacob Sakhnini, Hadis Karimipour, Ali Dehghantanha, Reza M. Parizi, and Gautam Srivastava. 2019. Security aspects of Internet of Things aided smart grids: A bibliometric survey. *Internet of Things* (September 2019), 100111. https://doi.org/10.1016/j.iot.2019.100111
12. Ian Turnipseed. A new SCADA dataset for intrusion detection system research. 69.
13. Amir Namavar Jahromi, Sattar Hashemi, Ali Dehghantanha, Kim-Kwang Raymond Choo, Hadis Karimipour, David Ellis Newton, and Reza M. Parizi. 2020. An improved two-hidden-layer extreme learning machine for malware hunting. *Computers & Security* 89, (February 2020), 101655. https://doi.org/10.1016/j.cose.2019.101655
14. Antoine Delplace, Sheryl Hermoso, and Kristofer Anandita. 2020. Cyber Attack Detection thanks to Machine Learning Algorithms. *arXiv:2001.06309 [cs, stat]* (January 2020). Retrieved August 15, 2020 from http://arxiv.org/abs/2001.06309
15. ICS Architecture|EPOS. Retrieved August 15, 2020 from https://www.epos-ip.org/data-services/ict-architecture/ics-architecture
16. Sara Mohammadi, Hamid Mirvaziri, Mostafa Ghazizadeh-Ahsaee, and Hadis Karimipour. 2019. Cyber intrusion detection by combined feature selection algorithm. *Journal of Information Security and Applications* 44, (February 2019), 80–88. https://doi.org/10.1016/j.jisa.2018.11.007

17. Amir Namavar Jahromi, Hadis Karimpour, Jacob Sakhnini, and Ali Dehghantanha. A Deep Unsupervised Representation Learning Approach for Effective Cyber-Physical Attack Detection and Identification on Highly Imbalanced Data. 10.
18. 2018. Decision Tree Classification in Python. *DataCamp Community*. Retrieved July 11, 2020 from https://www.datacamp.com/community/tutorials/decision-tree-classification-python
19. A new scada dataset for intrusion detection research—ProQuest. Retrieved July 11, 2020 from https://search.proquest.com/openview/bf0b546cac9a109aecb94419f7ee65a3/1?pq-origsite=gscholar&cbl=18750&diss=y
20. Thomas H. Morris and Wei Gao. 2013. Industrial Control System Cyber Attacks. https://doi.org/10.14236/ewic/ICSCSR2013.3
21. Thomas H Morris, Zach Thornton, and Ian Turnipseed. Industrial Control System Simulation and Data Logging for Intrusion Detection System Research. 6.
22. Sasanka Potluri, Christian Diedrich, Sai Ram Roy Nanduru, and Kishore Vasamshetty. 2019. Development of Injection Attacks Toolbox in MATLAB/Simulink for Attacks Simulation in Industrial Control System Applications. In *2019 IEEE 17th International Conference on Industrial Informatics (INDIN)*, 1192–1198. https://doi.org/10.1109/INDIN41052.2019.8972171
23. Hamid Reza Ghaeini and Nils Ole Tippenhauer. 2016. HAMIDS: Hierarchical Monitoring Intrusion Detection System for Industrial Control Systems. In *Proceedings of the 2nd ACM Workshop on Cyber-Physical Systems Security and Privacy* (CPS-SPC '16), Association for Computing Machinery, Vienna, Austria, 103–111. https://doi.org/10.1145/2994487.2994492
24. Taeshik Shon and Jongsub Moon. 2007. A hybrid machine learning approach to network anomaly detection. *Information Sciences* 177, 18 (September 2007), 3799–3821. https://doi.org/10.1016/j.ins.2007.03.025
25. Ensieh Modiri Dovom, Amin Azmoodeh, Ali Dehghantanha, David Ellis Newton, Reza M. Parizi, and Hadis Karimipour. 2019. Fuzzy pattern tree for edge malware detection and categorization in IoT. *Journal of Systems Architecture* 97, (August 2019), 1–7. https://doi.org/10.1016/j.sysarc.2019.01.017
26. Ahmad Javaid, Quamar Niyaz, Weiqing Sun, and Mansoor Alam. 2016. A Deep Learning Approach for Network Intrusion Detection System. In *Proceedings of the 9th EAI International Conference on Bio-inspired Information and Communications Technologies (formerly BIONETICS)*, ACM, New York City, United States. https://doi.org/10.4108/eai.3-12-2015.2262516
27. Geoffrey E. Hinton, Terrence Joseph Sejnowski, Howard Hughes Medical Institute Computational Neurobiology Laboratory Terrence J. Sejnowski, and Tomaso A. Poggio. 1999. *Unsupervised Learning: Foundations of Neural Computation*. MIT Press.
28. M. Govindarajan and RM. Chandrasekaran. 2011. Intrusion detection using neural based hybrid classification methods. *Computer Networks* 55, 8 (June 2011), 1662–1671. https://doi.org/10.1016/j.comnet.2010.12.008
29. Shadi Aljawarneh, Monther Aldwairi, and Muneer Bani Yassein. 2018. Anomaly-based intrusion detection system through feature selection analysis and building hybrid efficient model. *Journal of Computational Science* 25, (March 2018), 152–160. https://doi.org/10.1016/j.jocs.2017.03.006
30. Wathiq Laftah Al-Yaseen, Zulaiha Ali Othman, and Mohd Zakree Ahmad Nazri. 2017. Multi-level hybrid support vector machine and extreme learning machine based on modified K-means for intrusion detection system. *Expert Systems with Applications* 67, (January 2017), 296–303. https://doi.org/10.1016/j.eswa.2016.09.041
31. Md Tanzim Khorshed, Neeraj Anand Sharma, Aaron Vinek Dutt, A B M Shawkat Ali, and Yang Xiang. 2015. Real time cyber attack analysis on Hadoop ecosystem using machine learning algorithms. In *2015 2nd Asia-Pacific World Congress on Computer Science and Engineering (APWC on CSE)*, 1–7. https://doi.org/10.1109/APWCCSE.2015.7476223
32. Weizhong Yan, Lalit K. Mestha, and Masoud Abbaszadeh. 2019. Attack Detection for Securing Cyber Physical Systems. *IEEE Internet of Things Journal* 6, 5 (October 2019), 8471–8481. https://doi.org/10.1109/JIOT.2019.2919635

33. Mohamad Syahir Abdullah, Anazida Zainal, Mohd Aizaini Maarof, and Mohamad Nizam Kassim. 2018. Cyber-Attack Features for Detecting Cyber Threat Incidents from Online News. In *2018 Cyber Resilience Conference (CRC)*, 1–4. https://doi.org/10.1109/CR.2018.8626866
34. Faezeh Farivar, Mohammad Sayad Haghighi, Alireza Jolfaei, and Mamoun Alazab. 2020. Artificial Intelligence for Detection, Estimation, and Compensation of Malicious Attacks in Nonlinear Cyber-Physical Systems and Industrial IoT. *IEEE Transactions on Industrial Informatics* 16, 4 (April 2020), 2716–2725. https://doi.org/10.1109/TII.2019.2956474
35. Shahrzad Hadayeghparast and Hadis Karimipour. 2020. Application of Machine Learning in State Estimation of Smart Cyber-Physical Grid. In *Security of Cyber-Physical Systems: Vulnerability and Impact*, Hadis Karimipour, Pirathayini Srikantha, Hany Farag and Jin Wei-Kocsis (eds.). Springer International Publishing, Cham, 169–194. https://doi.org/10.1007/978-3-030-45541-5_9
36. Yuqi Chen, Christopher M. Poskitt, and Jun Sun. 2016. Towards Learning and Verifying Invariants of Cyber-Physical Systems by Code Mutation. In *FM 2016: Formal Methods* (Lecture Notes in Computer Science), Springer International Publishing, Cham, 155–163. https://doi.org/10.1007/978-3-319-48989-6_10
37. Giuseppe Bernieri, Mauro Conti, and Federico Turrin. 2019. Evaluation of Machine Learning Algorithms for Anomaly Detection in Industrial Networks. In *2019 IEEE International Symposium on Measurements Networking (M N)*, 1–6. https://doi.org/10.1109/IWMN.2019.8805036
38. Jingxuan Wang, Wenting Tu, Lucas C.K. Hui, S.M. Yiu, and Eric Ke Wang. 2017. Detecting Time Synchronization Attacks in Cyber-Physical Systems with Machine Learning Techniques. In *2017 IEEE 37th International Conference on Distributed Computing Systems (ICDCS)*, 2246–2251. https://doi.org/10.1109/ICDCS.2017.25
39. Randy C. Paffenroth and Chong Zhou. 2019. Modern Machine Learning for Cyber-Defense and Distributed Denial-of-Service Attacks. *IEEE Engineering Management Review* 47, 4 (Fourthquarter 2019), 80–85. https://doi.org/10.1109/EMR.2019.2950183
40. R. Vinayakumar, Mamoun Alazab, K. P. Soman, Prabaharan Poornachandran, Ameer Al-Nemrat, and Sitalakshmi Venkatraman. 2019. Deep Learning Approach for Intelligent Intrusion Detection System. *IEEE Access* 7, (2019), 41525–41550. https://doi.org/10.1109/ACCESS.2019.2895334
41. Steve Watson and Ali Dehghantanha. Digital forensics: the missing piece of the Internet of Things promise. Computer Fraud & Security 2016, 6 (June 2016), 5–8. https://doi.org/10.1016/S1361-3723(15)30045-2
42. Hadis Karimipour and Venkata Dinavahi. 2018. Robust Massively Parallel Dynamic State Estimation of Power Systems Against Cyber-Attack. *IEEE Access* 6, (2018), 2984–2995. https://doi.org/10.1109/ACCESS.2017.2786584
43. Philipp Kreimel, Oliver Eigner, and Paul Tavolato. 2017. Anomaly-Based Detection and Classification of Attacks in Cyber-Physical Systems. In *Proceedings of the 12th International Conference on Availability, Reliability and Security*, ACM, Reggio Calabria Italy, 1–6. https://doi.org/10.1145/3098954.3103155
44. Shailendra Singh and Sanjay Silakari. 2014. An Ensemble Approach for Cyber Attack Detection System: A Generic Framework. *International Journal of Networked and Distributed Computing (IJNDC)* 2, 2 (April 2014). https://doi.org/10.2991/ijndc.2014.2.2.2
45. Hongyu Chen, Jingyu Wang, and Dongyuan Shi. 2018. A Data Preparation Method for Machine-Learning-Based Power System Cyber-Attack Detection. In *2018 International Conference on Power System Technology (POWERCON)*, 3003–3009. https://doi.org/10.1109/POWERCON.2018.8602194
46. Yaokai Feng, Hitoshi Akiyama, Liang Lu, and Kouichi Sakurai. 2018. Feature Selection for Machine Learning-Based Early Detection of Distributed Cyber Attacks. In *2018 IEEE 16th Intl Conf on Dependable, Autonomic and Secure Computing, 16th Intl Conf on Pervasive Intelligence and Computing, 4th Intl Conf on Big Data Intelligence and Computing and Cyber Science and Technology Congress(DASC/PiCom/DataCom/CyberSciTech)*, 173–180. https://doi.org/10.1109/DASC/PiCom/DataCom/CyberSciTec.2018.00040

47. Dogukan Aksu and M. Ali Aydin. 2018. Detecting Port Scan Attempts with Comparative Analysis of Deep Learning and Support Vector Machine Algorithms. In *2018 International Congress on Big Data, Deep Learning and Fighting Cyber Terrorism (IBIGDELFT)*, 77–80. https://doi.org/10.1109/IBIGDELFT.2018.8625370
48. Naiji Zhang, Fehmi Jaafar, and Yasir Malik. 2019. Low-Rate DoS Attack Detection Using PSD Based Entropy and Machine Learning. In *2019 6th IEEE International Conference on Cyber Security and Cloud Computing (CSCloud)/2019 5th IEEE International Conference on Edge Computing and Scalable Cloud (EdgeCom)*, 59–62. https://doi.org/10.1109/CSCloud/EdgeCom.2019.00020
49. Tooba Qasim, M. Hanif Durad, Asifullah Khan, Farhan Nazir, and Tehreem Qasim. 2018. Detection of signaling system 7 attack in network function virtualization using machine learning. In *2018 15th International Bhurban Conference on Applied Sciences and Technology (IBCAST)*, 484–488. https://doi.org/10.1109/IBCAST.2018.8312268
50. Sona Taheri, Iqbal Gondal, Adil Bagirov, Greg Harkness, Simon Brown, and Chihung Chi. 2019. Multi-Source Cyber-Attacks Detection using Machine Learning. In *2019 IEEE International Conference on Industrial Technology (ICIT)*, 1167–1172. https://doi.org/10.1109/ICIT.2019.8755006
51. Erik M. Ferragut, Jason Laska, Mohammed M. Olama, and Ozgur Ozmen. 2017. Real-Time Cyber-Physical False Data Attack Detection in Smart Grids Using Neural Networks. In *2017 International Conference on Computational Science and Computational Intelligence (CSCI)*, 1–6. https://doi.org/10.1109/CSCI.2017.1
52. Giovanni Apruzzese, Michele Colajanni, Luca Ferretti, and Mirco Marchetti. 2019. Addressing Adversarial Attacks Against Security Systems Based on Machine Learning. In *2019 11th International Conference on Cyber Conflict (CyCon)*, 1–18. https://doi.org/10.23919/CYCON.2019.8756865
53. Bisyron Wahyudi Masduki, Kalamullah Ramli, Ferry Astika Saputra, and Dedy Sugiarto. 2015. Study on implementation of machine learning methods combination for improving attacks detection accuracy on Intrusion Detection System (IDS). In *2015 International Conference on Quality in Research (QiR)*, 56–64. https://doi.org/10.1109/QiR.2015.7374895
54. Chuadhry Mujeeb Ahmed, Carlos Murguia, and Justin Ruths. 2017. Model-based Attack Detection Scheme for Smart Water Distribution Networks. In *Proceedings of the 2017 ACM on Asia Conference on Computer and Communications Security*, ACM, Abu Dhabi United Arab Emirates, 101–113. https://doi.org/10.1145/3052973.3053011
55. Aboul Ella Hassanien. 2003. Classification and Feature Selection of Breast Cancer Data Based on Decision Tree Algorithm. *Studies in Informatics and Control* (2003), 8.
56. P.G. Campos, E.M.J. Oliveira, T.B. Ludermir, and A.F.R. Araujo. 2004. MLP networks for classification and prediction with rule extraction mechanism. In *2004 IEEE International Joint Conference on Neural Networks (IEEE Cat. No.04CH37541)*, 1387–1392 vol. 2. https://doi.org/10.1109/IJCNN.2004.1380152
57. 3.2.4.3.3. sklearn.ensemble.ExtraTreesClassifier—scikit-learn 0.21.3 documentation. Retrieved December 1, 2019 from https://scikit-learn.org/stable/modules/generated/sklearn.ensemble.ExtraTreesClassifier.html
58. Shaikh, R.: Feature Selection Techniques in Machine Learning with Python, https://towardsdatascience.com/feature-selection-techniques-in-machine-learning-with-python-f24e7da3f36e, last accessed 2020/07/11.
59. Marina Sokolova, Nathalie Japkowicz, and Stan Szpakowicz. 2006. Beyond Accuracy, F-Score and ROC: A Family of Discriminant Measures for Performance Evaluation. In *AI 2006: Advances in Artificial Intelligence* (Lecture Notes in Computer Science), 1015–1021. https://doi.org/10.1007/11941439_114

A Recurrent Attention Model for Cyber Attack Classification

Naseem Alsadi, Hadis Karimipour, Ali Dehghantanha, and Gautam Srivastava

1 Introduction

With the proliferation and development of contemporary computer systems, the demand for reliable data protection has reached an all-time high. Traditional methods of malware detection no longer provide systems with the security they so vitally need [1]. Attackers have developed a novel and advanced methodologies to disguise malicious activity and avoid detection, leaving systems vulnerable to malicious data acquisition. Novel algorithms have been developed with aim of detecting and classifying malicious attacks [2–7]. There are various methods to go about developing such algorithms, including operational code analysis.

All software initiates interaction with hardware components, and at some point, software intentions will need to be processed as operational codes, whether the software is good ware or malware. Operational code data, therefore, presents us with valuable information about software and software intentions.

Specifically, polymorphic malware has become increasingly difficult to detect due to its dynamic nature. Polymorphic code refers to a species of code that uses a mutation engine to change its code whilst keeping the fundamental algorithm the

N. Alsadi · H. Karimipour (✉)
Department of Electrical and Software Engineering, University of Calgary, Calgary, AB, Canada
e-mail: nalsadi@uoguelph.ca; hadis.karimipour@ucalgary.ca

A. Dehghantanha
School of Computer Science, University of Guelph, Guelph, ON, Canada
e-mail: adehghan@uoguelph.ca

G. Srivastava
Department of Mathematics and Computer Science, Brandon University, Brandon, MB, Canada
e-mail: srivastavag@brandonu.ca

H. Karimipour, F. Derakhshan (eds.), *AI-Enabled Threat Detection and Security Analysis for Industrial IoT*, https://doi.org/10.1007/978-3-030-76613-9_13

same [8]. This presents a problem for traditional methods that do not account for mutations and rely on firm pattern matching to detect anomalies. When polymorphic malware mutates it becomes increasingly difficult to relate it to its initial state. This necessitates an algorithm capable of recognizing and detecting the overarching algorithm structure. Unfortunately, polymorphic malware is not alone [9, 10]. Several other malicious attacks were developed to subtly bypass detection systems. Therefore, detection systems must be capable of closely analyzing software data and extracting software intentions [11]. Visualization of software data, in specific operational code data, can provide detection systems with the ability to examine program data and instruction correlation.

The use of visualization exploits image texture analysis, a vital subsection of computer vision. Allowing us to examine the spatial arrangement of intensity in regions of an image [12]. This is very helpful when we are looking to analyze the overall structure of the algorithm and analyze the relationship between non-adjoining instructions. Additionally, visualization allows for simple data set augmentation, making it easier to expand datasets. Visualization will require an image classification architecture. However, neural network-based image classification architectures typically have considerably higher computational cost.

The Recurrent Attention Model (RAM) is an alternative to prominent image classification models [13]. The RAM draws inspiration from human perception. Human focus does not holistically analyze visual input, rather it selects various sections of the visual space. Over a duration of time, information is gathered from variant fixations allowing for a more complete internal representation to be developed. As a result, images can be analyzed with a higher degree of attention in, especially crucial regions.

In this chapter, we focus on malware binary classification using operational code data and RAM. Operational code data is initially processed into an image, using image processing techniques. This allows for the analysis of the relationship between non-adjoining instructions, which may reveal initially obscure software intentions. The efficiency of the proposed model is evaluated using several metrics, including accuracy, precision, recall, and F-score to evaluate the performance of the proposed threat hunting model as follows [14]:

- **True Positive (TP):** when a malicious sample is predicted as malware.
- **True Negative (TN):** when a malicious sample is predicted as goodware.
- **False Positive (FP):** when a benign sample is predicted as malicious.
- **False Negative (FN):** when a malicious sample is predicted as goodware.

Utilizing the aforementioned core metrics, the performance of machine learning systems can be measured using the following formulas:

Accuracy: indicates how the proposed model can accurately predict malware and benign samples.

$$Accuracy = \frac{TP + TN}{TP + TN + FN + FP} \tag{1}$$

Precision: Precision for a certain APT group is the number of samples in a class that is correctly predicted, divided by the total number of samples that are predicted.

$$Precision = \frac{TP}{TP + FP} \tag{2}$$

Recall: for a certain class, is the number of samples in a class that is correctly predicted, divided by the total number of samples in that class.

$$Recall = \frac{TP}{TP + FN} \tag{3}$$

F-Score (F_1): F-Score is the harmonic mean of Precision and Recall. It can be applied as a general classifier performance metric:

$$F_1 = 2 \times \frac{Precision \times Recall}{Precision + Recall} \tag{4}$$

2 Previous Work

There has been a lot of research conducted on utilizing opcode data for malware classification. Researchers such as Haddadpajouh et al. proposed a Multi-Kernel and Meta-heuristic Feature Selection Approach for cyber threat hunting in an IoT environment [1]. The results were promising, with a validation accuracy of 94%.

Researchers at Dongguk University-Seoul and Baewha Women's University in South Korea used operation code data to train a Recurrent Neural Network for the identification of malicious activity. The results showed an exceptional accuracy of 97.59%. They further investigated the use of word2vec encoding against the classical one-hot encoding. Increased accuracy of 0.5% was reported when using the word2vec encoding over the one-hot encoding. However, the researchers report that with their proposed method there was a large computational cost, due to having to analyze a large set of opcode data [2].

Yuxin et al. at the Harbin Institute of Technology used operation code data to identify malicious software using a Deep Belief Network (DBN). The proposed model used DBN's to extract feature vectors and detecting malware. The proposed model was compared against three baseline models, which utilized decision trees, support vector machines, and the k-nearest neighbor algorithm as core classifiers. The proposed method showed that BDN was capable of accurately detecting malware. However, more importantly to this report, operation code data was used to reliably identify malicious programs [3]. Researchers at the Vision Research Lab at the University of California used to signal and image processing to examine malware data. They showed that the visualization of malware data was pragmatic and could distinguish between variant malware families. Furthermore, examining program data before execution allows for identifying malicious intent before exposing systems to the risk of attack. Besides, the relationship between non-adjoining program instructions can be analyzed through the analysis of spatial arrangement. The results showed that the accuracy of the proposed model was exceptionally high at 98% [4].

Recent developments in the field of image classification have led to the introduction of novel architectures. Researchers at Google DeepMind have developed an algorithm for image analysis based on RNN and reinforcement learning. The proposed model, named the Recurrent Attention Model (RAM), draws inspiration from biological human perception. With human perception, the entire environment is not processed at once. Rather, focusing is conducted by selecting particular regions of attention. Information is acquired from regions of attention and amalgamated with information from other regions of attention. Over time an internal representation is developed of the environment. This allows for an in-depth analysis of regions of emphasis while allowing the amount of computation to be controlled independently of image size. The RAM was compared against various other models including a convolutional neural network and a fully connected network. The results show that the RAM outperformed the other models and accurately classified input image with an error of 1.29% [5].

In conclusion, the proposed models in the literature confirm the validity of utilizing operation code data to classify malicious [6–15]. Furthermore, the visualization of program data has been shown to increase the accuracy of models when identifying malware. Novel image classification architectures allow for unique image analysis. The RAM allows visualized operation code data to be analyzed, therein allowing the relationship between non-adjoining instructions to be further examined.

3 Proposed Approach

The proposed model utilizes a deep reinforcement learning architecture to implement a vision attention based binomial malware classification system. The system is fundamentally composed of multiple processes, each process vitally contributing to the whole. The core processes are broken down and novelties are described in detail below.

3.1 Data Processing and Visualization

The architecture is initially presented with operational code data, which are essentially sequences of operands that stem from a specific instruction set architecture. Before data visualization, it is vital to identify instruction set architecture and encode prominent operands into their numerical representations. With a set of N

Table 1 Prominent Operands

Operand
LDR, MOV, CMP, BL, STR,
ADD, B, BEQ, BNE, LDRB,
STRB, LDMIA, BX, BLE,
MOVNE, MOVEQ, TST, BLT,
SUB, LDMFD

operands, each operand will be assigned a unique numerical value of $0 \rightarrow N - 1$. A list of prominent operands obtained from surveying the dataset is displayed in Table 1. Numerical vectorization followed for each respective data sample.

Proceeding vectorization, each data sample will need to be visualized using data visualization techniques. To do such, all data samples will initially need to be set to the same length. Each data sample will need to be reshaped to accommodate a height, width, and depth. Essentially, all data samples become a numerical pixel map. Any visualization or imaging library can then be utilized to generate an image. Figure 1 showcases an abstraction of the visualization procedure.

Figure 2 shows two different visualizations of variant software using operational code visualization. It is empirically evident that the two different images clearly show variance in texture, consequently supporting the ability to use texture analysis in detecting and classifying software. In this paper, visualization is used to exploit texture analysis.

3.2 Recurrent Neural Network (RNN)

Recurrent neural networks (RNN) are a category of artificial neural networks where a sequential connection is established among nodes. Nodes will be connected to other nodes within variant layers in a unidirectional fashion. The flow of data is unidirectional, therefore the RNN's will contain input, hidden, and output layers. The temporal flow of data from node to node allows previous outputs to be utilized as input for successive nodes. As a result, information from prior input is compiled and transferred to subsequent nodes, allowing for the model to dynamically learn from the past.

Classical neural networks work well on the presumption that the input and out are directly independent of each other, however, this is not always the case. This is crucial to the implementation of the proposed method and will be discussed in greater detail below.

Figure 3 shows a labeled RNN diagram. At each time step t, a new input x_t is provided to the network. The hidden section of the network, which compiles and stores all previous data, is calculated as $h_t = f(W_x x_t + W_h h_{t-1})$. Conclusively, the output y_t is sequentially generated at each node with $y_t = softmax(W_y)$ [16].

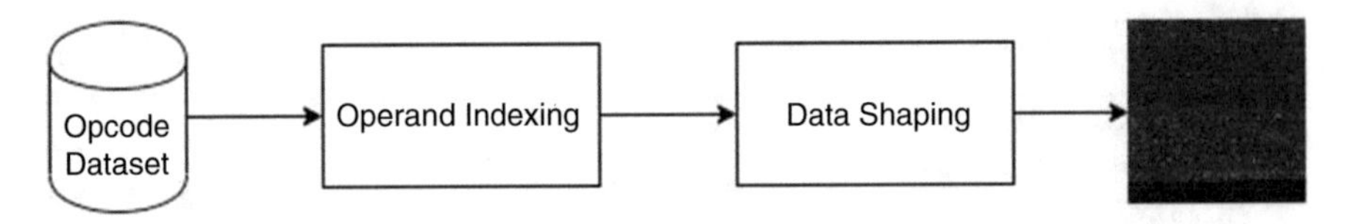

Fig. 1 Visualization Process

Fig. 2 Visualization of Program

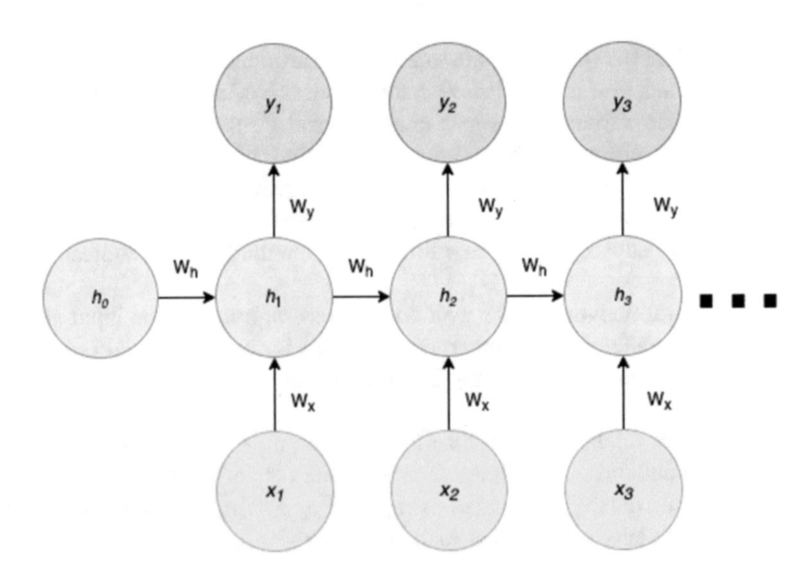

Fig. 3 Recurrent Neural Network Diagram

3.3 Reinforcement Learning (RL)

Reinforcement learning is a subsection of machine learning focused on the mapping of optimal actions to specific environment states, with aim of augmenting reward. An agent is placed within an environment and equipped with a finite set of possible actions. Through repetition, the agent learns optimal behavior within states. The foundation of reinforcement learning is the reward function, which drives the agent to pursue directed goals.

Reinforcement learning architectures are concerned with the identification of optimal action-selection policies for any given Markov Decision Process. A Markov Decision Process (MDP) is a mathematical framework for modeling decision-making in discrete time. The MDP can be formalized as a 4-tuple [17]:

$$(S, A, P_a, R_a)$$

- S: Defines a finite set of states.
- A: Defines a finite action space.
- P_a: The probability that an action, a, whilst in a current state, s, will lead to a future state, s'.
- R_a: Defines the reward from conducting the action, a.

The proposed model in this paper is an instance of the Partially Observable Markov Decision Process (POMDP). The POMDP is a generalization of the MDP, in which the actual state of the environment is never completely observed. Therefore, the POMP is required to keep a probability distribution over the set of possible states. The POMDP is widely accepted to be capable of being pragmatically implemented across various sequential decision-making applications. The POMDP can be formalized as a 7-tuple [18]:

$$(S, A, T, \Omega, R_a, O, \gamma)$$

- S: Defines a finite set of states.
- A: Defines a finite action space.
- T: Defines a finite set of conditional transition probabilities for state transition.
- Ω: Defines the observation space.
- R_a: Defines the reward from conducting the action a at state s.
- O: Defines the conditional observation probabilities.
- γ: Defines the discount factor.

At time step t the agent will select an action a from the action space A. Action a will create a change in the environment, causing the state of the environment to transition to state s'. This transition will occur with a probability of $T(s'|s, a)$. The agent is subsequently presented with an observation of the new state. The probability of being presented with observation o, provided the most recent action taken a and the new state s', is $O(o|s', a)$. Conclusively, the agent is provided a reward for selecting action a in state s, formalized as $R(s,a)$. The agent will aim to select actions at time step t, to maximize reward.

In this paper, the agent utilizes vision attention techniques to exploit regional texture analysis. The agent will attempt to perform a set of discrete sequential actions, which highlight the spatial arrangement of intensity with an image. The process aims at the discovery of optimal action sequences, conducted by the agent, which reveal an underlying relationship between adjoining and non-adjoining instructions, therein unveiling software intentions, which may have otherwise been initially obscured. Furthermore, Reinforcement Learning presents the ability to learn extensively about families of malware, specifically their polymorphic capabilities and their central entrenched algorithm structure.

3.4 Recurrent Attention Model (RAM)

The Recurrent Attention Model (RAM) is a recurrent neural network that takes inputs sequentially from various parts of a selected image to develop a dynamic representation of the environment. The RAM can do such by extracting information from various fixations, rather than the entire image.

At the time step t, an image of the environment will be provided to the agent, this image is named x_t. The agent will use a bandwidth-limited sensor p, which assists in focusing on a selected location l, to examine the image. The bandwidth-limited sensor utilizes a gradually lower resolution for pixels further away from l. As a result, the pixels in the region l will have a higher resolution than those elsewhere. This process is conducted within the glimpse sensor as seen in Fig. 4. The input to the glimpse sensor is the coordinates of the sensor and input image. The output is the various resolution patches, which form a retina-like representation.

Using an input image x with a selected location l, the retina encoder $p(x,l)$ will extract k overlapping square patches with variant sizes. The first patch will be of the size $g_w \times g_w$ pixels, subsequent patches will have twice the size of the previous. After

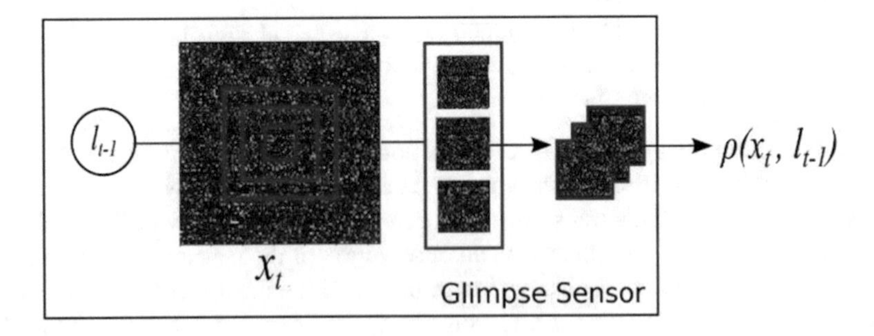

Fig. 4 Glimpse Sensor [5]

all the patches are extracted, they will undergo a uniform reshaping to the dimensionality of $g_w \times g_w$. The glimpse sensor process is completed when the patches have been reshaped and then subsequently concatenated.

Throughout the process, the agent keeps an internal state which compiles all information extracted from prior observations. The glimpse network utilizes the glimpse sensor to extract the retina representation. The retina representation and the glimpse location are both independently mapped into a hidden space utilizing linear layers. The detached outputs of the linear layers are combined with a final linear layer that amalgamates both streams of information. The process is depicted in Fig. 5. Fundamentally, the glimpse network $f_g(x, l)$ has two fully connected layers. The output of the glimpse network can be formalized as $g = Rect(Linear(h_g) + Linear(h_l))$, where $h_g = Rect(Linear(p(x, l)))$, $h_l = Rect(Linear(l))$, $Rect(x) = \max(x, 0)$ and $Linear(x) = Wx + b$.

At each time step, the bandwidth-limited sensors will be deployed. Along with the sensors, an environment action is conducted at every time step. For this specific application, environmental action is a SoftMax layer output. The environment action is produced by the action network, which makes a classification based on the internal state of the model. The action network can be defined using a linear SoftMax classifier as $f_a(h) = \exp(Linear(h))/Z$, where Z is a constant chosen for normalization. After the agent has completed the action, he is presented with a reward r_{t+1} and a novel visual observation of the environment x_{t+1}. The reward function is crucial in motivating the agent to choose instrumental actions. In the aim of achieving an accurate model, correct binary classification is our core aim. An agent receives a reward when a correct classification has been made, and none otherwise. The agent aims to perform a sequence of optimal actions to the input while maintaining the accuracy of the model. The reward function is formalized in Eq. 5.

$$r_t = \{0,\ y_{pred} != y_{label}\ 1,\ y_{pred} = y_{label} \tag{5}$$

The core network of the model uses the glimpse representation produced by the glimpse network and amalgamates it with the internal representation at the previous

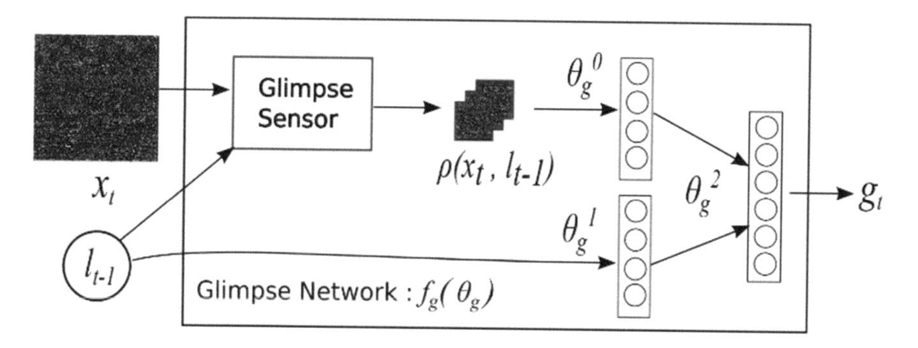

Fig. 5 Glimpse Network [5]

time step, which in turn yields a novel internal state of the model. The core network is formalized as $h_t = f_h(h_{t-1}) = Rect(Linear(h_{t-1}) + Linear(g_t))$. The location network utilizes the internal state of the model to select the next sensor location, as defined by $f_l(h) = Linear(h)$.

Keeping in mind the core aim of augmenting the reward achieved by the agent the training procedure can be laid out. The training parameters for the agent can be defined as $\theta = \{\theta_g, \theta_h, \theta_a\}$. The training procedure is then directed at learning these parameters to maximize the reward achieved by the agent. The action network f_a is trained via the utilization of backpropagation through the glimpse and core networks. The location network f_l is trained using the REINFORCE rule. The location network training procedure involves manipulating the value of θ to increase the log-probability of selected actions which leads to greater reward and decreases that of actions that do not. The cumulative diagram is displayed in Fig. 6, which showcases the methodology utilized from the data pre-processing phase to completion [5].

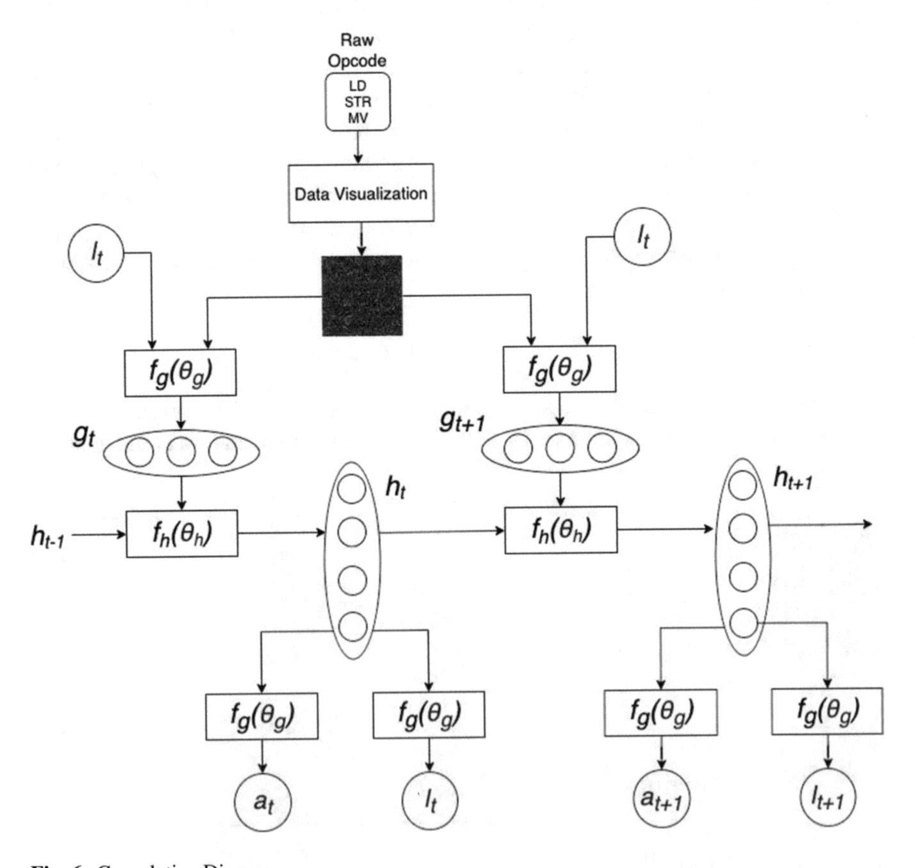

Fig. 6 Cumulative Diagram

4 Experimental Analysis

In this section, we give some descriptions into the two datasets that are used in our experimental analysis.

IoT Dataset: An IoT application dataset, which contains various samples of malware and goodware applications presented in operational code form is used for our experiments. Each data sample is obtained while using 32-bit ARM architecture. Normalization of data is critical to ensure uniformity of the model during both the training and validation phase. The vast majority of IoT platforms are ARM-based. Therefore, the utilization of an ARM-based IoT dataset prepares the model for real-world IoT attacks. The IoT dataset contained a total of 512 data samples, categorically divided into goodware and malware. Each data sample was a sequence of operational codes conducted by a specific application at a variant sequence span. The initial dataset provided an unbalanced split between goodware and malware, which was resolved with the use of image manipulation techniques.

BATADAL Dataset: The Battle of the Attack Detection Algorithms (BATADAL) is a dataset that was developed to evaluate and compare the performance of algorithms that detect cyber-attacks in water distribution systems. Contemporary water distribution systems utilize computers, sensors, and actuators to create a cyber-physical system. The evolution of these systems has improved the overall water distribution process; however, the introduction of cyber-physical systems has also increased the risk of cyberattacks. The dataset is partially labelled and was accumulated over 6 months. It contains 4177 data samples, with a mix of both normal and attack data [19].

5 Results

The proposed method was assessed using cross-validation. Each dataset was evaluated individually, and the proposed method was compared against various other models utilizing the performance metrics discussed prior.

5.1 IoT Dataset

The results for the IoT dataset, which contained 512 data samples are summarized in Table 2. As can be seen from the table, the proposed method outperforms all other models. The model was trained for 20 epochs over a cumulative period of 1 h. The results suggest that the adoption of visualization techniques and deep reinforcement learning when classifying operational code data increases model performance.

Table 2 Three Comparative Study of Models on IoT Dataset

Architecture	Accuracy	Precision	Recall	F-Score
Proposed model	**99.63%**	**99.59%**	**99.59%**	**99.59%**
Multi-layer perceptron	93.29%	93.29%	93.29%	93.29%
Haddadpajouh et al. [1]	94.0%	–	–	–
Yuxin et al. [3]	96.5%	–	–	–
Santos et al. [9]	95.91%	86.25%	81.55%	86.52%

Table 3 Comparative Study of Models on BATADAL Dataset

Architecture	Accuracy	Precision	Recall	F-Score
Proposed model	**99.95%**	**99.09%**	**99.54%**	**99.31%**
Aghashahi et al. [13]	98.6%	75.0%	95.2%	83.9%
Abokifa et al. [19]	–	–	–	88%
Chandy et al. [20]	71.3%	39.2%	85.8%	53.8%

5.2 BATADAL Dataset

The results for the BATADAL dataset, which contained 4177 data samples, is depicted in Table 3. The results indicate that the proposed method surpassed all other models in the selected criteria. The model was trained for 50 epochs over 8 h. The results further supplement the claim that an amalgamation of visualization and deep reinforcement learning augments model performance.

6 Discussion

In this paper, a method for malware classification was formulated and compared against various other models. The proposed method consisted of visualizing opcode data and using a RAM to perform a binary malware classification task. The proposed method was attractive for various reasons, including in-depth image analysis. This type of analysis is crucial for malware identification, where programs will be developed to subtly bypass detection algorithms. The results obtained from the experiment indicate that the proposed method performed excellently in accurately classifying malicious intention. An IoT dataset, obtained while using 32-bit ARM architecture, was utilized to test the accuracy of the proposed method on opcode data. The results show that the proposed method performed much better than other models. Emphasis must be drawn to the recall metric for malware classification, which estimates the rate at which true malware samples were correctly identified. Algorithms with low recall allow numerous malicious programs to pass undetected. The calculated recall from running the proposed method on the IoT dataset was 99.59%, which was higher than any of the other models which were analyzed. A total of only 1 malware sample went undetected out of a total of 244.

The proposed method was also tested on a dataset that did not include opcode data. The selected dataset was the BATADAL dataset which was a water distribution system dataset containing numerous system attack data. The aim was to test if the proposed method could be used across various dataset types with similar accuracy in malicious activity identification. The results show that the proposed method performed excellently, scoring a total accuracy of 99.95% and a recall of 99.54%. The proposed method was also able to surpass all other models implemented to classify malicious activity on this dataset.

Visualization of operational code data provided a reliable and accurate grounding for the model. As a means of protection against polymorphic code, the proposed architecture attempts to examine the global structure of the application rather than sequential operational code analysis. Successfully, through the use of image processing techniques, serial operational code data was manipulated into a visual representation that embodied the instructions of an application with numerical representations. An empirical examination of a collection of visualized data vividly shows the spatial arrangement of intensity across variant regions of an image.

The proposed model excels in its function of visualizing and emphasizing regions of operational code data, allowing for focused texture analysis. Moreover, the proposed model allows critical feature exposure, through its use of an RNN to emphasize and closely analyze distinct textural regions of an image. Future considerations include testing the proposed method across the variant dataset.

References

1. A. N. Jahromi, A. Dehghantanha, R. Choo, H. Karimipour, R. Parizi, "An Improved Two-Hidden-Layer Extreme Learning Machine for Malware Hunting", *Computer and Security*, vol. 89, pp. 1-11, Sept. 2019. https://doi.org/10.1016/j.cose.2019.101655
2. S. A. Roseline, S. Geetha, S. Kadry and Y. Nam, "Intelligent Vision-Based Malware Detection and Classification Using Deep Random Forest Paradigm," in *IEEE Access*, vol. 8, pp. 206303–206324, 2020, https://doi.org/10.1109/ACCESS.2020.3036491.
3. Alabasi, H. Karimipour, A. Dehghantanha, "An Ensemble Deep Learning-based Cyber-Attack Detection in Industrial Control System", *IEEE Access*, vol. 8, pp. 83965–83973, April. 2020. doi: https://doi.org/10.1109/ACCESS.2020.2992249.
4. S. Naval, V. Laxmi, M. Rajarajan, M. S. Gaur and M. Conti, "Employing Program Semantics for Malware Detection," in *IEEE Transactions on Information Forensics and Security*, vol. 10, no. 12, pp. 2591–2604, Dec. 2015, https://doi.org/10.1109/TIFS.2015.2469253.
5. M. Fan *et al.*, "Android Malware Familial Classification and Representative Sample Selection via Frequent Subgraph Analysis," in *IEEE Transactions on Information Forensics and Security*, vol. 13, no. 8, pp. 1890–1905, Aug. 2018, https://doi.org/10.1109/TIFS.2018.2806891.
6. S. Mohammadi, H. Mirvaziri, M. G. Ahsaee, H. Karimipour, "Cyber Intrusion Detection by Combined Feature Selection Algorithm", *Journal of Information Security & Applications—Elsevier* (IF: 2.6), pp. 80–88, vol. 44, Feb. 2018. https://doi.org/10.1016/j.jisa.2018.11.007
7. S. M. Tahsien, H. Karimipour, P. Spachos, "Machine Learning Based Solutions for Security of Internet of Things (IoT): A Survey", *Journal of Network and Computer Applications- Elsevier*, vol. 161, pp. 1–18, April. 2020. https://doi.org/10.1016/j.jnca.2020.102630

8. M. Nassiri, H. HaddadPajouh, A. Dehghantanha, H. Karimipour, R. M. Parizi, and G. Srivastava, "Malware Elimination Impact on Dynamic Analysis: An Experimental Machine Learning Approach", Handbook of Big Data and Privacy, Springer Books, pp. 1–39, Jan. 2020. https://doi.org/10.1007/978-3-030-38557-6_17
9. S. Cesare, Y. Xiang and W. Zhou, "Malwise—An Effective and Efficient Classification System for Packed and Polymorphic Malware," in *IEEE Transactions on Computers*, vol. 62, no. 6, pp. 1193–1206, June 2013, https://doi.org/10.1109/TC.2012.65.
10. R. Kaur and M. Singh, "A Survey on Zero-Day Polymorphic Worm Detection Techniques," in *IEEE Communications Surveys & Tutorials*, vol. 16, no. 3, pp. 1520–1549, Third Quarter 2014, https://doi.org/10.1109/SURV.2014.022714.00160.
11. H. H. Pajouh, A. Dehghantanha, R. Parizi, H. Karimipour, "A Survey on Internet of Things Security: Requirements, Challenges, and Solutions", *Internet of Things Journal*, pp. 1–16, Oct. 2019. https://doi.org/10.1016/j.iot.2019.100129
12. H. Shiravi, A. Shiravi and A. A. Ghorbani, "A Survey of Visualization Systems for Network Security," in *IEEE Transactions on Visualization and Computer Graphics*, vol. 18, no. 8, pp. 1313–1329, Aug. 2012, https://doi.org/10.1109/TVCG.2011.144.
13. J. Zheng and L. Zheng, "A Hybrid Bidirectional Recurrent Convolutional Neural Network Attention-Based Model for Text Classification," in *IEEE Access*, vol. 7, pp. 106673–106685, 2019, https://doi.org/10.1109/ACCESS.2019.2932619.
14. H. Karimipour, A. Dehghantanha, R. Choo, H. Leung, "A Deep and Scalable Unsupervised Machine Learning System for Cyber-Attack Detection in Large-scale Smart Grids", *IEEE Access*, vol. 7, pp. 80778–80788, May 2018. https://doi.org/10.1109/ACCESS.2019.2920326
15. H. H. Pajouh, A. Dehghantanha, H. Karimipour, X. Lin, R. Choo "A Multi-Kernel and Meta-heuristic Feature Selection Approach for IoT Malware Threat Hunting in the Edge Layer", IEEE Internet of Things Journal, pp. 1–13, Sept. 2020. doi: https://doi.org/10.1109/JIOT.2020.3026660
16. J. Kang, S. Jang, S. Li, Y.-S. Jeong and Y. Sung, "Long short-term memory-based Malware classification method for information security," *Computers & Electrical Engineering,* vol. 77, p. 366–375, 2019.
17. D. Yuxin and Z. Siyi, "Malware detection based on deep learning algorithm," *Neural Computing and Applications,* vol. 31, p. 461–472, 2017.
18. L. Nataraj, S. Karthikeyan, G. Jacob and B. S. Manjunath, "Malware images," *Proceedings of the 8th International Symposium on Visualization for Cyber Security—VizSec 11*, 2011.
19. V. Mnih, N. Heess, A. Graves and K. Kavukcuoglu, *Recurrent Models of Visual Attention*, 2014.
20. R. Doshi, N. Apthorpe and N. Feamster, "Machine Learning DDoS Detection for Consumer Internet of Things Devices," *2018 IEEE Security and Privacy Workshops (SPW),* 2018.

Printed in the United States
by Baker & Taylor Publisher Services